The Ecology of Tropical East Asia

The Ecology of Tropical East Asia

Richard T. Corlett

OXFORD
UNIVERSITY PRESS

Great Clarendon Street, Oxford OX2 6DP

Oxford University Press is a department of the University of Oxford. It furthers the University's objective of excellence in research, scholarship, and education by publishing worldwide in

Oxford New York

Auckland Cape Town Dar es Salaam Hong Kong Karachi Kuala Lumpur Madrid Melbourne Mexico City Nairobi New Delhi Shanghai Taipei Toronto

With offices in

Argentina Austria Brazil Chile Czech Republic France Greece Guatemala Hungary Italy Japan Poland Portugal Singapore South Korea Switzerland Thailand Turkey Ukraine Vietnam

Oxford is a registered trade mark of Oxford University Press in the UK and in certain other countries

Published in the United States by Oxford University Press Inc., New York

First published 2009

Reprinted 2010

British Library Cataloguing in Publication Data
Data available

Library of Congress Cataloging in Publication Data
Data available

Typeset by SPI Publisher Services, Pondicherry, India
Printed in Great Britain
on acidfree paper by
CPI Antony Rowe, Chippenham, Wiltshire

ISBN 978–0–19–953245–2 (Hbk)
978–0–19–953246–9 (Pbk)

3 5 7 9 10 8 6 4 2

Preface

Tropical East Asia is a region united by biology, but divided by history. The essential unity of the region has always been obvious to ecologists who have travelled through it, but, until very recently, few had done so. Arm-chair travel through the medium of books and journal articles has also been limited by the lack of a shared regional language. While knowledge of English, Spanish, and Portuguese gives you access to the entire ecological literature of the American tropics, and English and French suffice for tropical Africa, the ecologists of Tropical East Asia have written in at least ten languages. Language and other factors have also limited the development of regional journals—there are none in ecology—and conferences, with ecologists from different countries more likely to meet in the USA or Europe than in Asia. However, the traditional fragmentation of the region is changing. Economic and political developments, such as the rise of ASEAN (Association of Southeast Asian Nations) and the opening up of China, have created both opportunities and incentives for regional collaboration on environmental issues. There is no better illustration of this new attitude than the success of the first three annual meetings of the newly formed Asia–Pacific chapter of the Association for Tropical Biology and Conservation (ATBC).

The aim of this book is to provide an overview of the terrestrial ecology of Tropical East Asia. I hope to make it easier for people working in the region to acquire the understanding of its ecology necessary to put their own work into a broader context, while, at the same time, providing an accessible summary for people living outside the region. In order to maintain a reasonable degree of coherence, eastern Indonesia is omitted, because it is not biologically part of Tropical East Asia, and southern China is included, because it is. I must apologize, however, to my Indian friends for the abrupt western cut-off, at the Myanmar border. This hard boundary cannot be justified ecologically, but the inclusion of only the wetter parts of the subcontinent would have been confusing, so it seemed preferable to leave it all out.

Thanks to the Internet, I was able to write half this book on the non-tropical, non-eastern, non-Asian Greek island of Antiparos, whose inhabitants I must thank for their cheerful hospitality. This was the longest period I have spent outside the region for thirty years and I have no doubt that the distance helped, particularly with the early chapters. The other half was written largely at the University of Hong Kong, where I spent the middle 20 years of my career, and it was then completed in my new position at the National University of Singapore. So many people have helped with the book that any list must be incomplete, but I would particularly like to acknowledge the following ecologists who commented on chapters or other substantial sections: David Burslem, Kylie Chung, David Dudgeon, John Fellowes, Billy Hau, Nina Ingle, Michael Lau, Bill Laurance, Ng Sai-chit, Richard Primack, Yvonne Sadovy, Navjot Sodhi, I-Fang Sun, and Hugh Tan. Photographs and other graphics are acknowledged individually, but special thanks are due to Yeung Ka-ming (AFCD) and Lee Kwok Shing (KFBG) for offering me so much choice, and to Hugh Tan for filling in many gaps. Helen Eaton and the other staff at OUP were always a pleasure to work with. Finally, I am particularly grateful to Laura Wong, who not only drew all the maps and most of the other figures, but also checked facts, translated articles from Chinese, and acted as my Hong Kong agent while I was writing in Greece.

Contents

CHAPTER 1

Environmental history

1.1 Why 'Tropical East Asia'?

Tropical East Asia (TEA) is used in this book to refer to the eastern half of the Asian tropics and subtropics (Fig. 1.1). Politically, it consists of Myanmar, Laos, Cambodia, Vietnam, Thailand, Malaysia, Singapore, Brunei, the Philippines, western Indonesia, and southern China (north to 30°N), plus the Ryukyu Islands of Japan, and the Andaman and Nicobar Islands of India. Geographically, this is 'South-East Asia', but this term is nowadays most

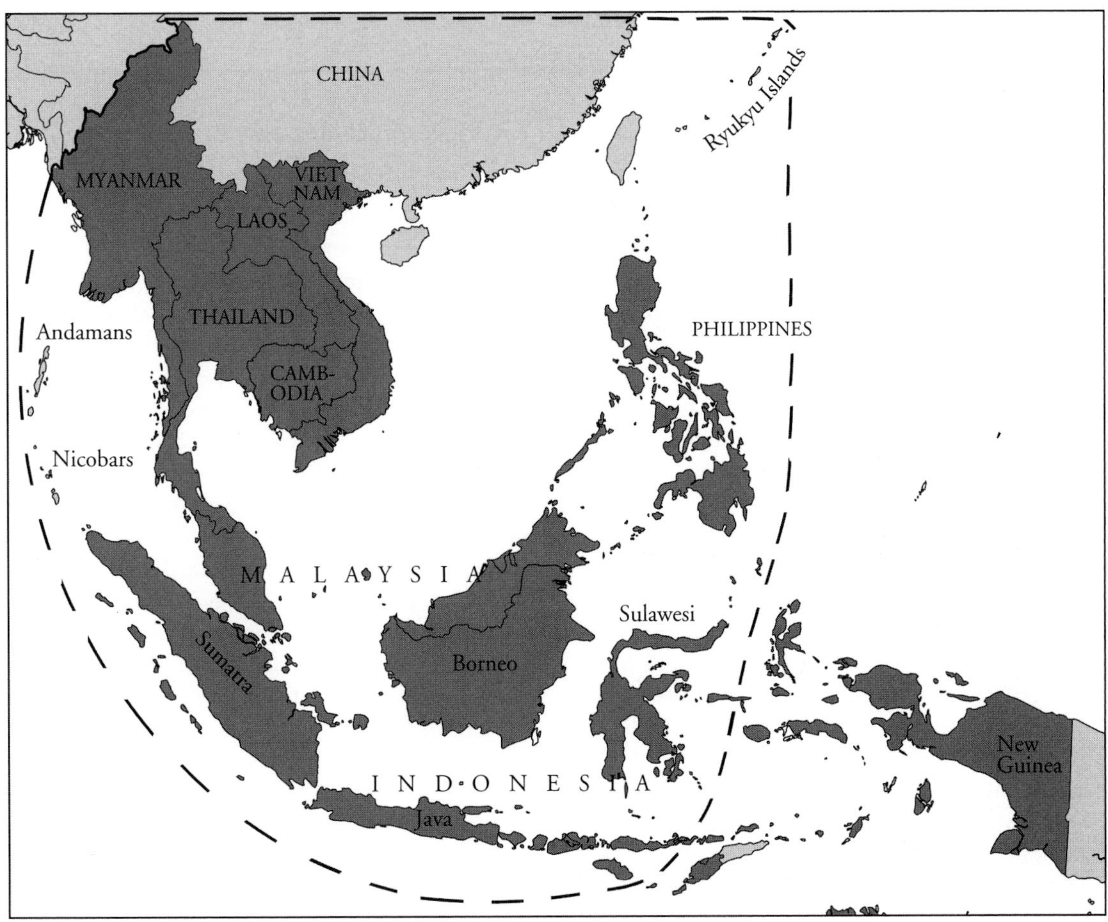

Figure 1.1 Map of the region showing Tropical East Asia as defined in this book (dashed line) and the member countries of ASEAN (the Association of Southeast Asian Nations) (dark grey).

often applied to the member countries of the Association of South-East Asian Nations (ASEAN), which excludes China, India, and Japan, and includes all of Indonesia, east to Papua. In this modern usage, South-East Asia is not a single, coherent biogeographical region, while Tropical East Asia is, even if the precise boundaries are somewhat arbitrary (see Chapter 3).

Tropical East Asia is almost all wet enough to support forest and these forests in turn support a flora and fauna that is distinctively tropical Asian (Corlett 2007c). Some of these distinctive elements drop out south of the northern border (30°N) chosen for this book, and some are missing from islands that have never had an overland connection with the Asian mainland, including most of the Philippines, Sulawesi, the Ryukyus, Andamans, and Nicobars. However, no significant part of TEA is dominated by plants and animals that are characteristic of other biogeographical regions.

The most arbitrary border is the western one. The western half of the Asian tropics is, in general, considerably drier than the east, and there are large areas that were never covered in closed forests. However, there are ecosystems that closely resemble those of TEA in parts of north-east India, Bangladesh, Nepal, and Bhutan, and as outliers in the Western Ghats of south-west India and in the wet zone of Sri Lanka. Inclusion of these areas would have made this book larger and geographically less coherent, so the India–Myanmar border was chosen as a convenient western limit.

1.2 Ecological vs. historical explanations

The primary goal of ecology is to explain the distribution and abundance of living organisms: to answer the question, 'Why is what where?'. Most such explanations can be grouped into two major categories: 'ecological' explanations relate the distributions of organisms to their present-day environments, while 'historical' explanations relate them to events that happened in the past. Each of these major types of explanation can, in turn, involve a wide range of possible factors. Ecological factors include climate, soils, and topography, as well as interactions with the other organisms present. Historical factors include the movements of tectonic plates, natural catastrophes, past changes in climates and sea-level, and a variety of human impacts.

Ecologists usually investigate ecological factors first and consider historical explanations only when the present-day environment proves insufficient to explain their observations. As a result of this approach, the importance of history in determining the present-day distributions and abundances of organisms may sometimes have been underestimated. Yet there are few other regions of the world that bear such a clear imprint of history on so many time-scales as TEA, so it makes sense to consider this history first. It is important to remember, however, that our knowledge of history is necessarily incomplete and that both the completeness and accuracy decrease the further back in time we go. The account given here is a simplified summary of the current consensus, but that does not mean that it is all true.

1.3 Plate tectonics and the origin of Tropical East Asia

Biologists have tended to view tectonic plates as 'rafts' that carry terrestrial and freshwater organisms across oceans, but this analogy can be misleading. Apart from the slow speed at which they move—typically 2–10 cm per year (about the speed fingernails grow)—tectonic plates float on the magma, not the ocean, and can become submerged under water, with fatal consequences for their biota. Excessive confidence in the ability of plate tectonics to explain modern distributions has caused some biogeographers to prefer tectonic explanations to alternatives, such as trans-oceanic dispersal. Dated molecular phylogenies, however, show that the modern distributions of many widespread taxa, particularly of plants, have arisen too recently for tectonic explanations to be possible (Pennington et al. 2006). Unfortunately, the extreme complexity of the plate-tectonic history of TEA creates a particularly strong temptation, since it can provide plausible explanations for almost any distribution pattern!

The whole of TEA, as defined in this book, is a giant jigsaw puzzle of continental fragments (Fig. 1.2) (Metcalfe 2005). In the Paleozoic (Table 1.1), 400 million years ago, the major pieces formed part

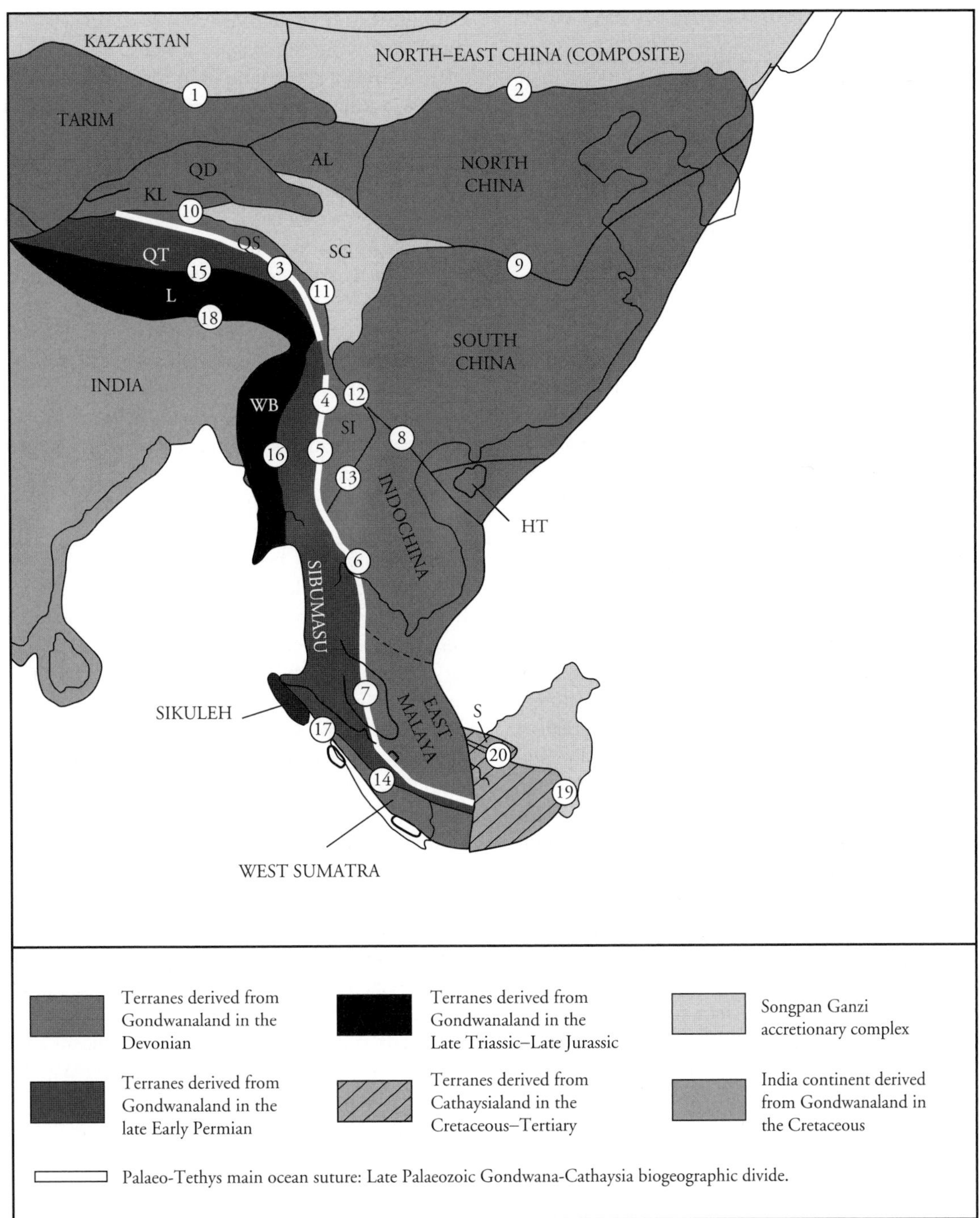

Figure 1.2 Tropical East Asia is a mosaic of continental fragments (terranes) rifted from the southern supercontinent of Gondwanaland over the last 400 million years. This diagram shows the major terranes, their approximate ages, and the boundaries (sutures) between them, with the numbers showing the order of assembly. Modified from Metcalfe (2005), which should be consulted for details.

Table 1.1 Simplified geological time-scale from the start of the Paleozoic era (ICS 2008).

Era	Period	Epoch	Age (Ma)
CENOZOIC	Quaternary	Holocene	0.012
		Pleistocene	1.8 (or 2.6)
	Neogene	Pliocene	5.3
		Miocene	23
	Paleogene	Oligocene	34
		Eocene	59
		Paleocene	65
MESOZOIC	Cretaceous		146
	Jurassic		200
	Triassic		251
PALEOZOIC	Permian		299
	Carboniferous		359
	Devonian		416
	Silurian		439
	Ordovician		488
	Cambrian		542

The 'Tertiary' comprises the Paleogene and the Neogene and has no official rank. The ages, in millions of years, are for the base of each unit. The base of the Quaternary has not yet been decided.

of the margins of the southern supercontinent of Gondwana. Between 350 and 140 million years ago, these fragments rifted from Gondwana and drifted northwards, as successive ocean basins opened between each fragment and Gondwana. The destruction and closure of these ocean basins by subduction then led to the progressive amalgamation of these continental fragments, so that by the end of the Cretaceous, 65 million years ago, the core of modern TEA had been formed, although there have been considerable changes in the shape since then. Fossil evidence shows that each fragment carried its own characteristic flora and fauna when it separated from Gondwana. However, few modern TEA lineages are old enough to have travelled on these fragments, so the Gondwanan geological origin of most of the region has had little direct influence on the modern biota. One possible exception is a South-East Asian endemic family of harvestmen (Stylocellidae, Opiliones), whose ancestors may have arrived from Gondwana on one of these fragments (Boyer et al. 2007).

While the major components of TEA were assembled by the end of the Cretaceous, two much larger Gondwanic fragments, India and Australia, were still heading north. Initially they were on separate plates and India moved much faster (*c.*21 cm/yr) than Australia. The timing of India's collision with Eurasia is still uncertain. The most widely accepted date is 55 million years ago. Aitchison et al. (2007) argue for a much later date, of only 34 million years ago, for the main India–Asia collision in Tibet, but suggest that a 'glancing contact' with Sumatra and then Burma may have occurred from around 57 million years ago, allowing for early biotic exchanges (Ali and Aitchison 2008). Although India is excluded from TEA as defined here, its collision with Eurasia had significant impacts on the shape, climate, and biogeography of the region.

Between 50 and 40 million years ago, the Indian and Australian plates became joined together and Australia began to move more rapidly, eventually colliding with the Philippine and Asian plates in the late Oligocene (*c.*25 million years ago). This ongoing collision has been responsible for much of the tectonic and volcanic activity in the region since then. Before the collision, East Asia and Australia were separated by a broad and deep ocean, allowing the largely unimpeded flow of warm waters from the Pacific to the Indian Ocean. The subsequent restriction of this flow coincided with, and may have caused, major changes in the regional and global climate systems (Kuhnt et al. 2004). Morley (2007) suggests that these changes included the origins of both the everwet climate of much of South-East Asia and the ENSO cycles (see 2.2.5) that periodically disrupt it. In contrast to the Gondwanic fragments that make up the Cretaceous core of TEA, both India and Australia carried modern plant and animal lineages on their northward journeys, and the biological interchanges that followed these plate collisions have made a significant contribution to the diversity of the regional biota at the family and genus level.

Although the core of TEA has changed relatively little since the Cretaceous, there have been big changes around the margins. What is now western Sulawesi was attached to eastern Borneo during the late Cretaceous and would have supported a typical Asian biota until the opening of the Makassar Straits in the late Eocene, 40 million years ago, created a deep-water barrier to further exchanges of organisms. Morley (1998) suggests that this Asian flora became a major source for the islands of eastern Indonesia, as they rose above sea-level in the Miocene. By the Miocene, the major pieces of Sulawesi had been assembled into approximately the arrangement we see today. Micro-continental fragments that joined on to eastern Sulawesi, from the Miocene to the Pleistocene, may have been above sea-level as they drifted towards Sulawesi, thus allowing island-hopping or rafting for plants and animals of Australian affinity (Moss and Wilson 1998) (see 3.9.10).

The Philippine islands have had an extremely complex history and there is still no fully accepted synthesis. The islands were formed by a variety of tectonic and volcanic processes in widely separated parts of the western Pacific, and were then progressively brought together over the last 25 million years (Hall 2002). Most have arisen from the sea and been colonized by overwater dispersal. Although they are now close together, many are still separated by deep channels that remained flooded during the lowest sea-levels (see below). Only Palawan and Mindoro contain continental crust, rifted from the margins of southern China by the opening of the South China Sea, and only Palawan may have been colonized across a terrestrial connection to Borneo (see 3.9.9).

By 10 million years ago, South-East Asia was largely recognizable in its present form and, from around 5 million years ago, it is reasonable to use the present-day geography of the region as a first approximation when discussing sea-level changes and the resulting bridges and barriers to the spread of organisms. However, it is important to remember that this continues to be a tectonically active region, particularly around the margins, so that vertical movements, large enough to have influenced the timing and duration of these bridges and barriers, are well within the bounds of possibility (e.g. Bird et al. 2006).

1.4 Sea-level changes

Because of the region's extensive shallow seas, changes in the global sea-level have been the major influence on both the total land area in TEA and the availability of dry-land connections between land masses over the last 3–5 million years. Fluctuations in global sea-level are caused by a variety of mechanisms, but changes in water volume as a result of ice-volume changes have dominated for the last few million years (Miller et al. 2005). From the Oligocene to the early Pliocene, global sea-levels probably varied within a relatively narrow range of 30–60 m, although brief wider fluctuations cannot be entirely ruled out. The global sea-level in the early Pliocene was around 25 m higher than today, submerging what are now coastal plains and deltas. It has been suggested that one or more sea-ways crossed the Isthmus of Kra, in southern Thailand, during this period, dividing Sundaland from mainland Asia (Woodruff 2003; see also 3.8), but this would probably have required a more extreme rise in sea-level than most evidence supports.

Large northern-hemisphere ice-sheets first appeared in the late Pliocene, around 2.5 million years ago, and subsequent fluctuations in their volume have resulted in variations in sea-level of up to 120 m. These changes in ice volume were driven by periodic variations in the Earth's orbit that change the amount, distribution, and timing of solar radiation reaching the Earth's surface. These orbital variations are, in turn, caused by gravitational interactions between the earth and the other planets in the solar system. The periodicity of the changes in ice volume and sea-level initially matched the 41,000-year cycle in the tilt of the Earth's axis with respect to the plane of the Earth's orbit (obliquity), but for last 800,000 years, a 100,000-year cycle has been dominant, for reasons that are still not fully understood (Claussen et al. 2007). Note that the complex interplay between orbital variations with different periodicities means that each glacial and interglacial period is unique.

These changes in sea-level periodically exposed large areas of land, almost doubling the land area of TEA at the lowest sea-levels (Fig. 1.3). The large Sunda Shelf islands of Borneo, Java, and Sumatra were connected to mainland Asia, along with many smaller islands, while Hainan and Taiwan were connected to southern China. To the east, large areas of the Sahul Shelf were also exposed, connecting Australia and New Guinea, but 'Sundaland' and 'Sahulland' were still separated by deep ocean water. Sulawesi and the other islands between the Sunda and Sahul Shelves remained isolated, as did

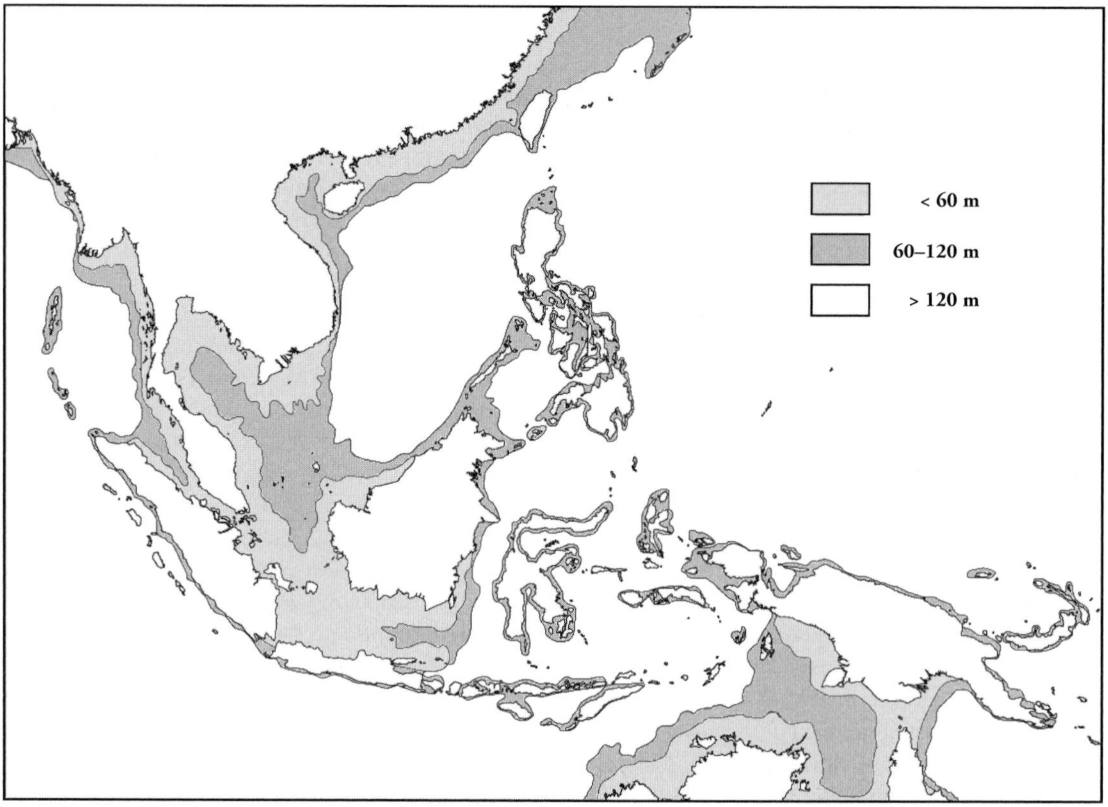

Figure 1.3 Additional land exposed by sea-levels 60 m below present (the average for the last 800,000 years) and 120 m below present (the lowest level reached at the last glacial maximum, 21,000 years ago). Note that the apparent dryland connection between Borneo and Palawan at low sea-levels was probably broken by deep-water channels too narrow to show on this map. From Hope (2005).

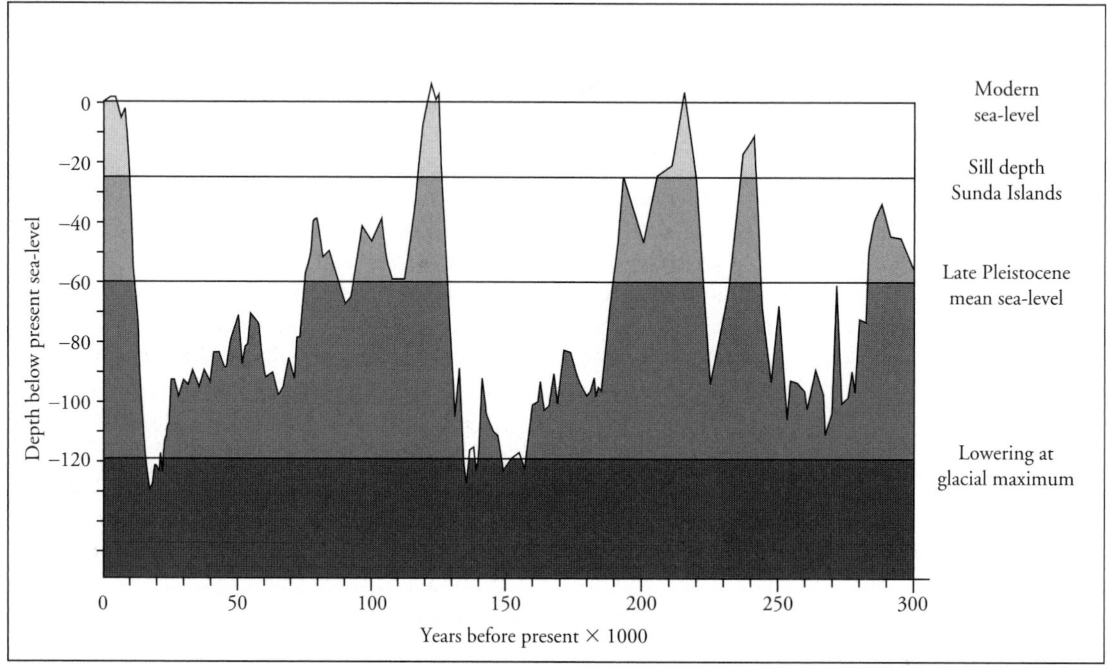

Figure 1.4 Sea-levels over the last 300,000 years relative to the present day. From Hope (2005).

those of the Philippines, although they coalesced into fewer, larger islands. The sea-level was at its lowest for only brief periods (Fig. 1.4), but the estimated average sea-level for the last 800,000 years was 60 m below that of today (Woodruff 2003), which would have been sufficient to connect all the major islands on the Sunda Shelf, as well as Hainan, to the Asian mainland.

The sea-level at the last glacial maximum, 21,000 years ago, was approximately 120 m below present. It rose to the present level by 6000 years ago, reached a high point of 4–5 m above present by 4200 years ago, and then fell back to the present level again by 1000 years ago (Sathiamurphy and Voris 2006). Most links to major islands were lost between 11,000 and 9000 years ago and the mid-Holocene 'highstand' would have flooded large areas of low-lying land, such as the Mekong and Chao Phraya deltas. Minimum sea-levels—lowstands—are important for explaining current plant and animal distributions (see 3.9), so the question of whether any of the previous Pleistocene glacial cycles produced lower sea-levels than at the last glacial maximum is an important one, on which there is no agreement at present. Currently, Marine Isotopic Stages (MIS) 10 (350,000 years ago) or 12 (430,000 years ago) seem most likely to have experienced significantly lower sea-levels (e.g. Rabineau et al. 2006).

1.5 Changes in climate and vegetation

The Earth's climate system is changing on all time-scales. On time-scales of tens of thousands to hundreds of thousands of years, much of the change is generated by the oscillations in the Earth's orbit mentioned above, leading to changes in the amount and distribution of solar radiation reaching the Earth's surface. These orbital oscillations cause climatic oscillations about a mean that varies on million-year time-scales, largely as a result of tectonic processes, such as changes in land distribution and topography, the opening and closing of oceanic gateways, and greenhouse-gas concentrations (Zachos et al. 2001).

Evidence for past climates comes largely from the archive formed by the gradual accumulation of marine, freshwater, and terrestrial sediments. A variety of physical indicators are available, including, in marine sediments, oxygen isotope ratios and Mg/Ca ratios in planktonic foraminifera (Wei et al. 2007), as well as unsaturated alkenones from marine phytoplankton (Bard 2001). The physical and chemical properties of lake sediments can provide information on past environmental conditions in the surrounding watershed, as well as trapping dust brought by wind from distant sources (e.g. Yancheva et al. 2007). The oxygen and carbon isotope records from stalagmites in caves are also valuable, especially because they can be accurately dated by Uranium–Thorium dating (Partin et al. 2007; Westaway et al. 2007; Wang et al. 2008).

For many ecological purposes, however, biological indicators can be more useful, since they integrate climatic variables in an ecologically meaningful way. For terrestrial ecology, the pollen and fern spores preserved in lake, swamp, and marine sediments are most important and can be used, with skill and experience, to reconstruct past vegetation and plant communities. Tree-ring records are another potential source of information for the last millennium (e.g. Buckley et al. 2007). The synthesis below is based on all the available types of data, but as the records become increasingly sparse with time back from the present, the inferences possible from them become increasingly broad and general.

A variety of evidence shows that there has been a long-term global cooling trend for the last 50 million years (Fig. 1.5). Everwet rainforest climates first became widespread in TEA about 23 million years ago, in the early Miocene, and extended furthest north about 15 million years ago, when tropical elements were found in fossil floras in Central Japan (Morley 1998, 2007). Coal formed in many areas of the Sunda region during this period, as peat-swamps became widespread. The appearance of humid forest vegetation across the south-eastern half of China in the early Miocene may mark the origin of the East Asian monsoon, which dominates the climate of the region today (Sun and Wang 2005). Before this, a broad band of arid, non-forest climates had stretched east–west across China, separating a narrow humid forest zone in the south from the temperate forests of the north-east.

This period also saw the first appearance of recognizably modern floras in the regional fossil record. Pollen records suggest a slow increase in

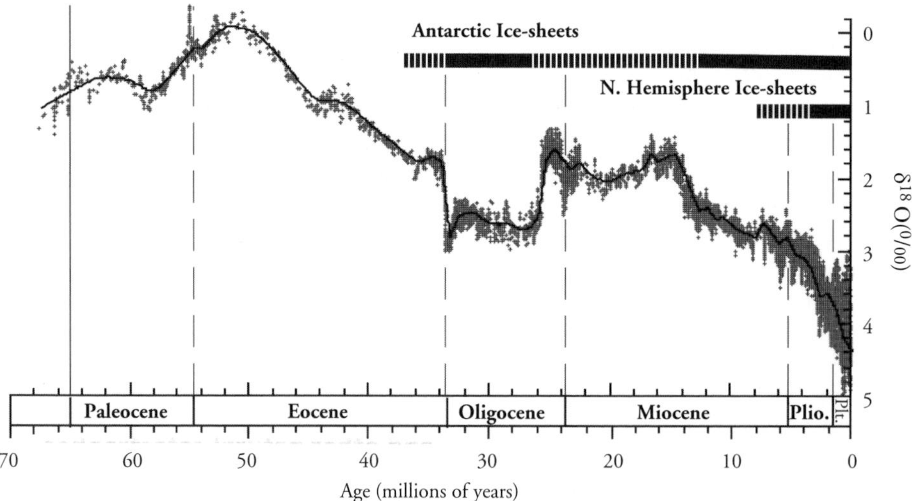

Figure 1.5 This composite oxygen isotope record from deep-sea foraminifera shows a long-term cooling trend from the early Eocene, punctuated by warm episodes. Note that after the formation of large continental ice-sheets, the record shows the combined effects of ice-volume and temperature. From Zachos et al. (2001).

the floristic diversity of rainforests since the early Miocene, but few major losses or additions, except at higher altitudes and latitudes, in response to post-Miocene cooling (Morley 1998, 2007). Both fossil and molecular data suggest that long-lived woody plants evolve very slowly (Levin 2000; Morley 2007), so it is not unreasonable to suggest that the Miocene forests of TEA were both floristically and functionally similar to those of frost-free parts of the region today. The component plants have expanded, contracted, and shifted their ranges in response to climate change, but there has been little turnover at the generic level. The same cannot be said of regional faunas, however, which have changed considerably since the Miocene.

The cooling trend continued, and the early Pliocene, 5–3 million years ago, was the last period that was consistently warmer than the present. Global average temperatures were around 3°C warmer than today at this time, although the difference was smaller in the tropics. Part of this extra warmth may have been contributed by a state of permanent El Niño in the Pacific Ocean. Harrison and Chivers (2007) speculate that this may have resulted in greater and more reliable rainfall seasonality in South-East Asia, with potentially major implications for flowering and fruiting patterns in the lowland dipterocarp forests (see 2.6.2).

As noted in the previous section, the global cooling trend culminated in the appearance of large northern-hemisphere ice-sheets from about 2.5 million years ago. Subsequent changes in the climate of TEA have largely followed the changes in ice volume and sea-level, although the match is by no means perfect, since regional and local climates are influenced by a wider array of factors than those that control global ice-volume. When the northern-hemisphere glaciers were at their maximum extent, global sea-levels and temperatures were at their lowest. In TEA, the replacement of large areas of shallow, warm, wet sea by cool, dry land also resulted in a general decline in rainfall, while the winter monsoon was strengthened and the summer monsoon weakened. This glacial–interglacial cycle was complicated by other cycles with different causes and periodicities. While ice volume may be the dominant influence on the strength of the Asian winter monsoon, high-resolution stalagmite records from China show that the strength of the summer monsoon follows 23,000-year cycles that closely match the precession-dominated cycles in northern-hemisphere summer solar radiation (Wang et al. 2008).

Most information is available for the most recent glacial cycle and, in particular, the period from the last glacial maximum, 21,000 years ago, to the present day. Most studies suggest that tropical sea surface temperatures were only 2–3°C cooler at the glacial maximum, while pollen data suggest up to 6°C cooling at lower altitudes on land and up to 8°C above 3000 m (Kershaw et al. 2006). There are also indications of lower rainfall from many sites. In general, the decline in temperature had most influence at high altitude, where the treeline was depressed and montane forest expanded downwards (e.g. Hope 2001), and towards the northern limits of the region, where tropical trees retreated and temperate deciduous species spread south (e.g. Mingram et al. 2004; Xiao et al. 2007). At lower altitudes and latitudes, the decline in rainfall seems to have had a more significant impact. Atmospheric carbon dioxide levels were also considerably lower (*c*.200 ppm) at the last glacial maximum than during the Holocene (*c*.280 ppm), but the consequences of this for tropical vegetation are not well understood (Cowling 2007).

There is pollen and other evidence for the persistence of areas of lowland tropical rainforest in Sundaland and Sulawesi at the glacial maximum (Gathorne-Hardy et al. 2002; Meijaard 2003; Hope et al. 2004), Indeed, it is possible that the total area was greater than in the Holocene, if there were substantial areas of rainforest on the exposed Sunda Shelf. However, a growing amount of evidence from genetic studies shows that free movement across the Sunda Shelf between the major islands has not been possible for rainforest organisms during most of the Pleistocene, and possibly not since the early Pliocene, three or more million years ago (e.g. Gorog et al. 2004; Harrison et al. 2006; Wilting et al. 2007). Although the flat terrain and large river systems of the exposed shelf could have acted as a barrier for some species, this evidence strongly suggests that the rainforest was fragmented by non-forest vegetation throughout the times of low sea-level. Borneo appears to have been particularly isolated, while movements between the Malay Peninsula, Sumatra, and Java may have been easier.

After a long period of warming, climate and vegetation approached their modern state by about 9500 years ago, soon after the rising sea-level had severed land connections between most of the major islands. There is evidence from many areas that the period between approximately 8000 and 5000 years ago was significantly warmer than today (Hope et al. 2004; Thompson et al. 2006; Liew et al. 2006), while in southern China, the earliest Holocene (*c*.9500–8000 years ago) seems to have been the warmest and wettest period (Wang et al. 2007a). Fluctuations in temperature and/or rainfall continued throughout the Holocene (e.g. Mingram et al. 2004; Li et al. 2006; Liew et al. 2006; Wang et al. 2007b; Xiao et al. 2007a), but, at many sites across the region, the interpretation of Holocene variability is complicated by human impacts.

1.6 Extraterrestrial impacts, volcanoes, and other natural catastrophes

In addition to the gradual changes described above, TEA has been subject to a number of catastrophic events, since its formation in the late Cretaceous, that have caused widespread destruction within a brief period of time. The biggest such event was undoubtedly the one that caused the global extinction of the dinosaurs and numerous other species of animals and plants at the end of the Cretaceous, 65 million years ago. Most scientists now believe that this mass-extinction episode was caused by the impact of a 10-km diameter asteroid at Chicxulub, Mexico, creating a crater 180 km in diameter (Kring 2007). Unfortunately, there is no known geological record for this period from TEA, so the regional impacts of this global catastrophe cannot be assessed. The global extinction of all large terrestrial vertebrates is well-documented—nothing bigger than a cat survived—but evidence from other sites remote from the impact suggests that plant extinctions were not as severe (Green and Hickey 2005; McLoughlin et al. 2008).

Extraterrestrial impacts have been blamed for other, smaller, global extinction episodes since the Cretaceous, but none of the correlations are, as yet, very convincing. Much more recently, a smaller impact in TEA itself appears to have caused regional devastation. High-velocity impact events produce glass bodies known as tektites, which are strewn over a large area. Four tektite-strewn fields are currently known, of which the Australasian one, from an impact 800,000 years ago, is the largest and youngest. Tektites and other debris from this impact were spread over 10–20% of the Earth's surface,

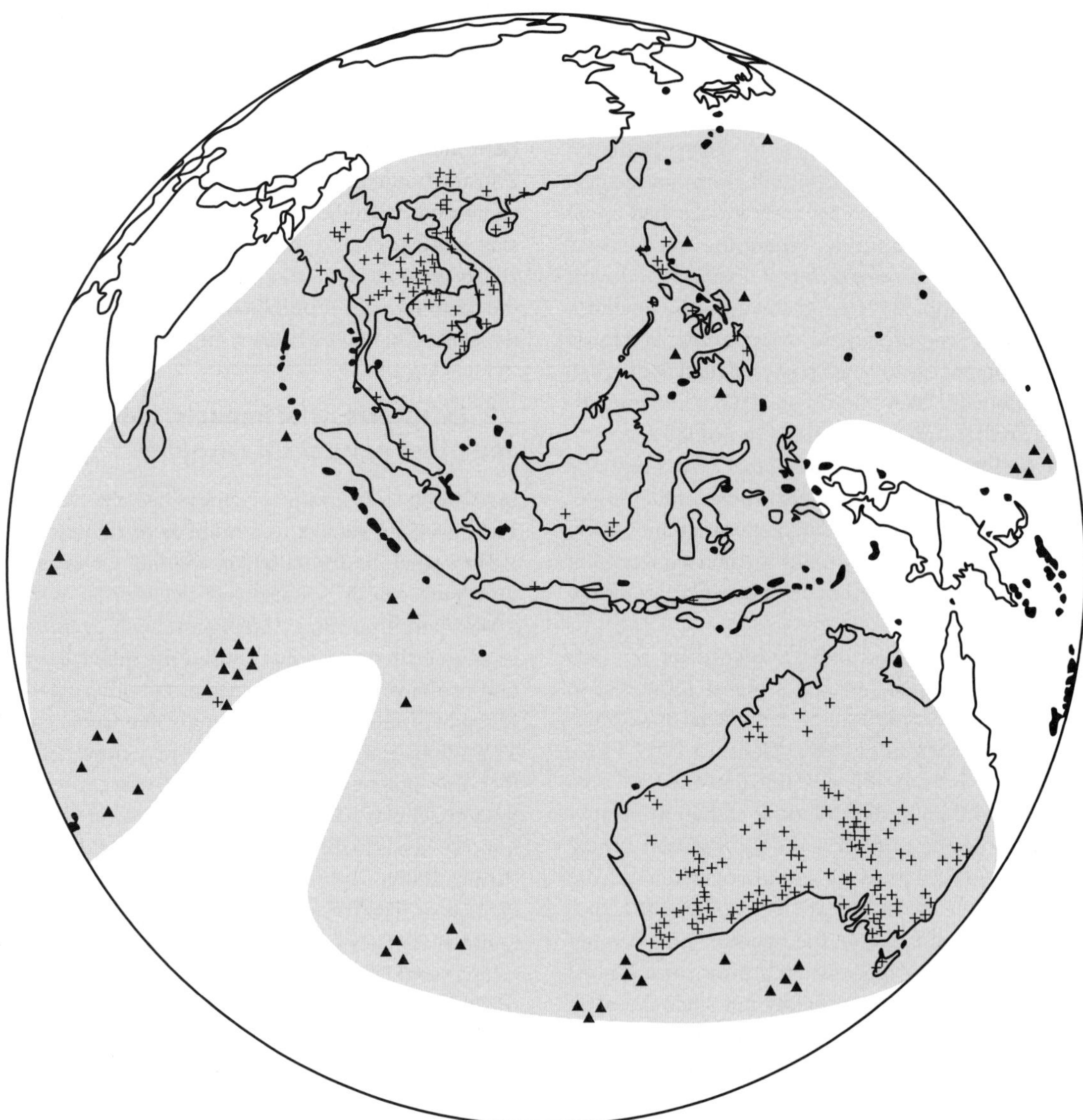

Figure 1.6 Map of the distribution of tektites (+) and microtektites (▲) produced by the low-angle impact of an asteroid somewhere in the Indochinese peninsula around 800,000 years ago. Courtesy of Aubrey Whymark.

from the Indian Ocean to Australia, South-East Asia, and South China (Fig. 1.6). These were apparently produced by the low-angle impact of an asteroid or comet, more than a kilometre in diameter, probably in Vietnam, Laos, Cambodia, or Thailand (Glass and Koeberl 2006; Prasad et al. 2007). No source crater has been found yet, but it is predicted to be around 40 km in diameter and may be buried under marine sediments. Haines et al. (2004) suggest that flood deposits with tektites in north-east Thailand reflect catastrophic forest destruction by the impact and, in Guangxi, tektites have been found in apparent association with stone tools and evidence of forest fire (Hou et al. 2000). Although an impact of this magnitude must have had massive regional consequences, it was probably too small to have a major global impact.

Large-impact events are extremely rare, but the destructive power of all but the largest impacts is matched by large volcanic eruptions—super-eruptions—which are much more frequent than impact events of similar energy (Mason et al. 2004). The complex tectonics of South-East Asia is reflected in the large number of active volcanoes. Indonesia has more than any other country in the world, mostly in the Sunda Volcanic Arc, stretching from north-west Sumatra to the Banda Sea, which results from the subduction of Indian Ocean crust beneath the Asian plate (Fig. 1.7).

The largest eruption in human history was that of Tambora in 1815, on the island of Sumbawa, which was heard 2600 km away in Borneo and killed at least 50,000 people. Other big eruptions in historical times in TEA were Krakatau (1883) and Agung (1963–64), in Indonesia, and Pinatubo (1991), in the Philippines, which was the largest eruption in the twentieth century. An even larger eruption of the Kikai caldera in the Ryukyus, *c*.6300 years ago, destroyed forest and disrupted human societies in southern Kyushu (Machida and Sugiyama 2002). It may also have caused extinctions on Yakushima Island, 40 km to the south-east, as well as a genetic bottleneck in the surviving macaques (Hayaishi and Kawamoto 2006). An earlier eruption on Flores may have wiped out both the dwarf hominids, *Homo floresiensis* (if this is what they were—see below) and the last of the elephant-like *Stegodon* in South-East Asia, around 12,000 years ago, although climate change and the arrival of modern humans (see below) provide alternative explanations (van den Bergh et al. 2008a, b). The disappearance of an earlier species of stegodont on Flores, around 900,000 years ago, also coincided with both a volcanic eruption and the earliest appearance of stone artefacts (van den Bergh et al. 2008b).

The largest historical eruptions devastated hundreds of square kilometres and had measurable impacts on global climate (D' Arrigo et al. 2009), but none were really large by geological standards. The last two really big super-eruptions were the Oruanui eruption in New Zealand, 27,000 years ago, and the Toba eruption on Sumatra, 74,000 years ago. This latter eruption—known as the Younger Toba eruption, to distinguish it from earlier eruptions of the same volcano—lasted perhaps two weeks and deposited metres of ash over large areas. Simulations with climate models suggest that emissions of sulphate aerosols from the eruption may have caused global cooling by as much as 10°C, with temperatures remaining several degrees below normal for more than a decade, as well as large reductions in rainfall (Jones et al. 2005). However, there is also evidence against a massive regional impact, including the lack of extinctions in the fossil record (Louys 2007), the survival of nine, rainforest-dependent, endemic mammals, and a diverse termite fauna on the isolated Mentawai Islands, 350 km to the south (Gathorne-Hardy and Harcourt-Smith 2003) (see 3.9.8), and the persistence of human populations in an area of southern India that was blanketed with volcanic ash (Petraglia et al. 2007). A sediment core from the southern South China Sea, north of Toba, shows an abrupt drop of 1°C in the sea-surface temperature immediately above the volcanic ash layer, which then persisted for *c*.1000 years (Huang et al. 2001). A variety of other, less well-dated, records suggest that the eruption may have switched the global climate into a cooler state, but with the maximum climatic impact at higher latitudes. Toba has had previous super-eruptions in the last few million years, as have several other volcanoes around the globe, so these events are not unusual on a geological time-scale.

Other types of natural catastrophe, including large earthquakes and major cyclones, can cause local or regional devastation, but few species are completely eliminated from the affected sites, so recovery is rapid. Most of the largest earthquakes occur at 'megathrust faults', where one tectonic plate slides beneath another, and the Sunda megathrust, where the Indo-Australian plate is subducted beneath the Eurasian plate, is particularly active. The Sumatran–Andaman earthquake of 26 December 2004 was the largest worldwide in the last 40 years and produced the most devastating tsunami in recorded history, killing >200,000 people; but ecological damage was localized. Further north, the Sichuan earthquake of 12 May 2008 killed >80,000 people and caused

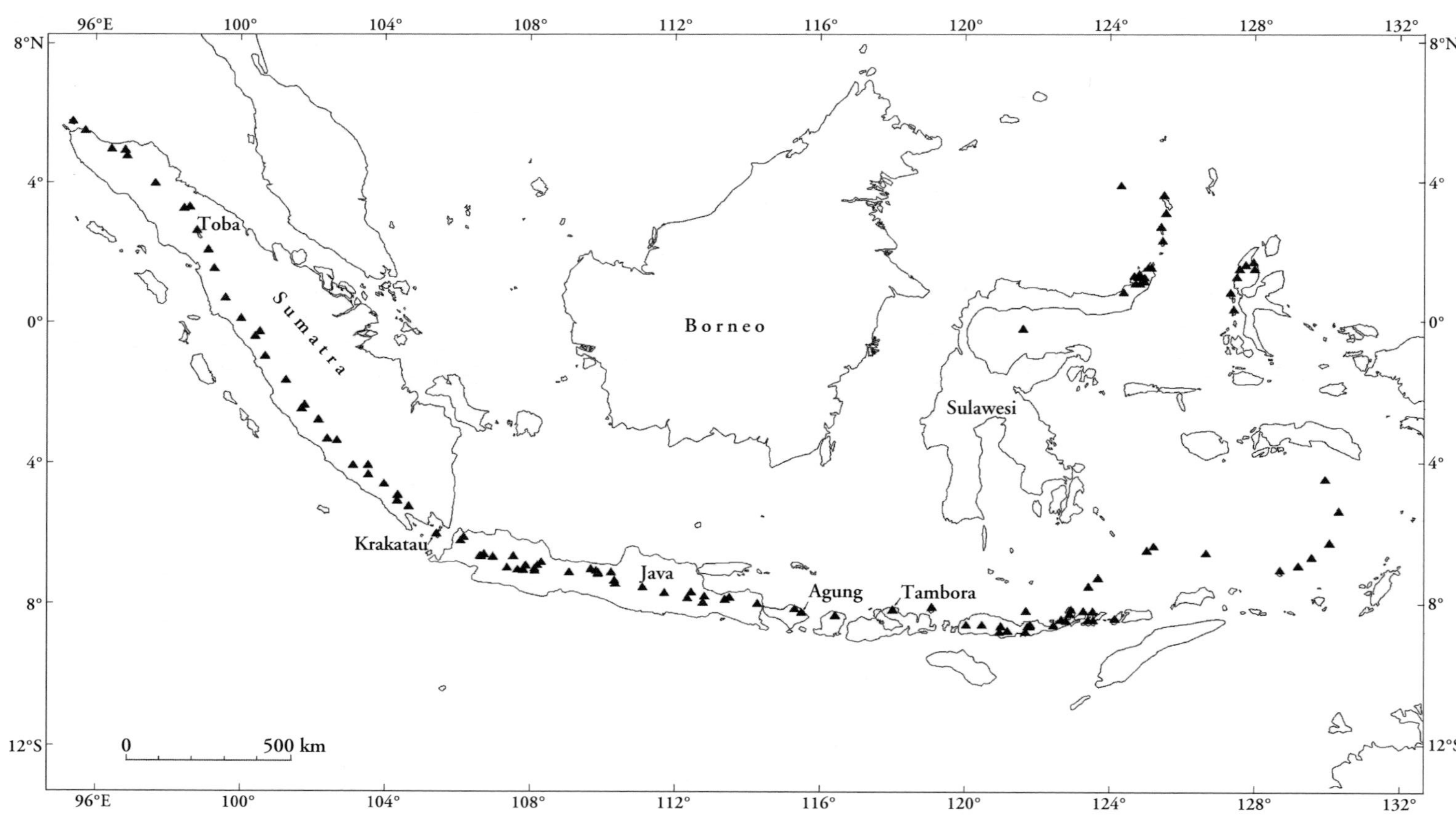

Figure 1.7 Indonesia has more active volcanoes than any other country in the world. Most of these are in the Sunda volcanic arc, stretching from north-west Sumatra to the Banda Sea, which results from the subduction of the Indo-Australian plate under the Eurasian plate—the Sunda megathrust. From Nossin (2005).

extensive forest loss on the steepest slopes (Ouyang et al. 2008).

1.7 The first humans

The best evidence for the presence of prehistoric people in an area is human bones or tools in a securely dateable context. Unfortunately, the archaeological record for TEA is scattered thinly in time and space, and many of the dates are disputed. Alternative indicators of human presence, such as evidence for fire, changes in vegetation, or the extinction of the animals that were most vulnerable to hunting, are therefore useful, but these indicators also have their problems. More recently, genetic evidence has become useful for distinguishing between alternative theories of human origins and dispersal. When all this evidence is assembled, a fairly coherent picture emerges, but there still remain scattered pieces of evidence that do not (yet?) fit in. As with other aspects of the history of TEA, the current majority view is not necessarily correct.

The earliest human remains in TEA are assigned to the species *Homo erectus*. Stone tools found without human bones, from the early and middle Pleistocene, are also usually assigned to the same species, but only because there is no evidence for any other human species in the region at that time. The archaeological record (bones and tools) is sparse and both the earliest (1.8–1.9 Ma) and latest (*c*.50,000 years ago; Yokoyama et al. 2008) dates claimed for *H. erectus* in TEA are disputed. There is virtually no information on the subsistence patterns of this species, and thus the potential impacts on the environment and on other species. Most skeletal remains are from Java, 1.0–1.8 million years ago, but the earliest securely dated stone tools are from Mata Menge on the island of Flores (Morwood et al. 1998) and the Bose Basin in South China (Hou et al. 2000), only 800,000 years ago. A recent study identified cut marks made by clamshell flakes on 1.5–1.6 million-year-old large mammal bones in Java as the work of *H. erectus*, and the use of shells or bamboo might explain why evidence for stone tools is lacking at this site (Choi and Driwantoro 2007). Most of the faunal remains associated with *H. erectus* finds suggest open woodland or savanna, and it is not clear if the species could live in closed forest.

Another unknown is the ability of *H. erectus* to manipulate fire. If, as has been suggested for East Asian populations, *H. erectus* had this ability by 400,000 years ago, then this would have greatly increased the power of this species to modify its environment. There was an increase in charcoal in a marine core from a site to the south of the Lesser Sunda islands from 225,000 years ago, which would be consistent with this, although there are alternative explanations (Kershaw et al. 2006). There is also the question of how *H. erectus* reached Flores, assuming that this is the species responsible for the tools there. Even at the lowest of Pleistocene sea-levels, reaching this island required three sea-crossings, the longest of at least 19 km (Morwood et al. 1998). Some authors have attributed these crossings to groups of humans accidentally washed out to sea on floating rafts of vegetation (as happened to several individuals during the Indian Ocean tsunami in 2004), while others have found deliberate seafaring more plausible. Along the chain of islands from Bali to Timor, the target island can always be seen from the likely point of departure on the previous one, while the dangers of crossing these treacherous straits may not have been so obvious. If these early humans were sea-farers, what else were they capable of? (Bednarik 2001).

Later Asian populations of *H. erectus* appear to have diverged increasingly from their presumed African ancestors, and a minority of experts believes that one or more of these populations ultimately gave rise to the *H. sapiens* populations of Asia (the 'Regional Continuity' model), rather than, as most evidence suggests, that they were displaced by modern humans who evolved in Africa (the 'Out of Africa' theory). Whichever view is correct, mainland East Asia was occupied between around 300,000 and 70,000 years ago by people who may have represented evolutionary developments from local *H. erectus* ancestors and/or archaic forms of *H. sapiens* and/or additional *Homo* species (Bacon et al. 2006).

Homo erectus in TEA was associated, at various times and places, with large vertebrates, such as the elephant-like *Stegodon* species (Fig. 1.8), which no longer survive in the region. Early Pleistocene extinctions of large vertebrates on Flores are associated with the first evidence for the arrival of humans, presumably *H. erectus* (van den Bergh et al. 2008b),

Figure 1.8 Skeleton of *Stegodon trigonocephalus* from the mid Pleistocene of central Java in the Bandung Geology Museum. Photograph courtesy of Gert van den Bergh.

and there are similar—but less clear—associations between extinct megafauna (including elephants, *Stegodon*, and rhinoceroses) and early stone tools in the Philippines (Bautista 1991). Human fossils and/or stone tools from Middle and Late Pleistocene cave-sites in southern China and Indochina are also often associated with the remains of extinct, large-bodied mammals, including *Stegodon* and the giant tapir *Megatapirus augustus*, (Ciochon and Olsen 1991; Bekken et al. 2004; Schepartz et al. 2005). Early *Homo* was also associated at some sites with *Gigantopithecus*, a giant terrestrial ape (Ciochon et al. 1996; Harrison et al. 2002). These associations suggest some form of interaction, but there is no direct evidence for hunting. It is hard to avoid speculation that early human populations had some role in the extinctions of these large mammals, but more evidence is needed before this can be taken further. Indeed, *Stegodon* and *Megatapirus* apparently survived in South China into the Holocene, when their final demise can be blamed on our own species (Tong and Liu 2004; Louys et al. 2007; Louys 2008).

On the island of Flores, just east of Wallace's line, a remarkable dwarf (*c.*1 m) hominid, with a brain the size of a chimpanzee's, inhabited a cave from, perhaps, 95,000–74,000 to around 12,000 years ago (Morwood et al. 2005). These remains are associated with evidence for hunting and/or scavenging of stegodonts, Komodo dragons, and a range of smaller species (van den Bergh 2008b).The widely divergent opinions on these hominid fossils range from them representing a distinct human species, *Homo floresiensis*, to their being diseased individuals of *Homo sapiens* (e.g. Obendorf et al. 2008). Further discussion of the Flores 'hobbits' must await new evidence.

1.8 The arrival of modern humans

Despite a number of claims to the contrary, there is no clear archaeological evidence for anatomically and culturally modern humans—*Homo sapiens* in the narrow sense—in Eurasia before around 45,000–50,000 years ago. This fits with the growing amount of archaeological, genetic, pollen, and charcoal evidence in support of a single dispersal event that brought modern humans from Africa, where they had emerged 150,000–200,000 years ago, along the coastlines of South and South-East Asia, to New Guinea and Australia, within a period of a few thousand years (Kershaw et al. 2006; Mellars 2006; Pope and Terrell 2008) (Fig. 1.9). Human arrival in Australia and New Guinea occurred by 45,000 years ago, and the only significantly earlier date from South-East Asia comes from an initial peak in charcoal around 51,000 years ago from a marine core in the Sulu Sea between Borneo and the Philippines (Beaufort et al. 2003). This 'coastal express train' model of modern human dispersal assumes that the pioneer populations lived initially on coastal resources, moving on as they became depleted, and only later moved inland. It is important to note, however, that alternative theories, involving two or more migrations out of Africa and different routes, can also be made compatible with much of the same evidence, as can the Regional Continuity model, which requires no such migration.

Human arrival in Australia coincided with a continent-wide disappearance of large vertebrate species (>45 kg), with a human role in these extinctions

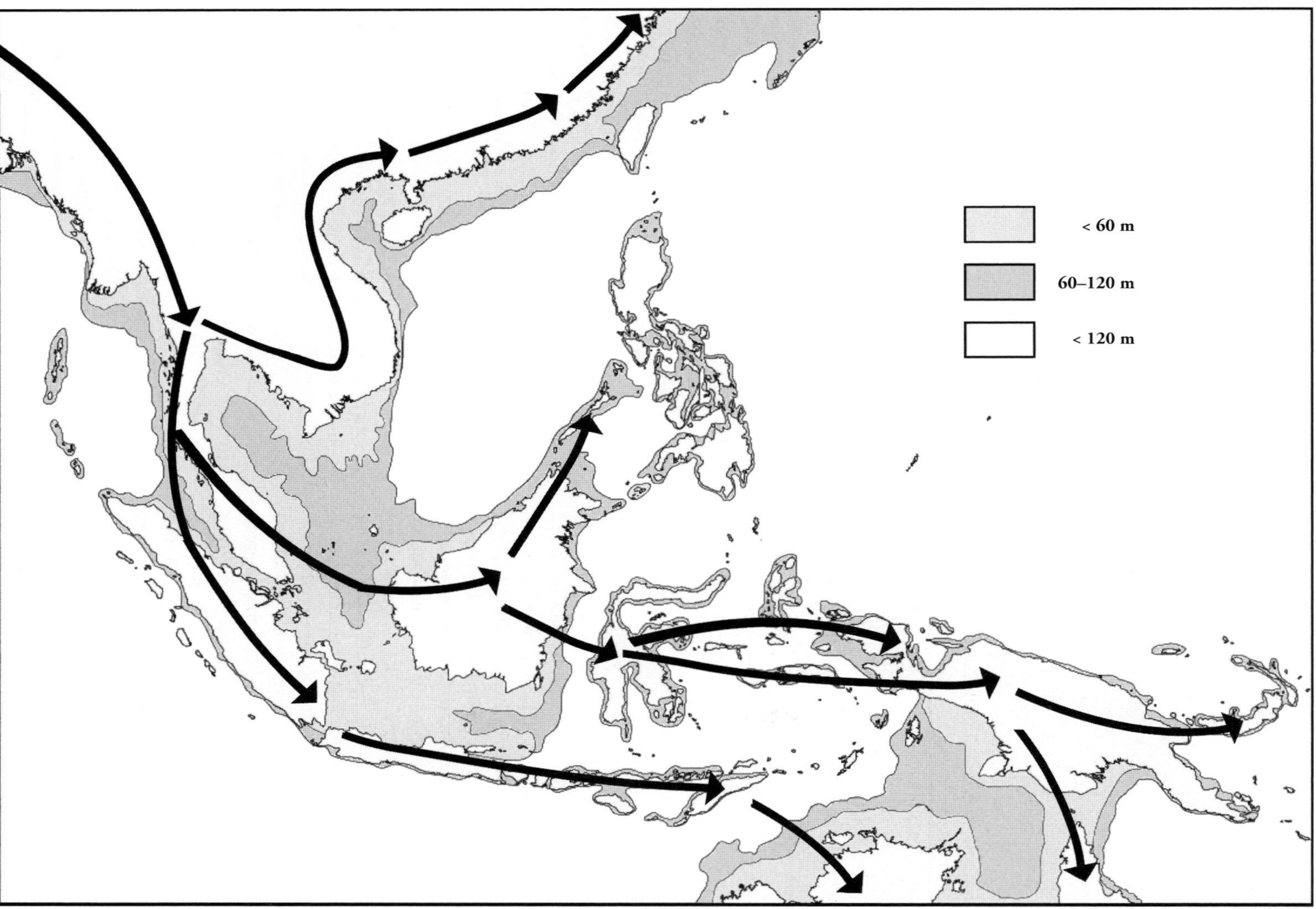

Figure 1.9 Possible routes by which modern humans, *Homo sapiens sapiens*, spread through Tropical East Asia between 50,000 and 40,000 years ago. The shading shows the land exposed by sea levels 60m and 120m below the present day. From various sources.

likely (Roberts et al. 2001). There was no such mass extinction event in tropical Asia in the 60,000–40,000 BP period during which modern humans probably arrived, perhaps because one or more species of *Homo* had been living in the region for >1 million years already. However, the last 40,000 years has seen at least four global extinctions in TEA: the giant pangolin, *Manis palaeojavanica*; the giant tapir, *Megatapirus augustus*; and at least two late-surviving species of *Stegodon*; as well as progressive range-restrictions for several other large mammals, including orangutans, rhinoceroses, tapirs, and elephants (Corlett 2007a; Louys et al. 2007; Louys 2008; Morwood et al. 2008). The extinction of the giant pangolin may reflect the loss of open habitats with a high density of accessible ant and termite nests (Medway 1972), but a slow-moving, 2-m long mammal, whose only defence was to roll up, must also have been extremely vulnerable to hunting. A similar argument can be made for the slow-moving, slow-breeding orangutans, which ranged widely over Tropical East Asia in the late Pleistocene, from southern China to Java, but were confined to the sparsely populated everwet rainforests of Borneo and Sumatra by historical times (Delgado and Van Schaik 2000). Others, however, doubt that human population densities, prior to the Holocene, were high enough to lead to faunal extinctions (Boomgaard 2007).

Both the earliest modern human skull and the longest record of human occupation in the region come from Niah Cave, Sarawak, on the island of Borneo, which was occupied from at least 45,000 years ago—possibly as early as 50,000 years ago—to around 8000 years ago (Krigbaum 2005; Barker et al. 2007; Hunt et al. 2007). The bones of at least 30 species of mammals, brought into the cave by people, have been identified (Medway 1972). Bearded pigs (*Sus barbatus*) dominate at all depths, suggesting that this was the most common prey. Monkeys of various species (macaques and several langurs) are the second most abundant bones overall, but show an abrupt increase from around 20,000 years ago. An increase in tree squirrel bones at the same time suggests that this reflects an increased ability to target arboreal prey, although terrestrial prey continues to dominate. Other mammalian prey include porcupines and other rodents, orangutans, deer, cattle, a variety of small carnivores, pangolins, and the occasional gibbon, Sumatran rhinoceros, tapir, and sun bear (Medway 1979). Non-mammal remains include monitor lizards, snakes, turtles, tortoises, birds, fish, and molluscs.

Other cave sites in Tropical East Asia present a broadly similar picture, with pigs often the commonest prey, although deer, cattle, or monkeys dominate at some sites (e.g. Cranbrook 1988; Higham 1989; Bellwood 1997; Shoocongdej 2000). On Palawan, deer dominated until their decline and eventual extinction in the mid-Holocene, when they were replaced by pigs (Lewis et al. 2008). At some Javan sites, a focus on hunting macaques is evident from around 12,000 years ago (Morwood et al. 2008). There is also evidence for exploitation of a wide range of plant foods, including some that are toxic without careful preparation (Barker et al. 2007; Barton and Paz 2007).

Several groups of nomadic hunter-gatherers survived into the twentieth century in TEA, including some that appear to be survivors from pre-agricultural times and others who appear to have given up the agricultural component of an earlier mixed economy (Bellwood 1999; Endicott 1999a; Oota et al. 2005). Some hunter-gatherer populations in the Andaman Islands and the Peninsula Malaysia harbour mtDNA lineages that suggest they can trace their maternal ancestry back to the first modern humans in the region (Macaulay et al. 2005; Hill et al. 2007).

The most widespread hunting technique among recent hunter-gatherers is hunting pigs and deer with dogs and spears. Domesticated dogs are absent from pre-agricultural archaeological records in the region, but were widespread in Neolithic China and may have spread through South-East Asia 6–5000 years ago (Savolainen et al. 2004). Some groups use blowpipes and poisoned darts to hunt arboreal monkeys, gibbons, squirrels, and civets (e.g. Kuchikura 1988, Endicott 1999b). Porcupines, bamboo rats, and other rodents, as well as pangolins, are smoked or dug out of burrows. A variety of traps and snares are also used by different groups. Although there have been suggestions that hunter-gatherers cannot survive in closed tropical forests without trading for carbohydrates (Bailey et al. 1989), the evidence from South-East Asia contradicts this (Endicott 1999a). However, almost all hunter-gatherer groups in historical times have maintained trade relationships with the farmers who colonized most of the lowland, coastal, and riverine areas of the region over the last 5000–3000 years (Morrison and Junker 2002).

1.9 The spread of agriculture

The Neolithic revolution has traditionally been defined by the presence of both pottery and agriculture, but in East Asia, pottery first appeared much earlier than agriculture—at least 13,000 years ago (Kuzmin 2006). Pottery implies a more sedentary human lifestyle, but the consequences for the ecological impacts of human populations are unknown. Agriculture started in East Asia soon after the climate improved at the start of the Holocene, between 9000 and 8000 years ago. Rice appears to have been cultivated first in the lower and middle Yangtze Valley, in eastern China, on the northern margins of TEA (Zong et al. 2007). The archaeological record suggests a gradual transition from the intensive harvesting of wild rice, through cultivation of wild rice varieties, to full domestication, over a period of several thousand years; but it is not always clear which stage is represented at each site.

The earliest well-documented record of rice cultivation comes from Kuahuqiao (30°N), in the lower Yangtze region, where coastal wetland scrub, dominated by alder (*Alnus*), was cleared by fire 7700 years ago for the cultivation of wild rice (Zong et al. 2007). These early farmers still collected wild plants and hunted wild animals, but agriculture requires a fixed base and allows higher population densities, so prey are more likely to be overhunted. There are also bones of domestic dogs and pigs from the same site and time period. Dogs were first domesticated in the late Pleistocene by hunter-gatherers (Savolainen et al. 2002), but the domestication of pigs presumably required a sedentary population and so is unlikely to have preceded the start of agriculture (Jing and Flad 2002).

Pre-historians have used the evidence from archaeology, linguistics, and the pollen record in sediments to put forward models for the spread of agriculture into southern China and South-East Asia (Bellwood 2005). For island South-East Asia, these models have taken the form of a 'two-layered' settlement process, whereby the original hunter-gatherer settlers of the region were largely absorbed and replaced by a later influx of Austronesians, around 4000 years ago, who brought pottery, rice cultivation, domesticated pigs and dogs, and other Neolithic innovations, and are the ancestors of most of the indigenous inhabitants today. Linguistic evidence points to Taiwan as the possible home of the Austronesian languages, so the 'Out of Taiwan' model envisages rice cultivation—and rice-cultivating human populations—spreading from south-eastern China to Taiwan and then to the Philippines, throughout island South-East Asia, and on into the Pacific (Bellwood 2005).

The first appearances of pottery on various islands are, in general, supportive of this model, with dates ranging from around 5300 years ago in Taiwan, 5000–4500 in the Philippines, and 4000–3000 in the rest of island South-East Asia (Spriggs 2003). However, human genetics do not support a simple two-layered model for the populations of the region (Hill et al. 2007a), the linguistic evidence can be interpreted in other ways, and pig genetics do not support an 'Out of Taiwan' model for the dispersal of domesticated pigs (Larson et al. 2007). Despite the attractiveness of this and similar models for the spread of agriculture in the region, the scattered evidence for early agriculture in TEA suggests that the final picture will be much less tidy.

The timing of the spread of agriculture south from the Yangtze Valley is still unclear. Rice cultivation had spread to the margins of the tropics in southern China by around 5000 years ago (Bellwood 2005; Dodson et al. 2006; Xiao et al. 2007a), but contemporary sites from further south lack clear evidence for either rice or other forms of agriculture. Most evidence from continental South-East Asia comes from the last 4000 years, but there are suggestions of widespread clearance from around 6400 years ago in north-east Thailand, which could be agricultural (White et al. 2004a). In island South-East Asia, the earliest rice remains from cave deposits date from around 4000 years ago in both Borneo and Sulawesi (Paz 2005). Rice is less suitable for the equatorial zone, so root crops (including taro, *Colocasia esculenta*, and yams, *Dioscorea* spp.) are likely to have been important from the start (Boomgaard 2007). Note also that genetic evidence suggests rice was domesticated from wild *Oryza rufipogon* at least twice, in separate areas, with the *japonica* form originating in south-central China and the *indica* form in India, Myanmar, or Thailand (Londo et al. 2006). If rice cultivation diffused from two or more centres of origin, it would further complicate the picture given above.

A striking feature of non-rice agriculture in TEA in historical times is the importance of American crops, both for subsistence (especially maize and sweet potato) and as cash crops (such as tobacco and later rubber). These became established in the region from the sixteenth century onwards and were particularly important in extending intensive agriculture on to land that was unsuitable for rice (Ho 1955; Boomgaard 2007).

More important ecologically than the first appearance of agriculture is its intensification. Rice was grown, initially, in naturally marshy areas, leaving the dry, upland landscape untouched; but other crops and other cultivation systems, including shifting cultivation, encouraged the use of all but the steepest slopes. From an ecological viewpoint, the most important transition is that from a forest landscape with small pockets of agriculture to an agricultural landscape in which any surviving natural ecosystems are all heavily exploited. In TEA, this transition is best documented in China, where it has spread south and west through the subtropics and tropics over the last 2000 years. The same process happened elsewhere in TEA, in areas of high human-population density, particularly the plains and deltas of the major rivers, including the Irrawaddy, Chao Phraya, Tonle Sap, Mekong, and Sông Hóng (Red River). Even today, these areas stand out on satellite photographs because of the completeness of forest clearance. However, the transformation was nowhere as widespread or as well-documented as in China. Indeed, for much of the region outside China, as well as for parts of tropical China, this transition from a largely forested to largely agricultural landscape has occurred only in the last 150 years, and in many places, only in the last 25 years. South-East Asia was still around 80% forested in 1900 (Flint 1994), and both Thailand and the Philippines, the countries with the least forest cover today, were still more than half covered in forest in 1950.

1.10 Hunting

Forest cover can, however, be a misleading indicator of the pervasiveness of human impacts. Hunting (and burning: see below) can spread the impact of relatively small farming communities over a much larger, uncultivated area. Traditional hunting practices by low-density human populations were sustainable, in that they provided a continued supply of protein, but this does not mean that they did not have drastic effects on the density and species composition of prey communities. Where hunter preferences and prey vulnerabilities coincide, as is often the case with large, terrestrial, slow-breeding mammals, then local extinction is likely (Corlett 2007a). Increasing human-population density, and/or the demand from trade networks, can expand and merge these local extinctions into regional and finally global losses. The possible impacts of pre-agricultural populations on the large mammals of the region have been considered already, although the discussion was necessarily speculative. More recent impacts are much better documented.

In China, the period of agricultural expansion was marked, in both archaeological and written records, by the disappearance of vulnerable large mammals. Elephants (Sun et al. 1998; Elvin 2004), rhinoceroses (Liu 1998), gibbons (Van Gulik 1967), and snub-nosed monkeys (Li et al. 2002) progressively disappeared from central and southern China over the last 2000–3000 years, as a result of a combination of hunting and habitat destruction. Rhinoceroses (one or two species; Fig. 1.10) were particularly sensitive, with the northern boundary of their distribution retreating southwards at a rate of around 0.5 km per year (Liu 1998). Historical data suggest a crude threshold value of around 4 people km^{-2} above which rhinoceros populations did not persist, compared with around 20 people km^{-2} for elephants (Sun et al. 1998). While the relative importance of habitat destruction and hunting is unclear, rhinoceroses were undoubtedly hunted, both for their horns and for their hides, which were used to

B60B1+

Figure 1.10 Rhinoceroses (one or two species) were once widespread in central and southern China. This bronze ritual vessel in the shape of a Sumatran rhinoceros is from Shandong Province, probably 1100–1050 BCE. The Avery Brundage Collection, B60B1 © Asian Art Museum of San Francisco. Used with permission.

make the armour worn by early Chinese soldiers (Elvin 2004), while almost all parts of an elephant are of value. Rhinoceroses are now extinct in China, while elephants, gibbons, and snub-nosed monkeys are reduced to a few tiny populations.

No other part of the region has as good historical records as China, but the records that do exist suggest that, while the most vulnerable species (elephants and rhinoceroses, in particular) already had restricted distributions more than a century ago, the decline of most species of large mammals has occurred largely within the last 50–100 years (Corlett 2007a). Today, there is probably no site in TEA that still supports an intact fauna of large mammals living at natural densities (see 7.4.8).

1.11 Burning

A rise in charcoal levels is one of the most consistent indicators for the spread of modern human populations through TEA (Kershaw et al. 2006) and the use of fire continues to be one of the major ways in which people expand their impact beyond the area they actually inhabit and cultivate. Fires can occur naturally, as shown by the presence of charcoal in sediments long before human arrival, but human-controlled fires are distinguished by their much higher frequency (see 2.3). Fire is used for a variety of purposes and these cannot usually be distinguished in sediment charcoal records because changes in the frequency, intensity, or area burnt can all lead to similar increases in charcoal levels. Farmers use fire to clear land for cultivation (Fig. 1.11)—a practice that has expanded to disastrous levels when used to clear rainforest for plantations of cash crops over the last two decades, particularly in Indonesia (see 7.4.9).

In the drier parts of TEA, much larger areas of relatively open forest are burned annually by hunters and collectors of non-timber forest products in order to remove dead plant material and stimulate growth of grasses. These ground-layer fires consume largely grass and fallen leaves, but undoubtedly help maintain large areas of forest in a more open and more deciduous state than would occur naturally (see 2.5.1). If, as was suggested above, much

Figure 1.11 Cutting and then burning is the main method of forest clearance for agriculture, in this case for rice in Central Kalimantan. Photograph © Natalie Behring-Chisholm, Greenpeace.

of TEA was covered in non-forest or open-forest vegetation at the last glacial maximum, then similar burning practices may have been used by hunter-gatherer populations with similar consequences.

1.12 Urbanization

There is no standard definition of an urban area, but the key elements from an ecological viewpoint are size, permanence, and a clear urban–rural distinction. Small or temporary human settlements are localized disturbances for the rural flora and fauna, but larger and longer lived ones can support independent populations of tolerant species that can then evolve adaptations specifically to the urban environment. Since all urban areas are built to meet the needs of the same one species, they tend to have similar structures and environments—a tendency that has greatly increased in recent times, as urban areas worldwide are now also constructed of very similar materials. This means that, once a species population has adapted to one urban area, it is likely to do well in other urban areas in the region, and often further away, if it can reach them. Human movements provide an ideal means of dispersal between urban centres for many urban-adapted species. Urban biotas thus tend to become increasingly distinct over time from the biotas of their surrounding rural areas, while becoming increasingly similar to each other.

China's first cities were built in the Yellow River Valley, on the northern margins of TEA, more than 4000 years ago, and there were cities in tropical southern China by 2200 years ago. In continental South-East Asia, the earliest cities emerged within the last 2000 years, on the coast or in major river valleys and deltas (Higham 2002); but large cities were largely absent from island South-East Asia until the last 500 years (Boomgaard 2007). The characteristics of these cities have changed over the centuries, but most dramatically in the last 200 years, with growing population sizes and densities, the use of modern building materials, and increasing proportion of the urban area covered in impermeable artificial surfaces. Today, more than 40% of the human population of TEA lives in urban areas and the region's biggest city, Jakarta, Indonesia, supports 13.2 million people (UN Population Division 2006).

Until recently, the direct negative impacts of urbanization on regional ecosystems would have been small, because the settlements themselves were relatively small, and the main impact of cities would have been through their consumption of agricultural and forest products, including luxury items, such as ivory and rhinoceros horn. Within the last few decades, however, urban areas in TEA have grown so big as to be a major direct threat to natural ecosystems over large areas, eliminating natural habitat diversity and replacing it with an increasingly uniform urban sprawl. Cities are also having an expanding influence through their contribution to regional and global air pollution, and by acting as a gateway for the introduction of exotic species (see Chapter 7).

CHAPTER 2

Physical geography

2.1 Introduction

A major unifying feature of TEA is that—with a few small exceptions—the whole region has forest climates and, without human impact, would be covered in some sort of forest. This fact is one of the keys to understanding the ecology of the region today. Human populations have expanded non-forest vegetation at the expense of forest in many parts of the world, but in TEA this has not meant the expansion of pre-existing non-forest communities, but rather the creation of such communities from scratch. More open vegetation was widespread in glacial times, however, and it is possible that some anthropogenic vegetation types today are similar to these. Pleistocene faunas from the region also included vertebrates that were adapted to more open habitats, such as hyenas, but most of these are now extinct and most of the Holocene fauna is unable or unwilling to live outside forest.

2.2 Weather and climate

Weather is the state of the atmosphere at a particular place and time, or over a short period. The climate of a site is not just the average of its weather, but the aggregate of its weather conditions over time, including their extremes and variability. Before human impacts became the dominant influence, climate was the major control on the distribution of vegetation types in TEA. Soils were also an important influence, but soil properties are, to a varying extent, determined by climate. Climate also has a major influence on the distributions of individual species of plants and animals, and on regional patterns of diversity (sees Chapter 3). Past changes in regional climate drove the long-term changes in vegetation described in Chapter 1, and future, anthropogenic, changes are a major conservation concern (see 7.4.13). On a shorter time-scale, seasonal and interannual changes in the weather are the major influence on plant phenology (see 2.6) and, both directly and through plant phenology, on animal phenology.

The most important climatic variables are temperature, rainfall, and their seasonality. Three major types of lowland climate can be recognized within TEA, along with all possible intermediates (Fig. 2.1): the equatorial climate, as in Singapore, where temperature and rainfall are adequate for year-round plant growth; the monsoonal or wet/dry climate, as in Chiang Mai, Thailand, where temperature is adequate year round, but drought limits plant growth in the dry season; and the subtropical climate, as in Changsha, China, where both temperature and rainfall are seasonal, and plant growth is concentrated in the hot, wet summer. Montane climates differ from lowland climates in the same region, in their lower temperatures and often less seasonal rainfall.

Figure 2.1 Climate diagrams for towns and cities in Tropical East Asia, illustrating the major lowland climate types. The horizontal axis runs from January to December and the graphs show mean monthly temperature and rainfall. On the vertical scale, 20 mm of rainfall is equivalent to 10°C. When the rainfall curve is below the temperature curve, the area between them is dotted, indicating a dry season, while when it is above, there are vertical lines indicating a wet season. The numbers at the top and bottom of the left axis are the mean daily maximum temperature of the hottest month and the mean daily minimum of the coldest month. The line under the place name includes, in square brackets, the number of years of observations for temperature and rainfall, respectively, then the mean annual temperature and rainfall. From Leith (1999), which should be consulted for additional details.

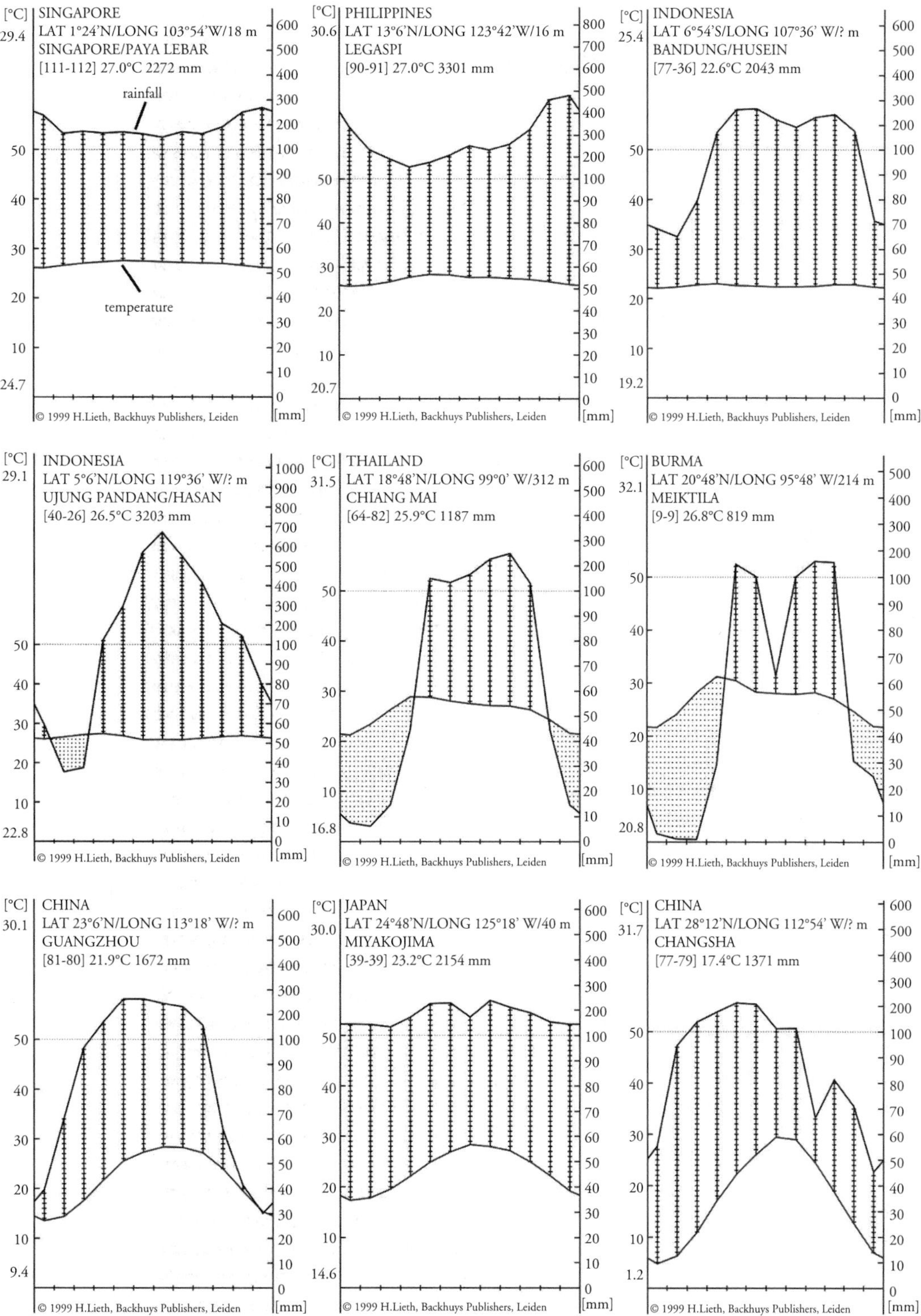
[°C]
SINGAPORE
LAT 1°24'N/LONG 103°54'W/18 m
SINGAPORE/PAYA LEBAR
[111-112] 27.0°C 2272 mm
29.4
24.7
rainfall
temperature
[mm]
© 1999 H.Lieth, Backhuys Publishers, Leiden
[°C]
PHILIPPINES
LAT 13°6'N/LONG 123°42'W/16 m
LEGASPI
[90-91] 27.0°C 3301 mm
30.6
20.7
[mm]
© 1999 H.Lieth, Backhuys Publishers, Leiden
[°C]
INDONESIA
LAT 6°54'S/LONG 107°36' W/? m
BANDUNG/HUSEIN
[77-36] 22.6°C 2043 mm
25.4
19.2
[mm]
© 1999 H.Lieth, Backhuys Publishers, Leiden
[°C]
INDONESIA
LAT 5°6'N/LONG 119°36' W/? m
UJUNG PANDANG/HASAN
[40-26] 26.5°C 3203 mm
29.1
22.8
[mm]
© 1999 H.Lieth, Backhuys Publishers, Leiden
[°C]
THAILAND
LAT 18°48'N/LONG 99°0' W/312 m
CHIANG MAI
[64-82] 25.9°C 1187 mm
31.5
16.8
[mm]
© 1999 H.Lieth, Backhuys Publishers, Leiden
[°C]
BURMA
LAT 20°48'N/LONG 95°48' W/214 m
MEIKTILA
[9-9] 26.8°C 819 mm
32.1
20.8
[mm]
© 1999 H.Lieth, Backhuys Publishers, Leiden
[°C]
CHINA
LAT 23°6'N/LONG 113°18' W/? m
GUANGZHOU
[81-80] 21.9°C 1672 mm
30.1
9.4
[mm]
© 1999 H.Lieth, Backhuys Publishers, Leiden
[°C]
JAPAN
LAT 24°48'N/LONG 125°18' W/40 m
MIYAKOJIMA
[39-39] 23.2°C 2154 mm
30.0
14.6
[mm]
© 1999 H.Lieth, Backhuys Publishers, Leiden
[°C]
CHINA
LAT 28°12'N/LONG 112°54' W/? m
CHANGSHA
[77-79] 17.4°C 1371 mm
31.7
1.2
[mm]
© 1999 H.Lieth, Backhuys Publishers, Leiden

2.2.1 Temperatures

The lowland equatorial tropics is characterized by continuous high temperatures or, perhaps more significantly, the absence of low ones (<18°C). Very high temperatures (>*c*.36°C) are also usually absent and the diurnal temperature range is much greater than the annual range in mean monthly temperature. Temperatures decline with altitude above sea-level. The rate of decline (the 'environmental lapse rate') averages around 0.6°C per 100 m increase in altitude, although it can vary between 0.4°C and almost 1.0°C depending on weather conditions. Absolute minima (the lowest temperatures ever reached) decline more rapidly with altitude and occasional frost (<0°C) has been recorded down to 1000–1500 m in some parts of Java. Temperature seasonality does not change with altitude, however. The diurnal range continues to exceed the annual range, so the thermal climate on near-equatorial mountains, such as Mt Kinabalu, can be described as 'summer every day and winter every night'.

Annual minimum temperatures also decline with increasing distance from the equator, while the seasonality increases. Maximum temperatures also tend to increase, as a result of reduced cloud cover in the later part of the dry season, when the sun is overhead. North of around 17°N, cold surges—outbursts of low-level, cold air towards the South China Sea—are a feature of the winter monsoon and can bring sub-zero temperatures to highland areas and, in extreme cases, down to sea-level, damaging tropical crops, such as rubber, in southern China (McGregor and Nieuwolt 1998). In Hong Kong (23°N), 130 km south of the Tropic of Cancer, temperatures fall below zero several times a decade above 700 m and, in January 1893, icicles hung from the rigging of ships in Victoria Harbour. Sub-zero temperatures become part of the normal annual cycle north of the Tropic of Cancer, and the temperature can fall below −10°C at 30°N.

2.2.2 Slope angle and aspect

The amount of solar radiation reaching a sloping site depends on the slope angle, the aspect (the direction it faces), and the latitude. North–south aspect differences increase with slope angle and with distance from the equator, so that at 20°N, a south-facing 30° slope receives about 45% more solar radiation over the year than a north-facing one (Fig. 2.2). Direct effects on vegetation at this latitude have not been reported, although they probably occur, but indirect effects, through the frequency of anthropogenic fires, are obvious (e.g. Dudgeon and Corlett 2004). The north–south difference is greatest around the winter solstice, when the sun is low in the southern sky. This coincides with the dry season, so the warmer, and thus drier, south-facing slopes are more fire-prone. East–west differences in solar radiation can also occur, even at the equator, if there is a consistent daily pattern of cloudiness. Even on cloudless days, east-facing slopes warm up and dry out earlier in the morning, increasing the chances of fire. The prevailing wind direction can also interact with slope angle and aspect to produce drier or wetter conditions, as well as having a direct impact on plant growth (see 2.2.7).

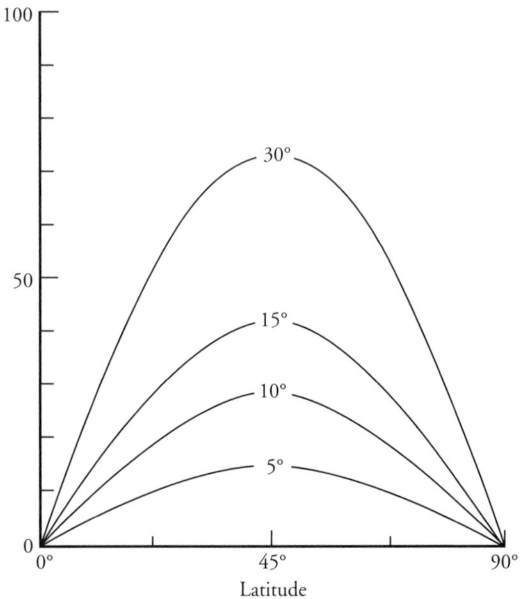

Figure 2.2 The variation with latitude of the difference in the amount of solar radiation received per year between equator-facing and pole-facing slopes for four slope angles (Holland and Steyn 1975). No allowance is made for cloudiness or for a slope shading its neighbours.

2.2.3 Rainfall

Mean annual rainfall within TEA ranges from only 500–800 mm, in the central dry zone of Myanmar and the Palu Valley of Central Sulawesi, to more than 3000 mm in many places, and nearly 8000 mm on Mount Bach Ma, in central Vietnam. For most organisms, however, the seasonal distribution of rainfall is at least as important as the mean, and, particularly in the 'everwet' Sunda Shelf region, interannual variation also has a significant impact on lifecycle events. Monsoon winds, movements of the inter-tropical convergence zone (ITCZ), and topography all contribute to the complex spatial pattern of seasonal rainfall regimes in the region, while the El Niño-Southern Oscillation (ENSO) and the Indian Ocean Dipole (IOD) have a major influence on the pattern of interannual variation.

The ITCZ is a band of low surface air pressure, cloudiness, and rainfall that moves north and south of the equator a month or two behind the overhead sun (McGregor and Nieuwolt 1998). The ITCZ is, therefore, furthest north during the northern hemisphere summer and furthest south during the winter. In TEA, the rainfall pattern produced by these movements of the ITCZ is complicated by interactions with the Asian monsoon system: the seasonal reversal of wind direction that results from the differential heating of the land and the sea. The Asian monsoon consists of two major subsystems, the Indian and the East Asian, with the East Asian monsoon dominating in TEA, with the exception of the north-west of the region, including northern Myanmar and western Yunnan, where the Indian monsoon prevails. A great deal of new information on the Asian monsoon has been gathered over the last decade or so, from field observations, palaeoclimate studies, and modelling, but this still leaves many unanswered questions about its origin, antiquity and variability (see also 1.5).

In the equatorial region (+/−10°C), the north-east and south-west monsoons bring similar masses of warm, humid air, but in most of TEA, one monsoon is wet and the other dry. In most of the region, the wet monsoon is in summer and the dry monsoon is in winter. This is clearest in the north, where the winter monsoon has travelled entirely over cool, dry land, while the summer monsoon has crossed a variable amount of warm, wet sea. Conversely, the same winter monsoon brings rainfall to coastal regions of Vietnam after it has travelled over the South China Sea. As a further complication, moisture-laden winds that blow towards the region are influenced by the local topography. Highlands oriented at a right angle cause uplift of the winds and greater rainfall on the windward side, and less on the other. This 'orographic uplift' accounts for much of the rainfall in the wettest places in the region, such as the west coasts of Myanmar, Sumatra, and Borneo.

2.2.4 Water from fog

Clouds, fog, and mist consist of tiny (mostly 5–50 μm) water droplets suspended in the air. In sites with frequent fog, such as montane cloud forests, the deposition of these droplets on to the vegetation can make a significant contribution to the water supply. Measuring this contribution in a standardized and ecological meaningful way is difficult, but various studies have suggested deposition rates equivalent to 0.2–5.5 mm of rainfall per day from cloud interception (Bruijnzeel 2005; Chang et al. 2006; McJannet et al. 2007). These inputs can be of particular significance in the dry season, where they may contribute up to two-thirds of the monthly water input to the forest. In areas that are frequently immersed in cloud, the amount of water captured by the vegetation appears to depend on both the canopy characteristics—complex woody canopies capture more than grassland—and the degree of exposure to the wind, since the main mechanism by which water is deposited is by impaction of droplets on the plant surfaces.

In the tropical seasonal forest of Xishuangbanna, south-west China, 'fog drip' contributes an estimated 5% of the total annual rainfall and more than a third of the dry season rainfall, as well as reducing dry season transpiration (Liu et al. 2008) (Fig. 2.3). This contribution from fog is probably all that enables tropical seasonal forest to survive at a site where the mean annual rainfall is less than 1500 mm, and less than 200 mm falls between November and April. The process here is rather different from that described above for montane cloud forests. The fog in Xishuangbanna is radiation fog, formed after sunset in calm, clear conditions, as a result of cooling by thermal radiation. Wind speeds are thus very low on foggy days and the dominant

Figure 2.3 Radiation fog forms after sunset and fills the valleys of Xishuangbanna, south-west China, in the dry season. 'Fog drip' contributes more than a third of dry-season rainfall and enables tropical evergreen forest to grow in a region where <200 mm of rain falls between November and April (Liu et al. 2008).

process of fog deposition is believed to be sedimentation, not impaction.

2.2.5 Interannual variation in rainfall

Differences between years in the amount and timing of rainfall occur throughout the region, but their impact is greatest in areas that have no annual dry season, or only a brief one. Even short, dry periods can have a major impact on plant communities that do not experience them every year. General flowering in the lowland dipterocarp forests of the Sunda Shelf is apparently triggered by a 30-day rainfall total of <40 mm (Sakai et al. 2006; see 2.6.2), while more severe dry periods lead to tree deaths (Nakagawa et al. 2000; Yoneda et al. 2000; Aiba and Kitayama 2002). The use of monthly rainfall averages to represent the climate of a site understates the true frequency of such droughts, and 30-day running totals show that there are dry periods, of a month or more, in many areas where the mean monthly rainfall is always >100 mm (Walsh 1996). On Mt Kinabalu, for example, weather stations at 1560 m, 2650 m, and 3270 m above sea-level, where there is no dry month in a normal year, all experienced 30-day running totals of <1 mm in March–April 1998 (Aiba and Kitayama 2002). Moreover, this was the third severe drought on Kinabalu in 30 years, so these are not rare events. Droughts of similar intensity have been recorded in tropical Asia as far back as historical and other records go (Grove 1998; D'Arrigo et al. 2006). The impacts of these droughts on tree growth and mortality vary considerably with topography and soil conditions.

Although the causes of interannual rainfall variation are not yet fully understood, most of the more extreme droughts in the aseasonal rainforest zone can be attributed to the El Niño-Southern Oscillation (ENSO) or, in the west of the region, the Indian Ocean Dipole (IOD). ENSO is a fluctuation between unusually warm (El Niño) and unusually cold (La Niña) sea surface temperatures in the tropical Pacific Ocean, which develop in association with changes in the atmospheric pressure pattern (the Southern Oscillation) that are related to the strength of Pacific trade winds (McPhaden et al. 2006). Under normal conditions, the trade winds pile up warm surface water in the western Pacific, while bringing up cold water from below the surface in the east. This results in an east–west contrast in sea surface temperature that reinforces the east–west difference in air pressure that drives the trade winds. During an El Niño episode, the trade winds weaken along the equator as pressure rises in the west and falls in the east, so warm water migrates eastwards and upwelling is reduced. The feedback is now reversed and the weakened trade winds and warming sea surface temperatures reinforce the El Niño. Eventually, however, other processes produce a delayed negative-feedback that ends the El Niño and, if strong enough, starts a La Niña.

The use of the term 'cycle' for ENSO is rather misleading, since the phenomenon is by no means regular. El Niño and La Niña typically recur at irregular intervals of 2–7 years (Fig. 2.4) and the strength, duration, rate of development, and spatial structure all vary between events, as does the impact on regional and global weather. Moreover, El Niños and La Niñas are also not simple opposites, with El Niños generally more important for the climate of TEA, as demonstrated by the consequences of strong events, such as those in 1982–83 and 1997–98.

A strong El Niño affects most of TEA, but the strongest and most predictable impacts of the ENSO cycle are on the rainfall in much of Indonesia and the surrounding countries, with El Niños bringing droughts, with all their consequences. One of these consequences is the mass-flowering and subsequent mast-fruiting in lowland rainforests across the region (see 2.6.2). Another is the anthropogenic forest fires, which during the 1997–98 El Niño—by some measures the strongest this century—released so much carbon dioxide into the atmosphere that they produced the largest annual increase in concentrations since records began in 1957 (McPhaden et al. 2006; see also 7.4.13). Even in Indonesia, however, the influence of ENSO on rainfall varies, and it is weak or absent in a region that covers northern Sumatra and south-west Kalimantan (Aldrian and Susanto 2003).

The importance of ENSO in recent decades has spurred a massive research effort aimed at deducing the history of the phenomenon and predicting future changes. The possible 'permanent El Niño' during the early Pliocene was mentioned in Chapter 1 (1.5) and there is also evidence for a lower amplitude, but higher frequency, during glacial times (Tudhope et al. 2001; Bush 2007). During the early to mid-Holocene, when the Asian monsoon was stronger, ENSO variability was apparently less and the average state was more like a modern La Niña (Gagan et al. 2004; Abram et al. 2007). There is, as yet, no agreement on how global warming will affect either the variability or the background state of ENSO, but the possibility of major adverse impacts has sparked vigorous debate.

The Indian Ocean Dipole (or Indian Ocean Zonal Mode) was first recognized in 1999 and is less well understood than ENSO. Like ENSO, the IOD arises from ocean–atmosphere interactions, but there are also major differences, reflecting both the presence of the Asian monsoon system and the smaller size of the Indian Ocean than the Pacific. In the positive phase of the IOD, sea surface temperatures are lower in the eastern Indian Ocean, off the coast of western Indonesia, and higher in the west, off the coast of Africa (Marchant et al. 2007). In the negative phase, the opposite occurs.

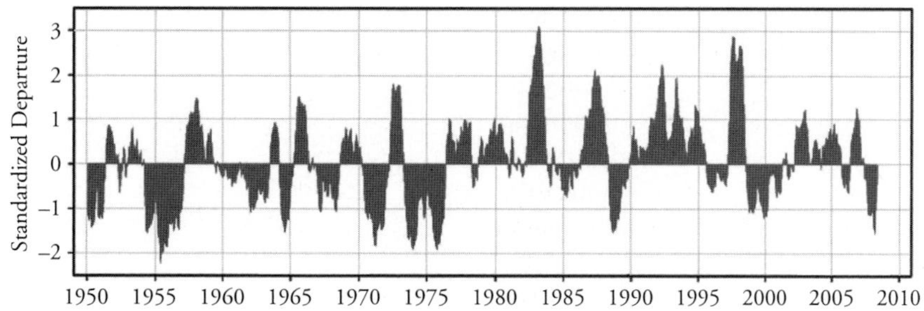

Figure 2.4 The El Niño-Southern Oscillation (ENSO) is a fluctuation between unusually warm (El Niño) and unusually cold (La Niña) sea surface temperatures in the tropical Pacific Ocean and has a large impact on rainfall patterns in Tropical East Asia. This graph shows the Multivariate ENSO Index (MEI) developed at NOAA: positive values represent El Niño events and negative values La Niña events (from the NOAA Multivariate ENSO Index web site).

Climate models suggest that IOD events can be triggered by El Niño conditions, but can also arise from processes internal to the Indian Ocean. A strong positive IOD in 1961, for example, was not linked to an El Niño, while the one in 1997 was. Whether or not they coincide with an El Niño, positive IODs bring droughts to western Indonesia, with the three strongest droughts in Padang, western Sumatra, coinciding with strong IOD events in 1961, 1994, and 1997 (Abram et al. 2007). Fossil corals from the Mentawai Islands, off the west coast of Sumatra, have been used to reconstruct the strength of the IOD over the past 6500 years. These suggest that positive IODs, and thus western Indonesian droughts, persisted longer in the mid-Holocene (6500–4000 years ago), when the monsoon was stronger, than they do today. Interactions between the IOD, ENSO, and the Asian monsoon system are currently a hot topic for research.

2.2.6 Snow and ice

North of the tropics, snow falls annually at high altitudes and occasionally lower down in the evergreen broadleaved forest, where heavy falls may break living branches. In early 2008, the longest cold spell for more than 40 years brought snow and ice-storms to a broad band of subtropical China, north of around 25°N, causing massive mechanical damage to native broad-leaved forests and devastation to many plantations (Stone 2008a). The main culprit seems to have been freezing rain—supercooled liquid water droplets that freeze on contact with a surface <0°C, leading to ice accumulation on branches and trunks. In the worst-affected forests, most trees were uprooted or their trunks snapped, while the few standing trees were stripped of most or all of their branches (personal observations) (Fig. 2.5). Three months after the storm, most damaged trees were re-sprouting vigorously, but some were dead and the canopy was still 60–90% open.

2.2.7 Wind

Winds strong enough to break or uproot large trees can occur anywhere in the region, but are commonest in the cyclone belt, north of around 10°N. Tropical cyclones—known as typhoons in TEA—are circular

Figure 2.5 This mixed broad-leaved forest at *c.*1000 m in Nanling National Nature Reserve (25°N), China, was severely damaged by ice accumulation in early 2008. Many trees were either uprooted or had the trunk snapped, and most of the standing trees were stripped of some or all branches. Three months after the event, most damaged trees were resprouting, but some standing trees were dead and the canopy was still very open. Photograph courtesy of Billy C.H. Hau.

storms into which winds spiral at great speed. In the western Pacific they form over warm (>26°C) ocean water, between approximately 5° and 10°N. They then typically move west, then north, before finally turning eastwards, although the actual paths are very variable (Fig. 2.6). Typhoons are classified on the basis of the maximum sustained wind speed at the surface: <63 km/h for a tropical depression, 63–117 km/h for a tropical storm, 118–240 km/h for a typhoon, and >240 km/h for a super-typhoon. Gust speeds (lasting a few seconds) can by considerably higher than this. Typhoons also vary widely in size, with gale-force winds (>63 km/h) extending from tens to hundreds of kilometres from the centre. Typhoons can occur in any month, but are most frequent from July to October.

The high wind-speeds associated with typhoons cause damage that can range from stripping foliage and damaging crowns, to completely flattening swathes of the forest (Whitmore and Burslem 1998; Lugo 2008; Turton 2008). Typhoons may also cause damage through heavy rainfall and, in coastal areas, storm surges. Strong winds also occur outside the typhoon belt, and there are anecdotal reports of damage ranging from the uprooting or snapping of single trees, to the flattening of forest over several square kilometres (Whitmore and Burslem 1998; Proctor et al. 2001; Baker et al. 2005). Forest fragmentation creates edges that increase the exposure of trees to wind and, in Central Amazonia, lead to increased mortality and damage within 100–300 m of the forest margin (Laurance 2005).

Chronic wind-stress from persistent winds of lower speeds can also have a dramatic impact on forest vegetation. At Nanjenshan (22°N) in southern Taiwan, the forest on rugged terrain shows a sharp differentiation into that on windward slopes, exposed to the strong and persistent winds of the north-east monsoon every winter, and that on protected leeward sites (Sun et al. 1996). The windward forest is very short (3–5 m), with a broken canopy, a high stem density, and a distinct floristic composition, while the heights of trees in the leeward forest range from 10 to 20 m. The direct impact of the wind is thought to be through increased mechanical stress and higher evapotranspiration. Sun et al. (1996) argue that these direct impacts lead to changes in tree stature, leaf morphology, and soil moisture that, in turn, trigger changes in microclimate and soil nutrients. High wind-speeds are also common

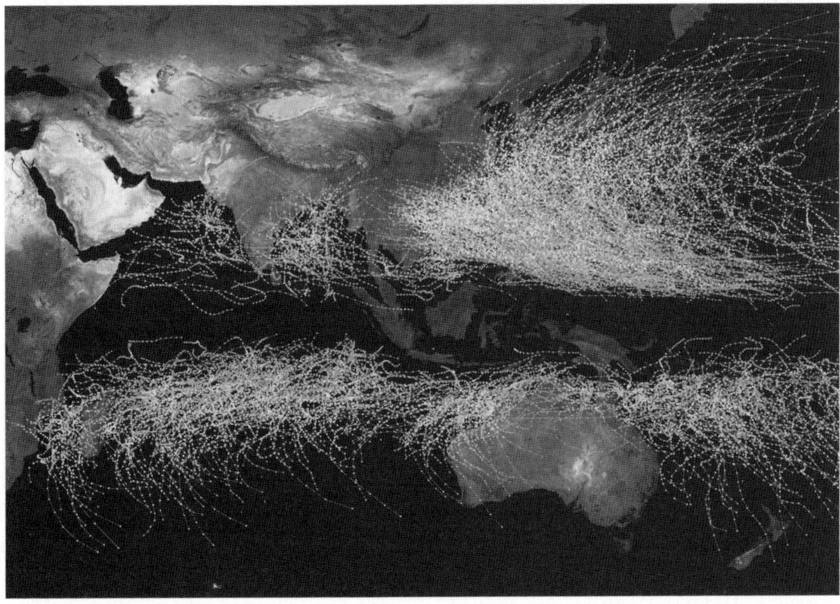

Figure 2.6 The tracks of all tropical cyclones from 1985 to 2005. Image created by Nilfanion, Wikimedia Commons.

in most montane regions and are likely to have a (currently under-studied) impact on montane forest communities (e.g. Noguchi et al. 2007).

2.2.8 Lightning

Lightning is the discharge of static electricity within and between thunderstorm clouds or from clouds to the ground, with only the latter having an ecological impact. Lightning is probably under-recognized as a cause of mortality in individual canopy trees because the initial symptoms of a strike are very variable, ranging from a tree explosion, to a visible scar, to the invisible death of roots. Lightning strikes may also kill groups of trees, presumably when the current flows through root connections (Whitmore and Burslem 1998). The role of lightning in starting natural fires has probably also been underestimated, with most authors assuming that all fires today are anthropogenic. Lightning is most frequent in the wet season, but it more likely to cause extensive fires during the transition between dry and wet seasons, when abundant dry fuels are available and ignition is less likely to be followed by heavy rain (Stott 2000). However, Cardoso et al. (2008) show that much of South-East Asia has a relatively low likelihood of lightning strikes during the dry/wet transition compared with Africa or the Neotropics. They then use this to explain the occurrence of a forest climax in parts of South-East Asia with a climate for which vegetation models predict savanna.

2.3 Fires

There may be areas of TEA that are consistently too wet to burn, but even at the wettest rainforest sites, studied for a decade or more, there have been irregular dry periods that increase litterfall and, thus, thin the canopy and increase the depth and dryness of the litter layer, making ignition possible. Vulnerability to fires in short, dry periods increases on topography (e.g. ridge tops) and soils (peat, sand, soils on limestone, etc.) that dry out easily and, conversely, is lower near streams. Even a brief annual dry season increases fire risk, as does the presence of deciduous trees in the canopy and a grassy understorey. Potential natural sources of ignition include lightning (see above), volcanoes, and burning coal seams (Goldammer 2007), while fires are started deliberately by people for reasons that range from stimulating growth of grasses, to forest clearance (see 1.11). Anthropogenic fires, once started, often burn out of control. Although the palaeo-ecological record provides evidence for fires before modern humans arrived in the region, the fire frequency has undoubtedly increased. More frequent fire has meant lower fire intensities for most of human history, since fuels do not accumulate as long, but modern commercial logging greatly increases the fuel supply and thus the fire intensity. There are also differences in the seasonal occurrence of natural and anthropogenic fires, with natural fires more likely at the end of the dry season (see 2.2.8).

Except in recently logged forests, and on peat, most forest fires in TEA burn only the litter layer and, if present, the grass layer. Flame heights and temperatures are low, and mortality is concentrated among smaller plants or near long-smouldering debris. In lowland rainforests, where most trees are thin-barked, most plants <5 cm diameter die, but mortality is more selective at larger diameters. The proportion of fire-tolerant species, with thick bark, the ability to heal fire scars, and to resprout after severe damage, increases in progressively drier forest-types; and in frequently burned deciduous forests, mortality may be confined to seedlings.

Grasslands are by far the most easily ignited of vegetation types in TEA, because of their abundant fine fuels. Annual anthropogenic fires are common in areas where seasonal drought causes the synchronous death and drying out of the stems and leaves. Grass fires spread rapidly and can reach high temperatures (>600°C), but these are short-lived, and lethal temperatures (>60°C) do not penetrate far into the soil (Zedler 2007). Many perennial grasses can recover rapidly from well-protected buds after such fires. Woody plants, in contrast, cannot avoid exposure of their buds and meristems to fires, and take longer to recover because of their investment in secondary tissues. Indeed, most grass-dominated vegetation in TEA depends for its existence on regular fires that prevent the invasion of woody species and its eventual replacement by forest.

2.4 Soils

Soils in the tropics are not only extremely variable, but these variations strongly influence plant distributions on a variety of spatial scales, from the local (e.g. Miyamoto et al. 2003; Davies et al. 2005), to the landscape (e.g. Cannon and Leighton 2004), to the regional (e.g. Potts et al. 2002). Soils are complex entities, however, and they co-vary with topography, vegetation, and climate, so it has so far proved difficult to relate the observed floristic variation to particular soil properties. Also, much of the evidence for edaphic control of floristics in TEA comes from Borneo, which, with its relatively wet and aseasonal climate, and limited volcanic activity, is not very representative of the region as a whole.

A major ecological function of soils is the supply of nutrients for plant growth. Nutrient supply, however, is best understood as a dynamic property of the whole ecosystem, rather than simply as a soil property. It is this dynamism that explains, for example, how tall, species-rich rainforest can grow on soils that have little or no agricultural potential. Nutrient cycling is, therefore, considered in a separate chapter (Chapter 6) and a basically descriptive approach to soil properties is taken here.

2.4.1 Soil classifications

Most countries in TEA have their own soil classifications systems and these vary considerably in both their principles and their nomenclature. Two international systems are also widely used in the region: the World Reference Base for Soil Resources (previously the FAO-UNESCO Soil Classification) (FAO 2007; Table 2.1) sponsored by the Food and Agriculture Organization (FAO) of the United Nations, and the USDA Soil Taxonomy developed by the United States Department of Agriculture (USDA 2006; Table 2.2). Both attempt to classify soils on the basis of characteristics that can be observed or measured in the field and which are also of relevance to soil management. Among other differences, the USDA system uses climate in its classification, while the WRB does not. Also, the USDA system is far more elaborate and detailed, while the WRB is intended more for international communication, than for soil

Table 2.1 The WRB reference soil groups found in Tropical East Asia (simplified from: FAO 2007).

Soils with a thick organic layer (peat):	Histosols
Soils with strong human influence	
Soils with long and intensive agricultural use:	Anthrosols
Soils dominated by human-made materials:	Technosols
Soils with limited rooting depth due to stoniness:	Leptosols
Soils influenced by water	
Clay-rich, shrink and swell with changes in moisture content:	Vertisols
Floodplains, tidal marshes:	Fluvisols
Soils influenced by groundwater:	Gleysols
Soils set by Fe/Al chemistry	
Allophanes or Al-humus complexes (on volcanic ash):	Andosols
Surface soil bleached by loss of organic matter and iron oxides over a dark layer where these accumulate:	Podzols
Accumulation of Fe under waterlogged conditions:	Plinthosols
Low-activity clay, P fixation, strongly structured:	Nitisols
Dominance of kaolinite and sesquioxides:	Ferralsols
Soils with stagnating water	
Abrupt textural discontinuity:	Planosols
Structural or moderate textural continuity:	Stagnosols
Soils with clay-enriched subsoil	
Low base saturation, high-activity clay:	Alisols
Low base saturation, low-activity clay:	Acrisols
High base saturation, high-activity clay:	Luvisols
High base saturation, low-activity clay:	Lixisols
Relatively young soils or soils with little or no profile development	
With an acidic, dark topsoil (in mountains):	Umbrisols
Sandy soils:	Arenosols
Moderately developed soils:	Cambisols
Soils with no significant development:	Regosols

Table 2.2 The soil orders found in Tropical East Asia (based on: USDA 2006).

Soils with a thick organic layer (peat):	Histosols
Surface soil bleached by loss of organic matter and iron oxides over a dark layer where these accumulate:	Spodosols
Soils formed on volcanic ash:	Andisols
High kaolinite and sesquioxides, low weatherable primary minerals:	Oxisols
Clay-rich soils that shrink and swell with changes in moisture content:	Vertisols
Strongly leached and weathered soils, with a clay-enriched subsoil:	Ultisols
Moderately leached and weathered forest soils:	Alfisols
Young soils with little profile development	Inceptisols
Soils of recent origin with no profile development:	Entisols

mapping at the local scale. Scientific publications in TEA usually use either the relevant national classification or the USDA system, but WRB has the advantage that it was designed from the start as a global overview, while the USDA system is optimized for the United States.

Soil mapping in the tropics is aimed at identifying agricultural potential. None of these systems was designed for ecological purposes and none makes direct use of the soil characteristics that are known to influence floristic composition. They, therefore, cannot be expected to have much predictive value in intact natural vegetation. However, in the absence of an ecological classification of tropical soils—or the funding to re-survey the vast areas already classified—ecologists must make use of existing soil classifications as a first approximation, when no detailed data on ecologically relevant soil properties is available.

Most soils in the humid tropics belong to two major soil groups: the Ultisols (USDA)/Acrisols (WRB) and the Oxisols (USDA)/Ferralsols (WRB) (Palm et al. 2007). The WRB and USDA definitions of these groups are similar, but not identical. Both groups consist of highly weathered, red or yellow soils, which are acid, low in nutrients, and with low cation exchange capacities (i.e. low capacity to hold nutrient cations). Ultisols/Acrisols are distinguished from Oxisols/Ferralsols mainly by the higher concentration of clays in the subsoil than the topsoil, as a result of clay migration. Oxisols/Ferralsols dominate on old, stable landforms in Africa and South America, but they are relatively rare in the tectonically more active TEA. In TEA, Ultisols/Acrisols dominate, covering more than half the region, including most of the non-volcanic areas (Dudal 2005). Two main suborders of Ultisols occur in the region: the humult Ultisols, low in nutrients and with a surface layer of raw humus; and the udult Ultisols, which are generally more fertile and lack the surface organic layer. In Sarawak, very different species associations occur on adjacent soils in these suborders (Potts et al. 2002; Baillie et al. 2006). Ultisols/Acrisols were traditionally farmed by shifting cultivation with a long, fallow period, and are now increasingly used for rubber and oil palm plantations.

The most fertile soils in TEA are relatively young (in terms of weathering, which does not necessarily correlate well with actual age) and still contain unweathered minerals from the parent material, which act as a nutrient reserve (see 6.3). Examples include: soils derived from recent fluvial deposits (Fluvents/Fluvisols), which are extensive in the deltas of the major rivers; soils derived from volcanic ash (Andisols/Andosols), which cover large areas in Java, Sumatra, and Bali; and young soils on steep slopes or recent colluvium (Inceptisols/Cambisols). Except on the steepest slopes, these soils have high agricultural potential and their natural vegetation has usually been cleared. Fluvents/Fluvisols are the most important soils for wet rice cultivation. There is also a range of other soil types, classified rather differently in the two systems, which are intermediate in age and fertility between these young soils and the highly weathered soils described in the previous paragraph (mostly called Alfisols in the USDA system, and Nitisols, Gleysols, or Luvisols in WRB).

At the opposite extreme, in terms of agricultural potential, are Histosols (both systems) and Spodosols/Podzols. Histosols are organic soils—peats—and are largely confined in TEA to areas with high rainfall and without a long dry-season. Around 20 million hectares of tropical forest peat border the Sunda shelf, largely in Sumatra and Borneo, mostly close to sea-level. Most coastal peatlands appear to be relatively young, with accumulation starting after sea-levels stabilized around 7000 years ago, but there is also evidence for peat formation in the late Pleistocene (*c.*26,000 years ago) at inland sites, despite the presumably drier climate of this time (Page et al. 2004). The huge amount of carbon stored in these peats (estimated at 42,000 megatonnes; Hooijer et al. 2006), which can exceed 20 m in depth, makes their fate of global concern. Shallow peat soils (<50 cm) can be successfully converted for agriculture but attempts to convert deep peats have usually failed (Wösten et al. 2008). Spodosols/Podzols are strongly acidic soils that have a surface layer of organic matter (<1 m thick) over an ash-grey sandy layer that has been bleached by the loss of organic matter and iron oxides. In the tropics they occur mostly on sand or sandstone and in TEA they are most extensive at low altitudes in Borneo. Other distinctive soil types occur on extreme geological substrates. Psamments/Arenosols form on unconsolidated sand and grade into the Spodosols/Podzols, described above. Soils on limestone vary greatly in thickness and chemistry, depending in part on the amount and type of the insoluble impurities in the rock, but are often shallow and relatively fertile. In contrast, soils formed on ultramafic rocks, although also very variable, tend to have both low concentrations of important plant nutrients and potentially toxic levels of nickel, magnesium, and other metals.

Clearance and cultivation of a soil do not necessarily change its classification, although with tropical soils there is often a big decline in fertility, associated with loss in soil organic matter. The WRB system does, however, recognize two groups of soils that have been greatly modified by human activities. Anthrosols are soils changed by prolonged agriculture, with the paddy soils (hydragric Anthrosols), formed by rice cultivation, the most widespread example in TEA. Paddy soils formed from a variety of different soil types tend to converge as a result of the rice-specific water-management regime and repeated wet cultivation, which promote the formation of a dense 'plough pan' beneath a structureless 'puddled layer'. Technosols are urban soils whose properties are dominated by their recent artificial origin, including those sealed by an impervious surface layer and those that contain a large proportion of human-made material (e.g. wastes, mine spoils). Technosols often contain toxic substances resulting from industrial processes.

2.4.2 Landslides and soil erosion

The rugged topography of much of TEA, coupled with high rainfall intensities and the inherent high erodibility of many tropical soils, promotes soil loss by surface erosion and landslides (Sidle et al. 2006). Surface erosion can remove the organic-rich topsoil, along with much of the nutrient capital, while landslides can remove the entire soil mantle, leaving the parent material exposed. On the other hand, the moderate rates of erosion in undisturbed landscapes can be important in rejuvenating the soil, giving plants access to rock-derived nutrients that would otherwise be beyond the reach of their roots (see 6.3). Landslides also increase environmental heterogeneity in forested landscapes, and may thus promote diversity. In both forested and deforested landscapes, landslides are associated with steep slopes and extreme rainfall (Dudgeon and Corlett 2004; Ohkubo et al. 2007). Landslides are also promoted by earthquakes and large areas of forest were lost from steep slopes in Sichuan, on the northern margins of TEA, following the massive (7.8 on the Richter scale) earthquake in 2008 (Ouyang et al. 2008).

Intact forest cover reduces surface erosion by promoting infiltration and protecting mineral soils from raindrop impact. Landslide frequencies are also reduced as a result of the removal of soil water by evapotranspiration and the stabilizing role of tree roots. Human impacts, from logging to forest clearance, disrupt this protection and can lead to large increases in both erosion and landslides. The impact on surface erosion is rapid and declines as soon as new vegetation is established, but landslide probability does not increase substantially until several years after forest clearance, when large roots have decayed, and may then stay high if the forest is replaced by weaker-rooted plants (Sidle et al. 2006). The impact of forestry operations is largely through road construction, rather than timber harvesting itself, and road construction can also be a big problem in agricultural landscapes. Agroforestry practices, which combine trees and low-growing cover crops, can greatly reduce surface erosion compared with monoculture plantations or crops that leave a lot of bare soil.

2.5 Vegetation

The broad patterns of natural vegetation across the Earth's surface can be predicted quite well from the mean annual temperature and mean annual rainfall, but a better fit to reality is obtained from models that incorporate information on temperature and rainfall seasonality (Prentice et al. 1992). In TEA, the most useful additional variables are the absolute minimum temperature (i.e. the lowest temperature ever reached) and some measure of the length and intensity of the dry season. Further improvements in precision then require information on the soil, particularly its water-holding capacity. The influence of soil on vegetation patterns is greater, the drier the climate, and soil may be the dominant factor at the landscape scale in the driest areas. In mountainous areas, altitude is often the best predictor of vegetation type, reflecting the decline in temperature with elevation, coupled with less regular, more-or-less correlated, changes in many other factors.

Human impacts are overlain on this natural pattern, but because human activities are partly controlled by the same factors that control natural vegetation, it can often be difficult to distinguish 'natural' from 'anthropogenic' vegetation. This is particularly true where the impacts have been continued over centuries or even millennia. Some people have argued that this makes the distinction meaningless, but this time-scale is short compared with that needed for significant evolutionary adaptation, so even old anthropogenic vegetation tends to have a relatively impoverished flora and fauna.

All systems for classifying and naming vegetation types are to some extent arbitrary and which is 'best' depends on the purpose. Traditional human communities classify the local vegetation in a way that reflects their uses of it (e.g. Delang and Wong 2006), and a production-forestry perspective is obvious in many national forest classifications. From an ecological perspective, the choice is between groupings based on features of the physical environment (climate, soil, and altitude), the appearance of the vegetation (structure, physiognomy, and seasonal changes), or the actual plant species present. The last of these—the floristic approach—undoubtedly gives the most precision, but is impractical on the scale considered here. An environment-based classification works well, if only natural, climax vegetation types are being considered, but human impact has broken the tight link between vegetation and environment on which it relies. I have, therefore, followed many other regional authors in using a combination of physical environment and appearance to divide the vegetation of the region into major vegetation types (Table 2.3). It should be remembered, however, that despite the prevalence of sharp boundaries between vegetation types at the landscape level, the vegetation of TEA as a whole forms a floristic continuum (i.e. most possible intermediates exist somewhere), so any classification is to some extent arbitrary. The following account of the vegetation of TEA borrows heavily from Corlett (2005).

2.5.1 Lowland vegetation

Tropical rainforests

Tropical, lowland, evergreen rainforests (Fig. 2.7) have the highest biomass, structural complexity, and both plant (Table 3.1) and animal diversity, of any vegetation type in the region. The main canopy is generally 30–40 m high, with scattered emergents rising to 50 m or more. Many of the larger trees have spreading buttresses, which function in support. Below the main canopy is an additional tree layer that includes both small, shade-tolerant understorey

Table 2.3 The major vegetation types of Tropical East Asia recognized in this chapter.

Lowland vegetation
Tropical rainforests
Tropical seasonal forests
Tropical deciduous forests
Subtropical evergreen broad-leaved forests
Forests on extreme soil types
  Heath forests
  Ultramafic forests
  Limestone forests
Secondary forests
Logged forests
Bamboo forests
Savannas and grasslands
Shrublands and thickets
Beach vegetation
Plantations, agroforestry, and other dryland crops

Montane vegetation
Lower montane forest
Upper montane forest
Subalpine forest
Alpine vegetation

Wetlands
Mangrove forests
Brackish water swamp forests
Freshwater swamp forests
Peat swamp forests
Herbaceous swamps
Rice fields

Urban vegetation

Figure 2.7 Tropical rain-forest canopy at Gunung Palung National Park, West Kalimantan, Borneo. Photograph © Tim Laman.

species and young individuals of the canopy and emergent species. True shrubs are relatively rare and the herbaceous ground layer is generally sparse and patchy. The dominance of the understorey of lowland rainforests in TEA by sterile saplings of canopy trees, contrasts with the diversity of small trees that flower and fruit in the understorey of Neotropical forests (LaFrankie et al. 2006). Woody lianas are common and include many species of rattans—spiny, climbing palms, which reach their greatest diversity in South-East Asia. The leaves of trees are mostly mesophyll (20–182 cm^2) and entire-margined, with an acuminate tip (Turner 2001). Some canopy trees may be deciduous, but the leafless period is brief and not synchronized between species.

Tropical rainforests are limited to areas without seasonal water or temperature stress, although less regular, dry periods may occur and a brief, annual, dry season (<2 months) may be tolerated on deep soils that hold water. In TEA, tropical rainforest occurs on the Sunda Shelf and in the wetter parts of the Philippines, Sulawesi, and Java. In the Thai-Malay Peninsula, the Kangar-Pattani Line (*c.*6°30′N), close to the Malaysian border, marks the northern boundary of continuous tropical rainforest, but dipterocarp-dominated lowland rainforest extends north to 18°N on Luzon Island in the Philippines, and there are also outlying areas of tropical rainforest in south-west Myanmar, and elsewhere in continental South-East Asia. Most such forests, however, probably fit better in the next category.

Tropical seasonal forests

In areas that experience a regular, annual, dry period of 1–4 months—up to 6 months on deep soils—the aseasonal rainforest described above is replaced by forests that, although still predominantly evergreen, exhibit regular, seasonal changes, synchronized by the annual drought. These changes are most striking in semi-evergreen rainforests, in which up to half the canopy trees are deciduous, although the lower storeys are largely evergreen. The proportion of deciduous trees does not show a simple relationship with the length of the dry season, and largely evergreen forests occur on deep soils in sheltered valleys and as a narrow 'gallery forest' along watercourses in the same climate as fully deciduous forests. These 'dry evergreen' forests exhibit annual cycles of growth and have a wilted and lifeless appearance at the peak of the dry season. In comparison with rainforests in areas without a dry season, seasonal forests are generally less tall and less diverse (Table 3.1), with a tendency to local dominance by one or a few species. Towards the north of the region, winter low temperatures reduce water stress, but absolute minima below 10°C probably exclude some tropical taxa. Despite this, dipterocarp-dominated tropical seasonal forests extend to 27°N in sheltered valleys in Myanmar (Kingdon-Ward 1945).

The dry season makes tropical seasonal forests susceptible to fire and thus replacement by more tolerant vegetation types, while their occurrence on the best soils makes them susceptible to clearance for agriculture. The limited extent of these forests in TEA today is thus probably a result of human impact, rather than the absence of suitable environments.

Tropical deciduous forests

Deciduousness in the tropics is a response to seasonal water stress, so one might expect to find a gradual change from fully evergreen to fully deciduous forest along a gradient of increasing length and severity of the dry season. This is rarely observed in modern South-East Asia, and boundaries between largely evergreen and largely deciduous forests are typically abrupt or through only a narrow ecotone. In most cases, this is because the boundary between tropical evergreen and deciduous forest is between a fire-sensitive, fire-excluding vegetation type and one that tolerates and often promotes fire. Much of the existing deciduous forest has probably replaced fire-sensitive, tropical seasonal forests in relatively wet areas while, at the other end of the spectrum, fire has degraded the drier deciduous forests into savanna and grassland.

Tropical deciduous forests typically occur in regions with a 3–7 month dry season and a mean annual rainfall of 700–1700 mm. However, as the length of the dry season increases, soil characteristics and topography have an increasingly important influence on the vegetation and, in combination with differences in human impact, can produce a complex mosaic of very different vegetation types in the same regional climate. Deciduous forests themselves are very varied in structure and floristic composition, as well as the degree of deciduousness and the length of the leafless period, which may be brief (see 2.6.1). Foresters often distinguish 'moist deciduous forest', with some trees more than 25 m tall and the lower storeys largely evergreen, and 'dry deciduous forests', with a lower canopy height and almost all species deciduous, but it is possible to find all intermediates between these types. Bamboos are common, but not universal, in the understorey, with grasses increasingly important as drought, fire, or other disturbance opens up the canopy. A dense undergrowth of bamboos can suppress tree recruitment, which is then concentrated in the period following the simultaneous death of the bamboos after gregarious flowering, which occurs at several-decade intervals (Marod et al. 1999; see 2.6.2).

One important deciduous forest type is recognized by both foresters and local people as distinct throughout its range in continental South-East Asia (Stott 1990). This is the 'deciduous dipterocarp forest' (Fig. 2.8). It varies considerably in stature and the openness of the canopy, but is dominated throughout its range by a characteristic group of tree species, including six deciduous (or semi-deciduous) dipterocarps. Deciduous dipterocarp forest is most widespread in areas with low rainfall (<1400 mm), a 4–7 month dry season, and poor sandy or gravelly soils. In some core areas it may be an edaphic climax, but its extent has been greatly increased by near-annual fires and cutting for timber and firewood. The dominants are thick-barked, fire-tolerant, and coppice well after cutting. The generally low diversity of both the flora and fauna, relative to other forest types, supports the idea that this is largely an anthropogenic formation. Other types of deciduous forest are distinguished by the dominance of a particular species, such as teak, *Tectona grandis*, in its native range in Myanmar, Thailand, and the Lao PDR, and where naturalized in seasonally dry parts of Indonesia. Most accessible teak has been logged, and in northern Thailand, the resulting forest has been termed 'bamboo-deciduous forest' because of the dominance of bamboos in the understorey (FORRU 2006).

In the central, dry zone of Myanmar, the Palu Valley of Central Sulawesi, and on some small islands of eastern Indonesia, where the total annual rainfall is only 500–800 mm and the dry season lasts 9 months or longer, the natural vegetation would probably be a 'thorn forest', dominated by low, thorny, deciduous trees, particularly in the genus *Acacia* (e.g. Stamp and Lord 1923). Such vegetation is extremely vulnerable to fire and other human

Figure 2.8 Dry deciduous dipterocarp forest at Huai Kha Khaeng, western Thailand, at the end of the dry season. Photograph courtesy of Tommaso Savini.

impacts, so little, if any, remains in anything like its natural state.

Subtropical evergreen broad-leaved forests

North of the tropics in TEA, there was, until recently, a broad belt of evergreen broad-leaved forest, stretching between around 24 and 34°N, which has been classified, by various authors, as 'temperate', 'warm-temperate', or—as done here—'subtropical' (Song 1995; Wang et al. 2007c). Mean annual rainfall in this belt ranges from around 900 to 2000 mm and frosts (temperatures <0°C) occur every winter. This forest once covered 25% of China's total land area, but very little now survives on the Chinese mainland, even in a degraded state, and some of the best-preserved examples are on Taiwan and the Ryukyu Islands. Most leaves are mesophytic, but the canopy is lower (typically 15–25 m) than in lowland forests further south, and both epiphytic angiosperms and large woody lianas are relatively rare. Common tree genera include: *Ilex* (Aquifoliaceae), *Beilschmidia, Cinnamomum, Cryptocarya, Lindera, Machilus, Neolitsea, Phoebe* (Lauraceae), *Castanopsis, Cyclobalanopsis* (*Quercus*), *Lithocarpus* (Fagaceae), *Schima* (Theaceae), *Elaeocarpus* (Elaeocarpaceae), and *Symplocos* (Symplocaceae). One or more species of *Castanopsis* is often dominant. There is often an admixture of conifers and deciduous, broad-leaved trees. In many ways, this subtropical lowland forest resembles the lower montane forest of the tropics (see 2.5.2), and it shares most plant families, many genera, and some species with these forests, despite the great difference in temperature seasonality.

The northern limits of dominance by broad-leaved evergreen trees seem to coincide with an absolute minimum temperature of around −15°C degrees (corresponding to a January mean of around 0°C). Although individual evergreen species penetrate further north, deciduous, broad-leaved trees become dominant.

Forests on extreme soil types

In general, the influence of soil type on vegetation appears to increase with declining rainfall and increasing seasonality, but, even in the wettest areas, three extreme types of substrate tend to produce distinctive vegetation types. The best studied of these are the 'heath forests' that are found on the Spodosols/Podzols and Psamments/Arenosols developed on sand or sandstone. They are most extensive in Borneo, but smaller areas occur on the Malay Peninsula, Sumatra, and in eastern Indonesia. Heath forests are both structurally and floristically distinct from the surrounding forest on other substrates. Typically, there is a relatively low, uniform, small-leaved canopy that is easily recognized on aerial photographs, and a much lower tree diversity (Table 3.1). Despite a considerable research effort, it is still not clear if the primary cause of this distinctiveness is a shortage of nutrients or a low water-holding capacity and thus high incidence of drought (Becker et al. 1999). The soils under heath forest are useless for agriculture and, although light, selective logging may be sustainable, clear-felling seems to lead to irreversible soil deterioration.

Ultramafic (or ultrabasic) rocks are extensive only on Sulawesi, but smaller areas occur throughout the region. Soils derived from these rocks are very variable, but tend to be relatively shallow and to have both low concentrations of important plant nutrients and potentially toxic levels of nickel, magnesium, and other metals (Proctor 2003). On some sites the vegetation is not distinct from the surrounding areas, while in others the forest is sparse or stunted, and contains species that are rare or absent on other substrates. On Mount Kinabalu, ultramafic forests are similar to non-ultramafic forests at lower altitudes, but become increasingly dissimilar with increasing altitude (Aiba and Kitayama 1999). There is evidence that it is the low water-holding capacity of the soil, rather than soil chemistry, that is responsible for the relatively low stature of many of these forests, but higher fire frequencies and slower rates of post-fire succession may be additional complications (Proctor 2003). Soils under these forests are infertile and not usually cultivated.

Exposed limestone has a very patchy distribution in the region, but covers large areas in total: around 410,000 km^2 in South-East Asia (Clements et al. 2006) and 430,000 km^2 in south-west China (Wang et al. 2004). Irregular weathering produces a wide variety of plant habitats, differing in slope, soil thickness, and soil chemistry. Part of the variation reflects differences in the purity of the limestone, since the soil is built up from the insoluble residues left after the carbonates have been dissolved. Forest on limestone is very variable but, in wet areas, tends to have a shorter stature and lower tree diversity (Proctor et al. 1983; Zhu 2008a) than forests on other substrates. The herbaceous flora, in contrast, may be rich and often includes species found nowhere else. In the seasonal tropics, limestone outcrops sometimes support more impressive forests than the surroundings (Fig. 2.9), although this may simply reflect reduced human impact, as a result of their inaccessibility and general unsuitability for agriculture. Cultivation can lead to rapid desertification as the thin soil layer is lost and the bedrock exposed (Wang et al. 2004).

Secondary forests

The term 'secondary forest' is often applied indiscriminately to all forests that have been disturbed, particularly by human impact. However, the major human impacts on forests in TEA—clearance for agriculture and selective logging for timber—have such different effects that it makes sense to distinguish them clearly (Corlett 1994). Secondary forest is, therefore, defined here as forest that has regrown after clearance. In contrast to logged forest, where recovery is dominated by species that survive on the site, secondary forest in this restricted sense consists largely of species that have dispersed to the site from elsewhere, although some may regenerate from root sprouts years after clearance. Young secondary forests are usually easily recognized by their low, uniform canopy and a tendency to dominance by one or a few species. Older secondary forests are more variable and may increasingly resemble primary forest in structure and species composition. When, or if, the differences finally disappear will depend on many factors, of which the most important are probably the size of the cleared area—and thus the proximity of primary forest seed sources—and the fertility of the soil (see 4.6).

One useful way of classifying anthropogenic secondary forests is by the degree of soil degradation, as shown by the capacity of the site to support fast-growing pioneer trees. At one extreme are undegraded soils, on which succession is dominated by fast-growing trees with low density wood, while at the other extreme are highly degraded soils, where succession is dominated by slow-growing shrubs

Figure 2.9 Semi-evergreen forest on limestone at Wangxia, Hainan Island. Photograph courtesy of Billy C.H. Hau.

and trees, often with high density wood (Corlett 1991) (Fig. 2.10). Natural catastrophes of various types and sizes provide analogues for anthropogenic secondary forests, but there is no natural equivalent of the repeated disturbances over large areas that characterize much of the landscape today.

Logged forests

Most lowland forests in the region have had some timber removed, but the logging intensity has varied greatly, from the harvesting of scattered individual trees by local people to the mechanized commercial logging of up to 72 trees per hectare, in those exceptional areas where valuable trees occur at such high densities (Johns 1997). More typically, the range in tropical rainforests is 8–24 trees per hectare, with only large individuals of commercially valuable species removed. Although the logging is highly selective, the damage caused to the remaining forest by the felling of huge-crowned emergents and their removal by heavy machinery along skid trails is not (see 7.4.6). Logging can kill more than half the trees in the forest, with the damage spread across species and size classes (Bischoff et al. 2005). Recently logged forest has an increased fuel load and is more open, and thus drier, than unlogged forest, making it more susceptible to fire. Moreover, the soil over much of the logged area is compacted, leading to decreased infiltration, increased erosion, and slow regeneration.

The botanical impacts of logging have received less attention than the impacts on animals (Meijaard et al. 2005; Cleary et al. 2007). Most botanical studies have been done only 1–6 years after logging, or in older forests that received post-logging management that is rarely carried out today, so the long-term impacts of present-day logging practices are largely unknown (Bischoff et al. 2005). In heavily logged forests there is generally a short-term increase in the recruitment of light-demanding species as a result of canopy opening, but there have been too few studies to generalize about what happens subsequently (Cannon et al. 1998; Slik et al. 2003; Bischoff et al. 2005). Around 200,000 km^2 of lowland forest in Borneo—half the remaining forest area—is in active logging concessions, as well as huge areas elsewhere in TEA, so improved management of these areas is of key conservation importance (Meijaard and Sheil 2007).

Figure 2.10 Secondary forest dominated by slow-growing *Adinandra dumosa* on a highly degraded site in Singapore. Photograph courtesy of Hugh T.W. Tan.

Bamboo forests

Bamboos are a component of most forest types in the region and some species can become abundant when the forest is disturbed by logging or shifting cultivation, particularly in seasonal areas. Although some bamboo-dominated vegetation is probably a self-perpetuating response to local environmental conditions, and other areas may result from catastrophic natural disturbances, the extensive, almost monospecific, bamboo forests that occur in parts of continental South-East Asia seem to be largely secondary in origin. Despite their importance, surprisingly little is known about the ecology of most bamboo species in TEA and few studies anywhere have followed the lifecycle from seed to seed. Some species appear to reproduce annually or more or less continuously, but most do so only at multi-year intervals. In some species, in the more seasonal parts of the region, reproduction is gregarious over large areas and followed by death (see 2.6.2).

Savannas and grasslands

In TEA, the term 'savanna' is usually applied to vegetation with a discontinuous tree layer over a more or less continuous grass layer. Some areas of lowland savanna, in the driest parts of eastern Indonesia, may be natural, owing their existence to some combination of drought, soil factors, lightning-caused fires, and/or seasonal inundation. However, all existing savannas are burned more or less regularly by people and there is little doubt that most have replaced forest as a result of human impact.

Stable, fire-maintained savanna depends on the occurrence in the local flora of tree species that can not only survive, but also regenerate, under such conditions. *Melaleuca* savannas occur throughout the region on seasonally waterlogged or inundated soils, while *Casuarina* and *Eucalyptus* savannas occupy huge areas in seasonal eastern Indonesia. Species of *Acacia* and the palms, *Borassus flabellifer* and *Corypha utan*, also form extensive savannas in this region. In contrast, savanna is much less extensive in lowland continental South-East Asia and, where it does occur, usually represents a transitional stage in the degradation or regeneration of forest, rather than a more or less stable, fire-maintained vegetation type. Upland pine savannas are considered in the montane vegetation section below.

Where the fire regime—or a combination of fire, cutting, and/or grazing by livestock—exceeds the tolerance of the local tree flora, treeless grassland replaces savanna. Grasslands also develop directly on soils exhausted by prolonged cultivation or short fallows, and are maintained by regular burning. Except in the drier parts of the region, the invasive and extremely fire-tolerant grass, *Imperata cylindrica* (alang alang, lalang, cogon, etc.), is characteristic of such situations. Garrity et al. (1996) estimated that 25 million hectares of TEA are covered in *Imperata* grassland, around 4% of the total land area.

Shrublands and thickets

Natural evergreen shrublands are found at the altitudinal limit of tree growth on some mountains. The extensive shrublands and thickets in the region today, however, are mostly on sites that have been deforested and abandoned. Deciduous and often thorny thickets are a common and persistent result of clearing forests in seasonally dry areas, while evergreen shrublands are a short-lived successional stage in wetter areas. Naturalized exotic plants, such as the tropical American composite, *Chromolaena* (*Eupatorium*) *odorata*, are often prominent or even dominant in such vegetation types (e.g. Laumonier 1997).

Beach vegetation

On accreting sandy beaches, a low community of creeping herbs, grasses, and sedges occupies the zone between high-water mark and the margin of the beach. On undisturbed beaches, there is then a belt of coastal forest, extending 5–50 m inland, sometimes with a seaward fringe of more or less pure *Casuarina equisetifolia*. In most of the region, however, this coastal forest has now been replaced by coconut plantations. In the Andaman Islands, there are still extensive coastal forests dominated by huge trees of *Manilkara littoralis* (Sapotaceae). Many beach communities are dominated by plant species with very wide geographical ranges: in some cases throughout the tropics.

Plantations

Huge and increasing areas in the region are planted with tree and shrub plantation crops, particularly rubber and oil palm (Fig. 2.11), with smaller areas of bananas, cashew nuts, coffee, coconuts, cocoa, tea, and other species (see also 7.4.2). The area planted with trees for pulp, plywood, timber, or fuel, in contrast, is relatively small, but also rapidly increasing. The total area of plantations in Malaysia now exceeds the area of natural forest. The most widely planted

Figure 2.11 A plantation of oil palm, *Elaeis guineensis*. Photograph © Peter Solness, Greenpeace.

tree species in the tropics, including *Acacia mangium*, *Eucalyptus urophylla*, *Gmelina arborea*, *Paraserianthes falcataria*, *Pinus merkusii*, *P. kesiya*, and *Tectona grandis*, are native somewhere in the Asia-Pacific region, although now planted far outside their natural ranges. Others, such as *Leucaena leucocephala* and *Swietenia macrophylla* (mahogony), are exotics. From the 1950s to the 1980s, huge areas of broad-leaved forest in Japan, including the subtropical Ryukyu Islands, were clear-cut and converted to conifers, such as *Cryptomeria japonica* (Agetsuma 2007). Almost all commercial plantations are monocultures and thus have a much simpler structure and lower diversity than any natural forest, although both plant and animal diversity increase if a native understorey is allowed to develop.

Agroforestry

Agroforestry is the term for a wide range of tree-dominated, multi-species cultivation systems used to produce both food and cash crops. They range from mixed-tree plantations to multi-layered systems that also include annual crops. The floristic diversity of agroforestry plots may be high and their structure is often more similar to secondary forest than to monoculture plantations. Such areas can provide an important habitat for wildlife in an otherwise deforested landscape (e.g. Beukema et al. 2007). Although each patch is usually small, their aggregate area in much of the region is huge. Homegardens—small, diverse, multi-storied gardens around houses—are a major land use in Indonesia, the Philippines, and other parts of TEA, and make up 63% of arable land in the Andaman and Nicobar Islands (Kumar and Nair 2004; Pandey et al. 2007). The main function of homegardens is usually to produce non-staple foods for family consumption, but surplus food and non-food crops (spices, coffee, cocoa, etc.) are also sold for cash income.

Other dry-land crops

Dry-land cultivation of non-woody crops ranges from the polycultures of some shifting cultivators to highly mechanized industrial monocultures. Shared features include the simplified structure and low species diversity compared with natural vegetation, and the incomplete soil coverage for at least part of each cropping cycle. In general—and in apparent contrast to wetland rice and dry-land tree crops—this type of agriculture requires either long, fallow periods or massive, external inputs in the form of fertilizer and pesticides. Numerous different dry-land crops are grown, but it is interesting to note the importance of the American crops, maize, cassava, and sweet potato, which were introduced to the region by Europeans in the sixteenth and seventeenth centuries (see 1.9). Soybeans and sugarcane, in contrast, have a much longer history in the region. Unlike shifting cultivation and agroforestry, dry-land crop monocultures, whether of shrubs, herbs, or grasses, support few native species of plants and animals.

2.5.2 Montane vegetation

Although TEA has extensive highland regions, few peaks exceed the climatic forest limit of 3800–4000 m. The highest mountain in the region, Hkakabo Razi (28°N) in northern Myanmar, reaches 5881 m and has a cap of permanent snow and ice. Mount Kinabalu (6°N), the highest peak in tropical Asia, attains 4094 m, but the summit is largely bare rock, scoured by Pleistocene ice, so there is no climatic treeline. Tambora, on the island of Sumbawa, may well have been higher than Kinabalu before the 1815 super-eruption, but the top third was lost in the explosion. Yushan (23°N), the highest peak in the north-eastern part of TEA, attains 3952 m and also has a summit region dominated by bare rock.

Temperature declines with altitude at an average rate of around 0.6°C per 100 m (see 2.2.1). All other environmental factors (except day length) also change as one ascends a mountain, but these changes are not necessarily unidirectional or correlated with each other. Flenley (2007) argues that the increase in UV-B (280–315 nm) radiation with altitude may have a significant influence on vegetation, but there is as yet no direct evidence for this. On Kinabalu, soil organic carbon content rapidly increases with altitude on sedimentary rocks, but not on ultrabasics (Kitayama and Aiba 2002b). Vegetation structure, physiognomy, and floristics all change with altitude. These changes may be gradual, but are often more or less abrupt, resulting in stepwise pattern of vegetation change along an apparently smooth environmental gradient. With increasing altitude, the forest becomes shorter, tree heights more even, the crowns and leaves smaller, rooting shallower, and cold-intolerant plant genera and families progressively drop out, while a smaller number are added.

The most widely used nomenclature refers to the forest zones above the lowlands as lower montane, upper montane, and subalpine, but only the tallest mountains have all three. In the alpine region above the altitudinal treeline at 3900–4000 m, the vegetation is dominated by grasses, low shrubs, or herbs. Permanent caps of snow and ice are only found on mountains that exceed 4650 m, although tongues from the glaciers on higher peaks extend considerably lower.

The details of the zonation vary considerably between the aseasonal and seasonal tropics, and between the tropics and subtropics. In the aseasonal tropics, the transition between the lowland rainforest and montane forest is typically gradual, and occurs between around 800 and 1300 m above sea-level. Most Dipterocarpaceae drop out and the new dominants usually include members of the families Fagaceae, Lauraceae, Myrtaceae, and Theaceae. Lower montane forest usually lacks emergents, the trees do not have prominent buttresses, and woody climbers are usually less diverse and abundant than in the lowlands. In many places this transition coincides with a change in soils, which become richer in organic matter and dominated by earthworms instead of termites (Ashton 2003). However, the causal relationships behind this correlation between declining temperature, changing flora, and changing soil and soil biota have not yet been worked out.

In areas with strong rainfall seasonality, where the lowland vegetation is deciduous and regularly burns, the lower montane forest is largely evergreen and fire-intolerant. The transition between the two may be sharp or occur through a band of mixed deciduous and evergreen tree species. A regular dry season makes montane forests very vulnerable to fire, however, and in areas with seasonal climates they have often been reduced to relict patches in moist or topographically protected sites, or eliminated altogether. Large areas of seasonal uplands are now covered in open forests dominated by fire-resistant pines or treeless grasslands. Fire-climax pine forests occur in lower montane areas in Sumatra, the Philippines, and mainland South-East Asia.

Above the lowland/lower montane transition, the most dramatic differences in the forest often coincide with the zone of persistent daytime cloud, where trunks and branches become contorted and bryophytes cover all surfaces. As well as being called upper montane forest, this vegetation is often referred to as 'cloud forest', 'elfin forest', or 'mossy forest', although the bryophytes are mostly liverworts rather than mosses. The altitude of the diurnal cloud base, and thus the lower montane/upper montane transition, is higher on large mountains than on isolated or coastal mountains, and lower on ridges. This transition can, therefore, be very sharp on any particular route up a mountain, while varying considerably between routes in the altitude at which it occurs. On the tallest tropical mountains, an additional zone, the subalpine forest, may be distinguishable just below the treeline, although the transition from upper montane forest is gradual.

No mountain in the Asian tropics has an altitudinal treeline, but in the subtropics and on tropical mountains elsewhere, the altitudinal tree limit occurs where the mean temperature during the growing season falls below 5–7°C (Körner and Paulsen 2004; Shi et al. 2008a). Studies in Sichuan and Tibet, on the extreme north-west of TEA (28–31°N), found that the non-structural carbohydrate concentrations in woody species at the end of the growing season generally increased with altitude, suggesting that the treeline is not determined by inadequate carbohydrate supply, but rather by a direct impact of low temperature on growth (Shi et al. 2008a).

On an equatorial mountain, frosts occur every night at the treeline and occasionally down to 2000 m or below in open 'frost hollows'. The diurnal temperature range is much greater than the annual range, so plants cannot adapt by deciduousness or seasonal dormancy. In the north of the region, in contrast, frosts are a predictable, annual event throughout the montane zone, and winter cold is a significant ecological factor. North of about 18–20° latitude, both winter-deciduous trees and evergreen conifers from temperate genera start to become an important part of the montane flora, particularly above 2000 m. On Yushan (23°N), in Taiwan, evergreen, broad-leaved trees dominate up to around 2500 m and conifers (*Abies*, *Picea*, *Tsuga*) dominate above (Su 1984). On the highest mountains in northern Myanmar (26–28°N), the broad-leaved evergreen forest of lower altitudes acquires an increasing admixture of deciduous species and conifers up to around 3000 m, from where coniferous forest dominated by *Abies* continues to an irregular treeline at about 4000 m (Kingdon-Ward 1945). Snow carpets the ground below the fir trees for several weeks every year. In the alpine region above the treeline, a *Rhododendron*-dominated scrub gradually gives way to an herbaceous turf, with stunted shrubs, below the line of permanent snow. A similar zonation occurs on the eastern slopes of the Gangga Mountains (30°N), on the eastern margins of the Tibetan Plateau (Luo et al. 2004), and on Mt Emei (29°N) in Sichuan (Tang 2006). On Yakushima Island (30°N, 1936 m), in contrast, conifers (*Abies*, *Cryptomeria*, *Tsuga*) dominate increasingly from 1200 m to near the summit, which is covered in a dwarf bamboo scrub (Hanya et al. 2004).

2.5.3 Wetlands

Wetlands can be defined as areas where flooding or saturation of the soil occurs with such frequency and duration that the organisms that live there need special adaptations. In the absence of human impacts, the natural vegetation of most wetland areas in South-East Asia would be forest. The existence of natural, non-forest wetlands is confirmed, however, by the presence of several animal species in the regional fauna that require such habitats (Dudgeon 2000). The forested wetlands of South-East Asia are extremely varied and the classification used here is necessarily both simplified and somewhat arbitrary.

Mangrove forests

Mangrove forests occupy the upper-half of the intertidal zone on muddy shores. Compared with other forest types in the region, they have a much simpler structure and much lower floristic diversity, although the mangrove forests of South-East Asia and northern Australia are more species-rich than those elsewhere in the tropics. Typically, there is a clear zonation of species, controlled by the frequency and duration of tidal flooding, the amount of freshwater input, and the characteristics of the substrate. Around a quarter of the world's mangrove forests were in South-East Asia, with the largest areas in Sumatra and Kalimantan. Huge areas have been lost or severely degraded in the last few decades, by conversion to shrimp or brackish fish ponds, reclamation for agriculture, salt ponds, or urban development, and logging for charcoal, woodchips, and pulp.

Brackish water swamp forests

Areas subject to tidal flooding by brackish water support a flora distinct from both the mangrove forest and the freshwater swamp forest. The nipa palm *Nypa fruticans* is characteristic of such sites, forming pure or mixed stands along the tidal section of rivers, as well as covering extensive low-lying areas in estuaries.

Freshwater swamp forests

Forests subject to flooding by freshwater are extremely varied within the region and are lumped together into one category purely for convenience. Much of this variability reflects differences in the periodicity and duration of flooding. Swamp forests near the coast may be flooded daily or a few days a month, when river water is backed up by the tides. This 'freshwater mangrove', as it has been called, shares many features with the true mangrove forest, including stilt roots and peg-like pneumatophores (Corlett 1986). Further inland, flooding may be semi-permanent, irregular, or seasonal, and vary in depth from a few centimetres to several metres. It is difficult to make generalizations about forests growing in such varied environments but the flora is generally less diverse than in adjacent dry-land forests and there is a tendency for dominance by one or a few tree species.

Freshwater swamp forests are not confined to the everwet parts of the region, but those in seasonally dry areas are easily degraded by cutting and fire. Large areas dominated by more or less pure stands of fire-resistant paperbark (*Melaleuca* species) are one result of this process (e.g. Chokkalingam et al. 2007). Elsewhere, swamp forests have been replaced by woody thickets or by grass-dominated vegetation. In addition, vast areas of freshwater swamp forest have been converted to wet rice cultivation.

Peat swamp forests

The freshwater swamp forests described above sometimes have a thin layer of peat on the soil surface, but in peat swamp forests this layer is deeper (0.5 m to more than 10 m) and the surface is above the highest limit of wet season flooding by mineral-rich river water. The water table is higher than the surrounding area, so the only external input of water and nutrients is from rainfall. The peat generally consists of partly decomposed woody material in an amorphous semi-liquid matrix. The organic matter content is 90–98%, while both pH and nutrient content are low (see Histosols in 2.4.1). Unlike freshwater swamp forest, forest on deep peat is confined to areas with high rainfall and without a long dry-season, and is most extensive on the islands of Sumatra and Borneo, where it blankets huge areas. Smaller areas occur in the Malay Peninsula and Southeast Thailand, and on Mindanao, Sulawesi, Halmahera, and Seram. It is found mostly in the coastal and sub-coastal lowlands, close to sea-level, but also extends up river valleys and occurs in isolated basins at higher altitudes.

The best-developed peat swamps have a characteristic convex surface and a sequence of forest types occurs from the margin towards the centre (Fig. 2.12). The outer zone, on the thinnest peat, is similar to dry-land forest in structure and floristics, while the successive zones on deeper peat have

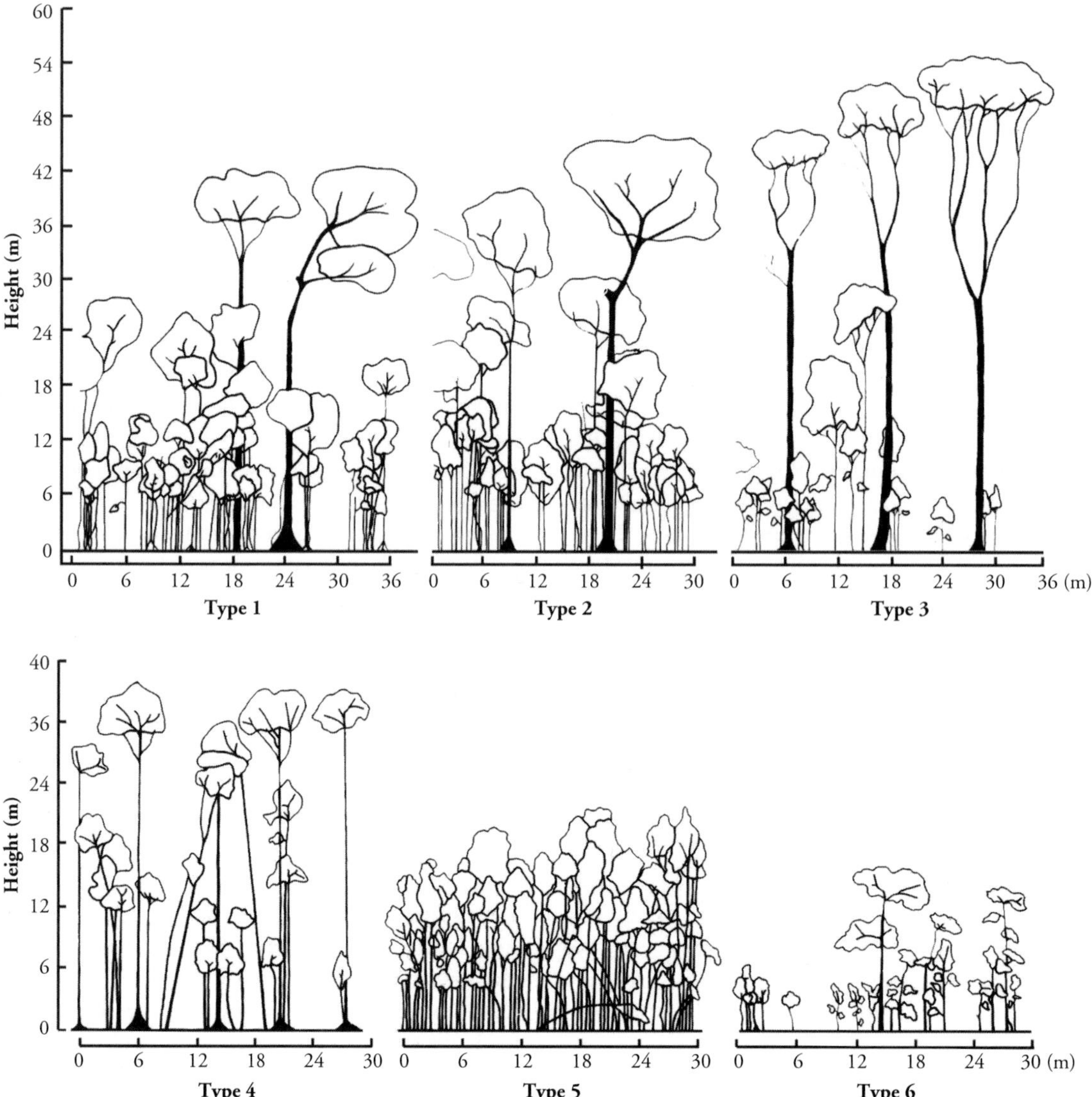

Figure 2.12 The sequence of forest types from the margins (Type 1) to the centre (Type 6) of a peat swamp in Sarawak. From Anderson (1983).

a progressively lower height, smaller tree girth, higher density of stems, and lower species diversity, presumably in response to declining nutrient supply. In some extreme cases, the vegetation at the centre of the largest swamps is like an open, savanna woodland (Anderson 1983). Peat swamp forests appear to have a lower diversity and density of wildlife than dry-land or freshwater swamp forests, presumably because of the relatively low primary productivity.

The less extreme types of peat swamp forest are a very important source of commercial timber, particularly ramin (*Gonystylus bancanus*). Swamp forest on shallow peat (<50 cm) can be successfully converted to rice, pineapple, coconut, or sago production *if* the subsoil beneath is fertile, but attempts to convert deep peats (>1 m) have usually failed (Wösten et al. 2008). The biggest and most catastrophic failure involved the infamous Mega Rice Project, which attempted to convert a million hectares of peat swamp forest in Central Kalimantan into rice paddies, but instead resulted in a fire-prone wasteland (Aldhous 2004; Wösten et al. 2008). Pulpwood plantations of *Acacia crassicarpa* appear to be financially viable on drained peat, although their long-term sustainability is in doubt. There is now very little peat swamp forest in the region that has not been either cleared or severely degraded by logging and/or drainage.

Herbaceous swamps

In South-East Asia, natural, non-forest wetlands seem to be confined, in the lowlands, to areas with seasonal rainfall. In these climates, swamp grasses, sedges, herbs, or ferns cover large areas on alluvial plains, where flooding is too deep, frequent, or prolonged, or the substrate too unstable, for the establishment and growth of trees. The major non-forest wetlands of mainland South-East Asia were in the lower basins of the major rivers, the Irrawaddy, Chao Phraya, Mekong, and Sông Hóng (Red River). Over the last century, these have been almost entirely converted to the cultivation of rice, and now support dense human populations. The animals that were dependent on this habitat have become endangered or are extinct (e.g. Schomburgk's deer in central Thailand and Père David's deer is subtropical China). Herbaceous swamps, dominated by sedges, grasses, or herbs, are also common in montane basins throughout the region and, at least at higher altitudes, some of these are apparently natural.

Rice fields

Rice was domesticated and first cultivated in natural wetlands (see 1.9), but the area used for wet rice cultivation has been greatly extended by the development of techniques for water retention and irrigation (Fig. 2.13). Today, rice is by far the most important crop in the region, occupying a total of around 600,000 km^2. Some areas may have been in continuous cultivation for millennia, demonstrating a sustainability not shown by any other farming system in the region. With the exception of the relatively small areas of upland rice, rice fields are flooded for part of the year, making them also the most extensive wetland type in the region. These fields and the associated dikes, ditches, and ponds, provide an important habitat for the more tolerant wetland birds, mammals, and other vertebrate and invertebrate animals, as well as a diverse weed flora. However, agricultural intensification, including heavy pesticide use, has greatly reduced the wildlife value of many areas.

2.5.4 Urban vegetation

All over the world, the proportion of the land surface covered in urban areas is increasing. Nowhere is this happening more rapidly than in TEA, where new urban areas now occupy land that was used for agriculture a few years ago. Although urban vegetation is diverse in terms of structure, it is relatively uniform floristically, not only within a particular city, but throughout TEA. Cities are a relatively new habitat—less than 4000 years old in TEA—so wildlife has had little time to adapt to them, but those species that have adapted have often been able to spread to other cities within the region, either spontaneously, as weeds, or through deliberate human agency as cultivated plants.

From an ecological viewpoint, the key feature of urbanization is the replacement of surfaces covered by natural soil and vegetation by surfaces

Figure 2.13 Rice is the most important crop in the region and also the most extensive wetland type. This photograph was taken in Guangxi. Photograph courtesy of Lee Kwok Shing, KFBG.

covered in impervious materials, such as concrete, stone, and asphalt. The proportion of land covered by such materials varies from <20% in some low-density suburbs, to 100% in many city centres and industrial areas. The vegetation of parks and gardens is a better habitat for wildlife, but it is usually heavily managed in order to prevent successional development and exclude most native plant species, while only a few wild species can tolerate the close-mown grassland of lawns and playing fields. The most common type of urban vegetation in the region is an 'urban savanna' with scattered trees, perhaps reflecting innate human preferences for a habitat similar to that in which our African ancestors evolved. There are also small patches of unmanaged vegetation within most urban areas, where wild plants are allowed to grow and animals to feed, rest, and nest, with little disturbance. Most of these patches are on steep slopes and they may be absent in cities that have spread over flat agricultural land. Where present, they increase the biodiversity of urban areas by providing refuges for many less tolerant wild species.

Urban areas are, in general, warmer, drier, noisier, and more polluted than other habitats. The increase in temperature—known as the 'urban heat island effect'—is a result of a variety of factors, including waste heat from buildings and vehicles, dark surfaces that absorb more heat from the sun, the absence of evaporative cooling from vegetation and soil, and streets lined by tall building that trap heat. Satellite thermal data shows dry-season urban-rural temperature differences of 5–8°C in Bangkok, Manila, and Ho Chi Minh City (Tran et al. 2006). Within an urban area, vegetated areas, such as parks, are several degrees cooler than their surroundings. The lower humidity of urban areas results from the rapid draining of water from impervious surfaces after rain, leaving none to evaporate. Motor vehicles, especially diesels, are the main sources of air pollution, particularly suspended particulates and nitrogen oxides, and the biggest sources of noise. Where soil is present, it is usually highly compacted, poorly drained and aerated, low in nutrients, and full of concrete, brick, mortar, and other urban debris. Another important characteristic of most urban habitats is their continual state of change, as buildings are constructed and demolished, roads and pavements dug up, and slopes re-surfaced. This disturbance favours short-lived, quick-growing, fast-breeding species.

2.6 Plant phenology

Phenology is the study of the timing of recurring biological events. The phenological patterns shown by plant communities are a major outcome of plant–climate relationships and, in turn, a major component of the environment of other organisms. Changes in plant phenology will be among the first impacts of climate change (Corlett and LaFrankie 1998) and phenological flexibility sets limits on the past, present, and future geographical ranges of plants (Chuine and Beuabien 2001). Most information available for TEA concerns the patterns of flowering and fruiting, production of new leaves, or leaf fall. There is also some information on the phenology of diameter growth. The phenology of below-ground processes is poorly known, although Green et al. (2005) showed that fine root growth and mortality are related to rainfall in the lowland rainforest at Danum, with fine root biomass lowest after dry periods. In contrast, fine root biomass was highest at the end of the dry season in Sulawesi (Harteveld et al. 2007) and in early spring, before the main period of active above-ground growth, in subtropical China (Yang et al. 2004).

Seasonal variation in water availability—reflecting rainfall, but also evaporation and the water storage capacity of the soil—has traditionally been seen as the main driver of plant phenology in the tropics, both as a physiological constraint and as a cue. More recently, it has become apparent that tropical trees are also constrained by the amount of light available (i.e. the solar irradiance), particularly during cloudy periods. Trees given extra light from high-intensity lamps during the cloudiest part of the wet season in Panama grew more than those getting only natural light (Graham et al. 2003). At Barro Colorado Island, also in Panama, community peaks of both leafing and reproduction follow the seasonality in irradiance, not rainfall, despite the 4–5 month dry season (Zimmerman et al. 2007). In theory, storing carbohydrate to even out supply should be more practical than storing water, because of the much smaller volume required, but there is little evidence that this happens on a large enough scale and a dipterocarp, *Dryobalanops aromatica*, used current photosynthate from the leaves to produce fruits, even during a mast year (Ichie et al. 2005). Seasonal low temperatures also limit plant growth at higher altitudes and latitudes, with plants generally unable to grow when the mean air temperature falls below zero. At the altitudinal treeline in the extreme north-west of TEA, the growing season is reduced by winter cold to around 120 days (Shi et al. 2008a).

2.6.1 Leaf phenology

The forests of TEA range from fully evergreen, with no seasonal changes in canopy density, to fully deciduous, with at least a brief period when all trees are leafless. In general, this gradient correlates with the predicted (but rarely measured) seasonal availability of water. Forests are largely evergreen where it rains year round, or the dry season is brief and/or cool, or fog is common in the dry season (Liu et al. 2008b), or ground water is available all year. They are mostly deciduous where there is a prolonged, annual period of low water-availability, unmitigated by low temperatures, fog, or ground water. However, because natural fires are more common at the drier end of this spectrum, fire-tolerant tree species tend to be deciduous and anthropogenic fires have consequently facilitated the spread of deciduous forests into areas that could support evergreen forest. At the species level, there are many ways of being evergreen or deciduous, and there can also be considerable within-species variation between individuals at the same or different sites. Leaf phenology—particularly leaf lifespan—is also generally correlated with other leaf and whole-plant traits, but these have been little investigated in the TEA.

Even at the least seasonal sites, continuous growth and leaf production is rare, at least in trees and shrubs. The great majority of tree species show synchronous flushing of new leaves at the branch level, most are more or less synchronous at the individual level, and many are at the species level. Even brief dry-periods (<15 mm rainfall in 14 days), of the type that occur several times a year, are enough to cause synchronous loss of old leaves and flushing of new leaves in canopy trees at aseasonal Lambir (Ichie et al. 2004), and a severe supra-annual drought caused a massive increase in leafing in most canopy species over the next 3 months (Itioka and Yamauti 2004). At Sepilok, on the east coast of Sabah, where no month receives less than 100 mm of rainfall on average, litterfall increased linearly with decreasing rainfall in the previous month in forest overlying well-drained sandstone, but not in alluvial or heath forests in the same watershed (Dent et al. 2006).

The advantages of synchronization of leaf production, even at the branch or the individual level, are not obvious, since this strategy is likely to incur additional storage costs over the alternative of continuous production of leaves, as resources become available. Synchronization at the branch, individual, and species level may be an anti-herbivore measure, since the populations of specialized herbivores that require young leaves will be depressed between leaf flushes and cannot instantaneous expand to take advantage of the post-flush bonanza (Aide 1992). Community-level synchronization could provide similar protection against more generalist herbivores, as long as they cannot switch to an alternative diet between flushes. However, the community-level increase in leafing after the severe drought at Lambir, described above, led to a massive *increase* in lepidopteran abundance and a large *increase* in average leaf damage. For such extreme events, the simplest explanation is that water stress causes the tree to abscise older leaves with poor control over water loss, thus improving tree water status and promoting flushing (Borchert 1994), but whether this passive mechanism applies in all cases is unclear.

An opportunistic leafing response to tree water-status, is also one possible strategy for surviving regular annual periods of water stress, where these vary significantly in length or severity. In species with this strategy, drought causes the abscission of old leaves and new leaves are produced as soon as the tree water status is sufficiently good. This could be more or less immediately ('leaf-exchanging') for plants with access to soil-water reserves, or after a leafless period that lasts until the first heavy rains (>20–30 mm), for plants without such access (Rivera et al. 2002). The timing for such species will, therefore, vary from year to year and site to site. Some tree species in dry forests are confined to moist sites where leaf-exchange is possible, but others are opportunists, which can be leaf-exchangers or deciduous depending on water availability (Elliott et al. 2006). In a *Shorea siamensis* stand at Huai Kha Khaeng, the largest trees expanded most of their new leaves before rain, but small trees—presumably with relatively shallow root systems—expanded their leaves before the first rains only near a stream bed.

The commonest leafing strategy among deciduous trees in TEA is not, however, the opportunism described above, but synchronous 'spring flushing' around the time of the spring equinox, apparently induced by increasing day length (Rivera et al. 2002; Elliott et al. 2006; Williams et al. 2008). These trees thus leaf at the hottest, driest time of the year, relying on soil water-storage. This could be a risky strategy if the start of the rainy season were delayed, but this risk is presumably outweighed by the advantages, in a normal year, of using soil water-reserves to extend the period of photosynthetic activity beyond the wet season and to have new leaves during the period of maximum solar irradiance. Leafing before the start of the wet season may also reduce the impact of herbivorous insects on the vulnerable young leaves, if insect populations are limited directly by moisture availability (Aide 1992). Whatever the proximate and ultimate causes of phenological patterns in seasonal tropical forests, the result at Huai Kha Khaeng, in western Thailand, is a continuum of deciduousness, from complete crown loss to no crown loss, with the duration of the leafless period ranging from 2 to 21 weeks (Williams et al. 2008).

Further north, regular, annual cycles of leaf production and fall predominate, with most species in the subtropical, broad-leaved evergreen forest exchanging old leaves for new between March and May (Dudgeon and Corlett 2004; Li et al. 2005). At these latitudes, temperature, rainfall, and day length are all strongly seasonal, and neither the proximal cues nor the ultimate adaptive significance of this phenological pattern are known. There are also deciduous species, mostly from tropical genera (e.g. *Cratoxylum*, *Sapium*) in the south of the broad-leaved evergreen forest belt and from temperate genera (e.g. *Betula*, *Populus*) in the north. Deciduousness is more common among the woody pioneers than the canopy species of climax forest.

Leaf phenology is clearly an understudied subject in TEA and there are numerous gaps in our knowledge and understanding of this complex and important phenomenon. A simple quantification of canopy deciduousness across local (soil, topography, fire) and regional (climate) environmental gradients in normal years and extreme droughts would be very valuable (Condit et al. 2000). One promising technique that could help fill in many gaps is the use of remote sensing for the regional mapping of leaf phenology (Chambers et al. 2007; Cleland et al. 2007; Huete et al. 2008; Ito et al. 2008). This should be particularly useful for distinguishing regular seasonal cycles from supra-annual, often ENSO-linked, changes, and both of these from the long-term impacts of global climate change. More research is also needed at the other end of the scale: the phenology of individual shoots. A regional network of phenological plots could provide a link between shoot-level and remote-sensing information.

2.6.2 Reproductive phenology

The life history of a flower bud is more complex than that of a leaf, and thus the opportunities for phenological diversity are much greater. At the individual level, there is also the relationship with leafing to be considered, since flowering often coincides with, or closely follows, leafing. Fruits are a morphological continuation of flowers, but although the phenologies of the two phases are obviously linked—a plant cannot fruit more often than it flowers—fruit development times can vary widely among related plants, suggesting that the timing of flowering and fruiting can respond to different selective pressures (Corlett 2007b). Fruiting typically requires more carbohydrate resources than flowering, so fruiting times may be more often constrained by seasonal variations in the leaf canopy and solar irradiance. On the other hand, fruiting times may also be influenced by the advantages of dispersing seeds at a time when they can germinate immediately.

The reproductive phenology of plants is far more diverse in the humid tropics than in the temperate zone, presumably because there are fewer abiotic constraints (Sakai 2001, 2002). In TEA, even in areas with a distinct dry and/or cool season, flowers and fruits can usually be found in all months. Newstrom et al. (1994) classified phenological patterns in tropical forests on the basis of frequency, regularity, duration, amplitude, synchrony, and date. They recognized four basic classes of patterns that can be applied at all levels from branches, through individuals, populations, guilds with shared pollinators or dispersal agents, to whole communities: continual (often with small breaks), subannual (more than once a year), annual, and supra-annual (every few years). Species with all these patterns and various intermediates can be found in all but the most extreme climates in TEA, but the predominant pattern at the community level varies with climate.

Even at the least seasonal sites in TEA, continual flowering at the individual level is even less common than continual leafing. Continuity at the population level is more widespread, but only in figs (*Ficus*), herbs (e.g. Zingiberaceae; Sakai 2000), and in some understorey and pioneer trees. In the figs, population-level continuity is necessary to maintain populations of species-specific pollinators, while prolonged or irregular flowering in the understorey may be a strategy to take advantage of traplining pollinators, or simply a consequence of the absence of synchronizing cues in this relatively constant environment (Sakai 2000). Continuous flowering, and thus fruiting at the individual level, would appear to be advantageous to light-demanding trees and shrubs because of the seasonal unpredictability of suitable gaps for establishment, but although some pioneers flower and fruit more or less continuously at equatorial sites (e.g. Corlett 1991), most do not (Davies and Ashton 1999), perhaps relying on a dormant seed bank to fill gaps in seed production.

Continual reproduction should be the most efficient strategy physiologically, since it requires no storage of resources, so its rarity must reflect the advantages of synchronizing reproduction within individuals and across populations. For canopy and emergent trees that make use of generalist pollinators, a highly synchronized 'big bang' reproductive strategy may be necessary to attract pollinators, and synchronization between individuals of their relatively brief flowering periods is then necessary to promote outcrossing (Sakai 2000, 2001). Large fruit crops may also attract disproportionately more seed dispersal agents and cause local satiation of seed predators.

Bamboos and Strobilanthes

The 'big bang' reproductive strategy is taken to extremes by the many species of bamboos in which, after many years of vegetative growth, all individuals flower simultaneously, set seed, and then die, to be replaced by a new generation. This phenological behaviour is most common in the more seasonal parts of the region and the bamboo lifecycle can have a dramatic impact on the whole community, where a single bamboo species dominates a large area. The seeds provide a sudden superabundance of food for rodents, jungle fowl (*Gallus* spp.), pheasants, and other granivores (Janzen 1976), while the simultaneous death of the parent plants reduces shade and allows a pulse of tree recruitment in the understorey (Marod et al. 1999). Satiation of seed predators must be one advantage of this strategy, since bamboo seeds are highly edible and would surely all be consumed if they were produced in small amounts. However, the interval between reproductive events can be several decades (<120 years), which is far longer than needed to prevent the build-up of seed predator populations. An alternative hypothesis is that this strategy evolved to encourage fire, which in turn favours the bamboo seedlings (Keeley and Bond 1999). This is conceivable in some areas with a long dry-season, but bamboos with this strategy are also common in the understorey of montane forests in the north of TEA that do not normally burn.

These subtropical forests were the home of the ultimate bamboo specialist, the giant panda (*Ailuropoda melanoleuca*), which once roamed over much of southern China. Pandas are now confined to a tiny fraction of their historical range, in the mountains of western China (28–36°N). Their fragmented habitat makes them vulnerable to the synchronous death of their preferred bamboo species, since migration is now difficult and it may be more than a decade before the next generation of bamboo plants can support pandas again.

A similar reproductive strategy has evolved in a number of shrub species in the genus *Strobilanthes* (Acanthaceae) and related genera, found mostly in montane forests, which are reported to flower, seed, and die at 5–15 year intervals (van Steenis 1942; Daniel 2006). The impact of these events on granivores and competing plants is also similar, where *Strobilanthes* is abundant in the understorey over large areas, except that the flowers are insect pollinated, unlike the wind-pollinated bamboos, so the mass flowering also provides a bonanza for bees, in some cases attracting colonies of the migratory giant honey bee, *Apis dorsata* (Janzen 1976).

General flowering in lowland dipterocarp forests

Synchronization of reproduction at the population level is usual in large trees across the tropics, and supra-annual cycles are also common in relatively aseasonal climates (e.g. Norden et al. 2007). The aseasonal dipterocarp-dominated forests of TEA, however, are unique in the phenomenon of 'general flowering'—synchronized, supra-annual flowering at the community level (Fig. 2.14). General flowering can occur at intervals ranging from less than 1 year to as many as 9 years, and typically involves most dipterocarp species in an area, as well as a wide range of non-dipterocarps, particularly canopy trees, but also many subcanopy trees, understorey treelets, herbs, climbers, and epiphytes (Corlett 1990; Sakai 2002). Moreover, most dipterocarps and many of the other species flower little or not at all in the years between general flowering episodes. General flowering is followed by community-wide mast fruiting, involving the full range of fruit types, from the giant, winged fruits of many dipterocarps to the dust-like seeds of orchids, and fleshy fruits of various sizes. In the dipterocarps, at least, related species often flower in a sequence that is consistent between general flowering episodes, which is most easily explained as a means of minimizing competition for pollinators (Ashton et al. 1988; LaFrankie and Chan 1991; Brearley et al. 2007a). Fruiting of the same related species, in contrast, appears to be more or less simultaneous, which is consistent with the hypothesis that mast fruiting has evolved as a means of satiating seed predators (see 4.3.4).

General flowering and mast fruiting may be synchronized over an area that varies from a few square kilometres to most of the Sunda Shelf. Years of debate over the nature of the cue that triggers flowering have apparently been resolved recently, with

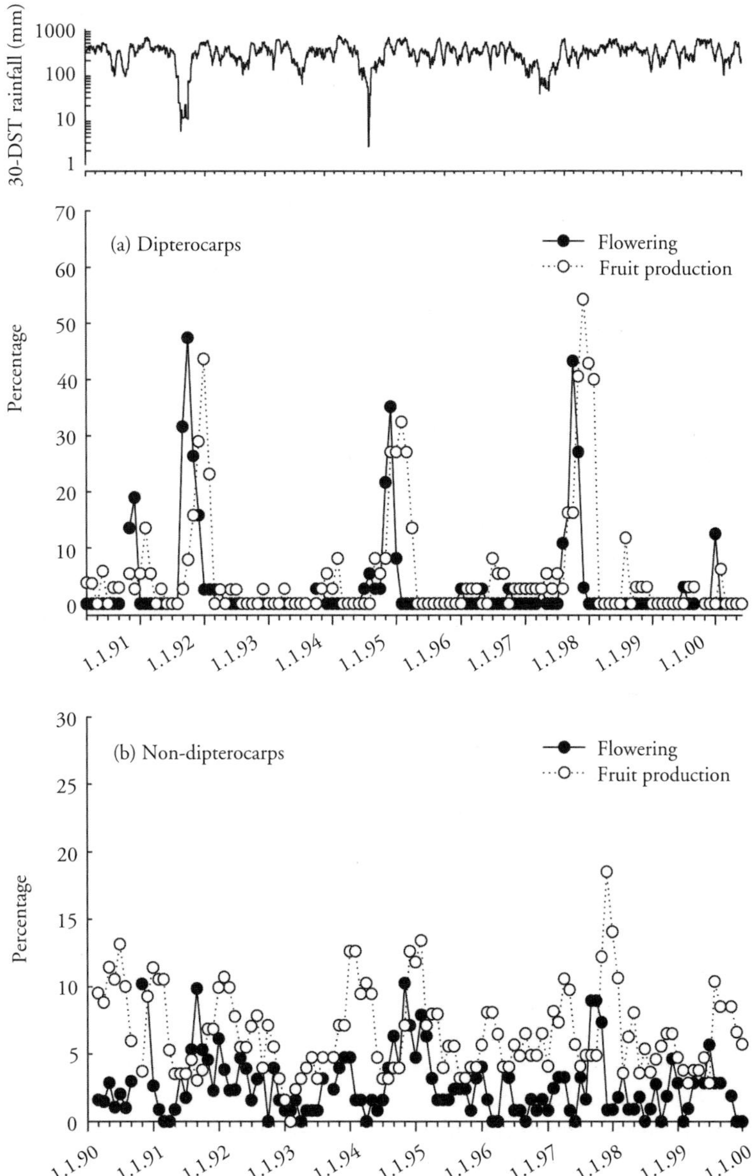

Figure 2.14 The percentage of flowering and fruiting dipterocarp trees (a) and non-dipterocarp trees (b) over a 10-year period in lowland dipterocarp forest at Barito Ulu, Central Kalimantan. Also shown is the 30-day sliding rainfall total. There were three 'general flowering' events, in 1991, 1994, and 1997, and each was preceded by a significant dry period. From Brearley et al. (2007).

widespread agreement that it is irregular periods of drought, rather than one of the other potential signals (a brief drop in night-time temperature or increased solar radiation) that co-occurs more or less regularly with drought (Sakai et al. 2006; Brearley et al. 2007a). At Lambir, the best correlation was with a 30-day rainfall total of <40 mm. The magnitude of the general flowering (i.e. the number of species and individuals that participate) does not have a clear relationship with the strength of the drought, but appears to depend more on the time for accumulation of resources since the last episode. Phosphorus has been suggested as the possible key resource (T. Ichie, in Sakai et al. 2006). In much of the Sunda Shelf region, the triggering droughts are associated with the ENSO cycle, occurring at the transition from La Niña to El Niño or at the start of the El Niño period (Sakai et al. 2006). These droughts can affect most of the Sunda Shelf region and thus synchronize flowering over vast areas. In northern Sumatra and north-west Kalimantan, however, ENSO has little or no effect on droughts and general flowering events may not be synchronized over such large areas (Wich and van Schaik 2000; Aldrian and Susanto 2003).

Reproductive success for many species appears to be greater in larger mass-flowering episodes, with individual trees producing larger fruit crops from similar numbers of flowers, reflecting either an improved quantity or quality of pollination, or reduced pre-dispersal seed predation (Sakai et al. 2006). Increased post-dispersal seed and seedling survival after larger masts have also been demonstrated in some cases (Curran and Leighton 2000). The alternation of brief 'feasts' with prolonged 'famines' for many flower-, fruit-, or seed-dependent animals has also had a major impact on the ecology of animals in the less seasonal parts of TEA (see 5.2.7, 5.2.8).

Numerous hypotheses have been put forward to explain the general flowering phenomenon. The simplest explanation would be the matching of reproductive effort to resource availability, but, although increased solar irradiance during the triggering droughts may increase the availability of photosynthate, the between-year differences in reproductive effort are very much greater than any variation in resources. More plausible is the suggestion that general flowering is in anticipation of good conditions for the establishment and growth of young seedlings, as a result of drought-induced thinning of the forest canopy (Williamson and Ickes 2002).

Other plausible hypotheses involve biotic interactions. The satiation of seed predators has already been mentioned and will be considered in more detail in Chapter 4. The facilitation of pollination is another possibility, but while this is a reason for synchronization at the population level, enhanced competition is a more likely general outcome of synchronization at the community level, as suggested by the evidence for sequential flowering in related dipterocarp species. A community-level advantage is only likely where pollinators are attracted from outside the community, as has been shown for the giant honey bee, *Apis dorsata*, at Lambir (Itioka et al. 2001), but this is unlikely to be true for the majority of plant species, particularly if general flowering extends over very large areas.

A final possibility is that general flowering simply reflects the independent evolution of the same flowering trigger for synchronization at the population level by many plant species. This could simply reflect the lack of any alternative cue in the aseasonal climate of the Sunda Shelf or could, conceivably, be a relic of the more seasonal climates of glacial times when droughts may have been annual (Sakai et al. 2006). The Holocene is only around 100 tree-generations old and the median climate of the last 2 million years was probably more seasonal than today's, so this idea is not as implausible as it might first appear to be. However, use of the same cue by different species is unlikely to have, by itself, led to the observed degree of community-level synchronization because of varied floral and fruit development times. An additional factor or factors must be selecting for strong synchronization.

Reproductive phenology in seasonal climates

How far the general flowering phenomenon extends beyond the Sunda shelf is currently unclear. While a big episode is unmistakeable, particularly at the fruiting stage, smaller episodes are only obvious when long-term records are available and there are few of these for most of TEA. Individual species mast in most forests, including some dipterocarps in deciduous forest in Thailand (e.g. Marod et al. 2002) and 'submontane' forest (at 1000 m a.s.l.) in the Philippines (Hamann 2004), and many Fagaceae (e.g. Du et al. 2007) and a variety of other woody species (Noma 1997) in the subtropics, but the cycles of different species are usually more or less independent. The little data from montane forest suggests that reproduction is more continuous at the community level than in the lowlands (Kimura et al. 2001; Hamann 2004) and the same is probably true of swamp forests (e.g. Singleton and van Schaik 2001), but there have been no long-term observations in either community.

In more seasonal tropical forests in TEA, while the phenologies of individual species are still diverse, the community-level pattern is clearly annual (Elliott et al. 1994; Corlett and LaFrankie 1998). There is a peak of flowering in the late dry-season, presumably controlled by the same factors as leafing, i.e. increasing day length or stem re-hydration following leaf fall or rain. Fruiting is more spread out, but the community-level peak usually starts just before the beginning of the wet season and continues into the early wet-season. This fruiting pattern is consistent with both a likely peak in photosynthetic activity resulting from a high sun, new leaves, and cloudless skies near the end of the dry season, and the dispersal of seeds at the best time of the year for germination and establishment before the start of the next dry season. In Chiang Mai, northern Thailand, peak production of wind-dispersed species

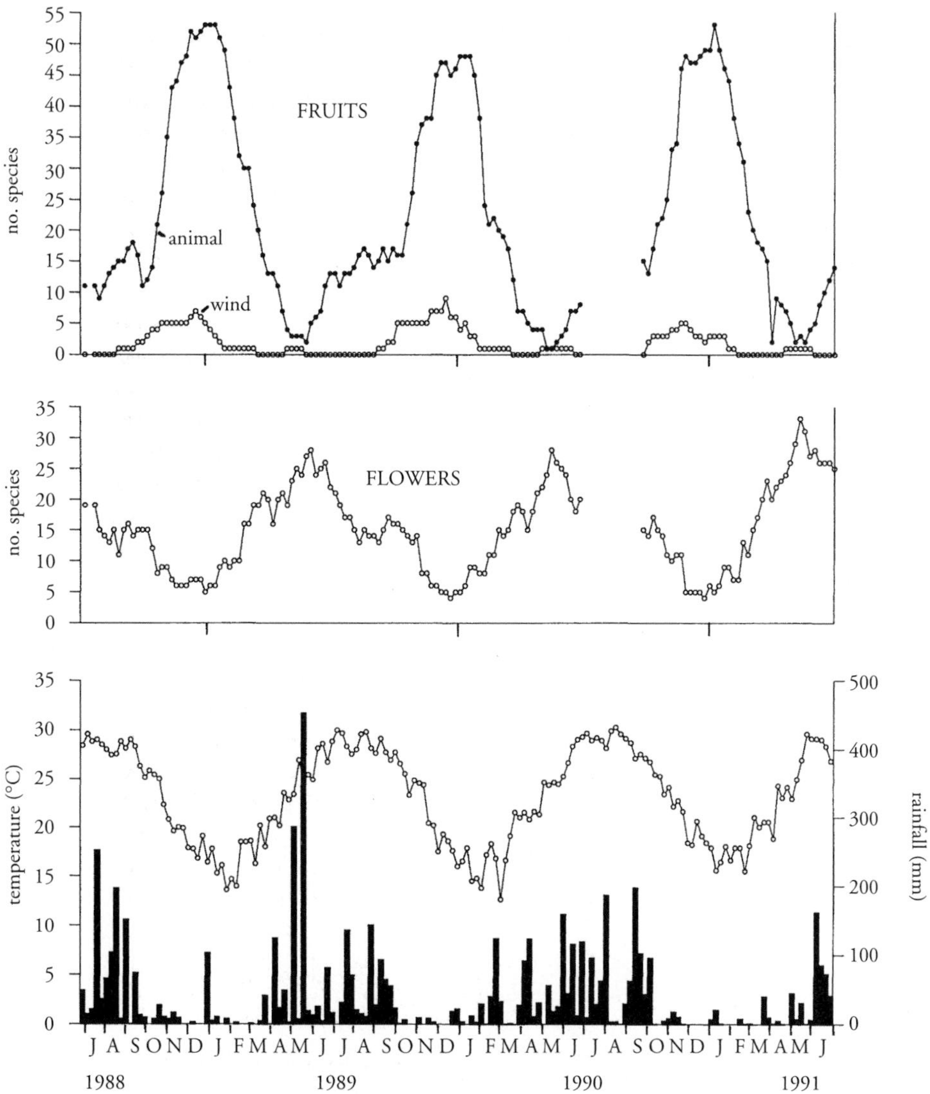

Figure 2.15 Numbers of woody plant species in fruit and flower over a 42-month period in Hong Kong, South China (22°N). Also shown are weekly rainfall totals and mean temperatures for the same period. From Corlett (1993).

also coincided with the April maximum in mean wind-speed (Elliott et al. 1994).

The subtropical evergreen broad-leaved forest in the north of the region also has an annual reproductive cycle at the community level, with a flowering maximum in April–June, coinciding in many species with new leaf production, and a fruiting maximum between September and January (Li and Wang 1984; Dudgeon and Corlett 2004; Shen et al. 2007; Su et al. 2007; Zang et al. 2007) (Fig. 2.15). Most species flower and fruit annually, although with varying intensity, and masting has been reported in many Fagaceae (Liu et al. 2002; Du et al. 2007). Fruiting data from different latitudes along the eastern margins of the region suggest a phenological correspondence between fleshy fruit availability and the abundance of fruit-eating birds, many of which migrate south for the winter (Corlett 1993; Noma and Yumoto 1997; Takanose and Kamitani 2003; Shen et al. 2007). Diet-switching by omnivorous birds in response to declining invertebrate availability may also favour winter fruiting.

Figs

In striking contrast to the annual reproductive cycles of most plant species in the strongly seasonal regions of TEA, is the phenology of the figs (*Ficus* spp.), which continue to have subannual cycles to the northern limits of their ranges. The specificity and short lifespans of the adult pollinating wasps (<2 days) constrain the phenologies of figs, since a phenological gap would lead to the local extinction of the only pollinator of a fig species. In the monoecious species in TEA, fig crops are synchronous at the individual level, ensuring out-crossing, but asynchronous at the population level, ensuring the maintenance of the pollinator population. Even in the most seasonal climates in the region, crops are apparently initiated, pollinated, ripened, and dispersed throughout the year. In the north of the region this may mean that figs are the only fruits available at certain times of the year. Phenologies are more complex in dioecious fig species, where seed production and the raising of pollinating wasps occur on separate plants. Avoiding self-pollination is no longer an issue so fig crops are not necessarily highly synchronized at the individual level. At the population level, reproduction appears to be generally asynchronous in aseasonal climates (Corlett 1987, 1993; Harrison et al. 2000), but at the northern margins of their range male and female subpopulations of some species have different seasonal peaks in fig production, apparently timed to allow wasps leaving male syconia to pollinate female syconia (e.g. Tzeng et al. 2006).

CHAPTER 3

Biogeography

3.1 Introduction

Biogeography is the science of describing and explaining the geographical patterns of distribution of species and of higher taxa, such as genera and families. A key task of biogeographers has been the biogeographic 'regionalization' of the planet: the identification of regions that are recognizably distinct in terms of their flora and fauna. Note that this is a very different task from recognizing biomes or vegetation types, such as those described in Chapter 2. There are lowland tropical rainforests in both tropical Asia and South America, but they share no species and only a proportion of their genera and families. Regionalization might seem to be a rather old-fashioned task—more appropriate for the nineteenth century, perhaps—but if regions can be defined in a way that has relevance to their ecology and conservation, then they can be very useful. This book, for example, is based on the assumption that TEA is different in important ways from other regions of the tropics (see 1.1).

The Asian tropics have been included in the Oriental or Indomalayan biogeographical region (or 'kingdom', 'realm', or 'ecozone') by vertebrate biogeographers since Wallace (1876) (Fig. 3.1). Plant geographers, in contrast, have generally recognized different, broader, boundaries for the region that includes tropical Asia, either putting it in a huge Palaeotropical region, stretching from Africa to New Guinea and the islands of the Pacific, or, if Africa is separated, recognizing an Indo-Pacific region (Cox 2001). The distributions of invertebrates groups are generally less well known, but are often more similar to those of the plants than the vertebrates. However, the key role of vertebrates in many ecological processes (see Chapters 4 and 5) and in biological conservation (Chapters 7 and 8) means that vertebrate distributions make a better basis for an ecological division of the Earth's surface. The term 'Oriental Region' will, therefore, be used in this book for the biogeographical unit incorporating the Asian tropics. It is defined to include all of TEA, plus Pakistan, India, Nepal, and Bhutan below the Himalayan treeline (at *c.*4000 m a.s.l.), Sri Lanka, and Bangladesh. 'Oriental' is preferred to the more widely used 'Indomalayan', because it is potentially misleading to name a region that includes all or part of 16 countries, after just two of them.

TEA (Fig. 3.2) is the eastern half of the Oriental Region and is not a purely biogeographical unit, since, as explained in the introduction, its western boundary is a political border that was chosen largely for convenience. Although the two halves of the Oriental region are different in many ways, most of these connected with the lower rainfall of the western half, the transition between them is gradual and there are exceptions on both sides to any generalizations. Many species, most genera, and almost all families of plants and animals occur on both sides of the boundary.

The number of major biogeographical regions recognized varies between authors, and the tropics have variously been divided between two (Neotropics and Palaeotropics) to five (Neotropics, Africa, Madagascar, Tropical Asia, Australia–New Guinea). Madagascar is usually considered to be part of the African region, because its biota is distinguished more by the many taxa that are missing, than by those that are present, but it is ecologically as distinct as any of the other four. Richard Primack and I have argued that each of these five major regions is a distinct ecological and biogeographical entity (Primack and Corlett 2005; Corlett

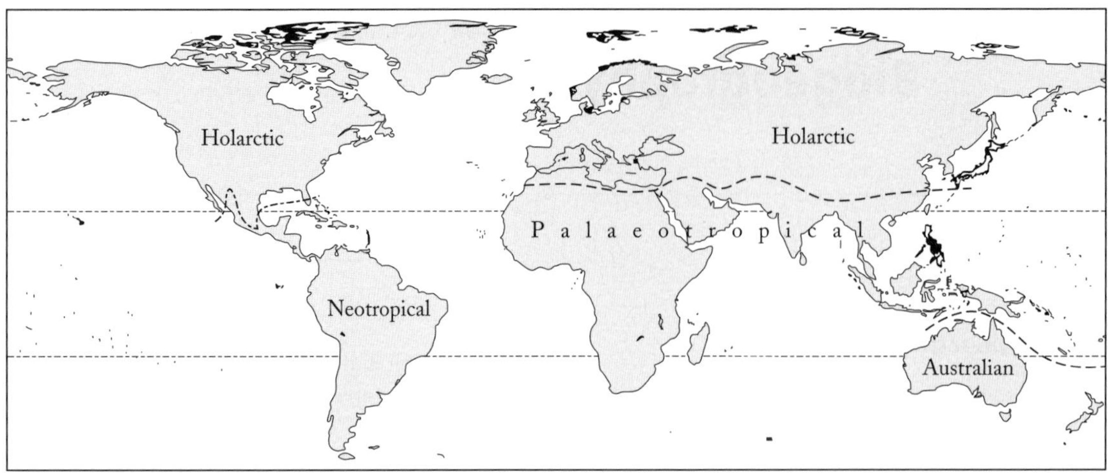

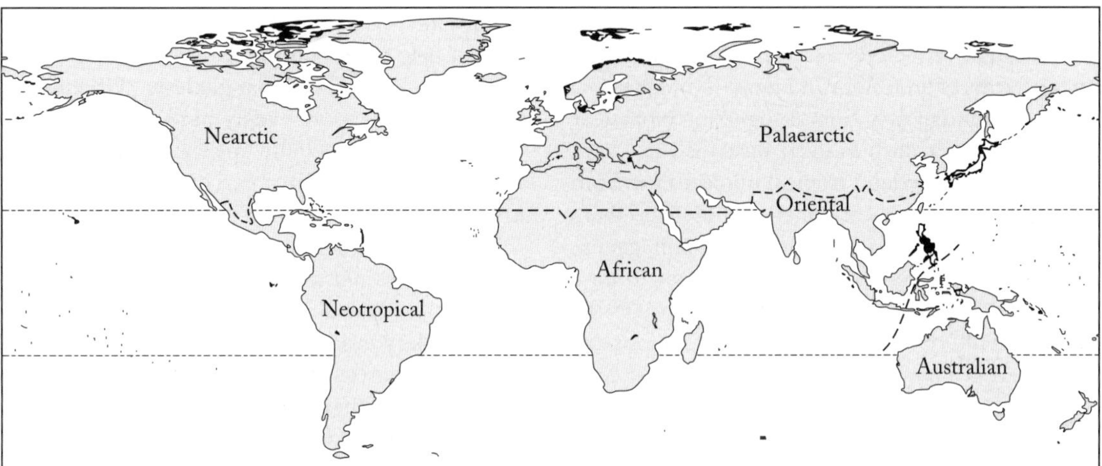

Figure 3.1 The biogeographical regions of the world as recognized by plant geographers (upper map) and animal geographers (lower map). Based on Cox (2001).

and Primack 2006). Although our arguments were confined to lowland rainforests, they apply equally well to other tropical forests.

3.2 Differences between the biogeographical regions

Most tropical forests today, like those of TEA, are on fragments of the ancient southern supercontinent of Gondwana (see 1.3). The precise age of the last terrestrial connections between the major fragments is still uncertain. Most geological models have these connections ending in the early Cretaceous (145–100 million years ago), but dated molecular phylogenies of organisms that cannot cross oceans, such as amphibians, suggest that some links may have survived into the late Cretaceous (e.g. Bocxlaer et al. 2006). In any case, most modern groups of plants and animals originated after the break-up was completed and radiated during the Tertiary, following the mass extinctions at the end of the Cretaceous (K/T), 65 million years ago. This coincided with the period of maximum

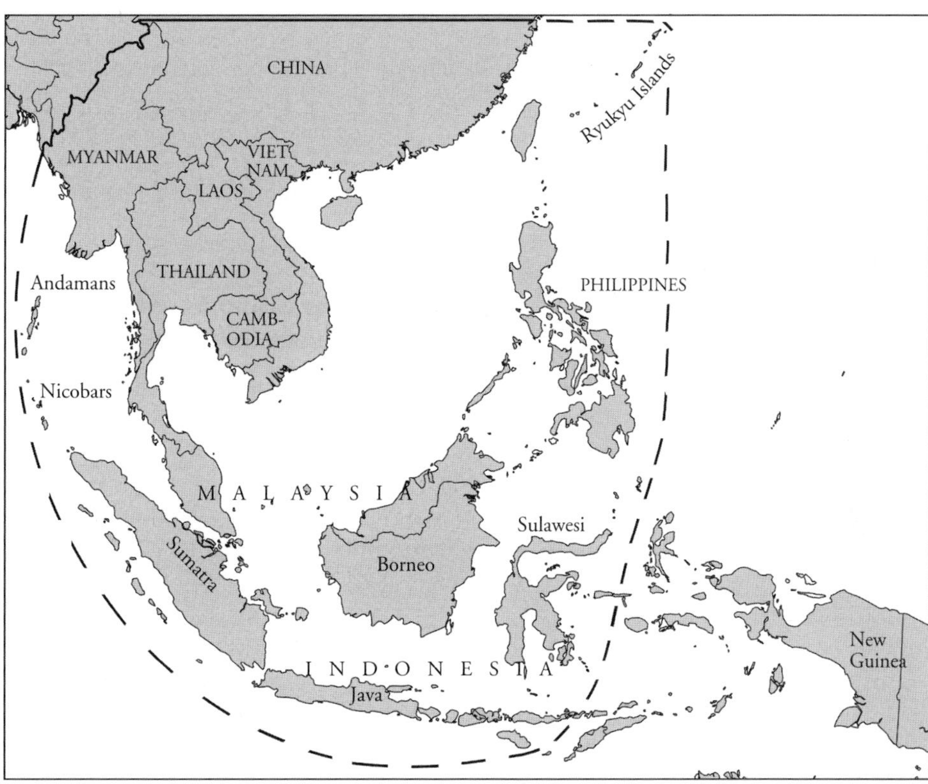

Figure 3.2 Tropical East Asia (TEA), as defined for this book, is the wetter, eastern half of the Oriental faunal region of biogeographers.

isolation among the five tropical forest regions. Dated molecular phylogenies have demonstrated the occurrence of long-distance trans-oceanic dispersal between regions in some plant families (e.g. Dick et al. 2007), but this appears to have been far less common in animals, and it is not yet clear how frequent it was in plants.

Madagascar has been an island for almost 90 million years, so the relatively few groups of vertebrates that were able to recolonize it after the K/T extinctions had to cross hundreds of kilometres of ocean (Yoder and Yang 2004). Africa was an island during the late Cretaceous and early Tertiary (90–24 million years ago), when it produced a spectacular mammalian radiation in an endemic group, the Afrotheria, represented today by the elephants, hyraxes, elephant-shrews, aardvark, and golden moles (Jaeger 2003). Africa has been physically connected to Asia for 24 million years, but a belt of dry climates has limited the exchange of tropical forest organisms for much of that period (Morley 2003), while strong African affinities in the biota of the semi-arid western Oriental Region suggest that non-forest exchanges have been much easier. Strikingly, the Afrotheria are still confined to Africa, with the exception of the Asian elephant (and its many extinct relatives) and the members of the aquatic order Sirenia (manatees and dugong).

Australia and New Guinea moved north towards Asia throughout the Tertiary, resulting in a major influx of Asian rainforest plants into lowland New Guinea (Morley 2003) and exchanges in the more mobile invertebrate groups (e.g. Schaefer and Renner 2008), birds (e.g. Jønsson et al. 2008), and reptiles (e.g. Rawlings et al. 2008), but no dry-land connection. Exchanges between Asia and Australia–New Guinea have become gradually easier as the collision progressed, but still involve multiple water-crossings, the longest of which was around

70 km, even at the lowest of Pleistocene sea-levels. Among the mammals, only the bats and rodents have managed to 'island-hop' from Asia to New Guinea and Australia, although the elephant-like stegodonts made it as far as Timor (and would have made a spectacular addition to the Australian fauna!).

Leaving aside long-distance oceanic dispersal, the last frost-free land routes between Asia and the Americas appear to have been across the north Atlantic in the early Eocene, around 50 million years ago (Morley 2003). South America was an island for most of the past 70 million years, giving rise to many endemic radiations. It became connected to North America through the Isthmus of Panama a mere 3–4 million years ago, enabling a dramatic intermingling of biotas known as the Great American Interchange (Webb 1997).

Many families of plants and animals occur in two or more of the five tropical forest regions, but molecular phylogenies, particularly for animals, usually show deep divisions between the taxa in different regions, with only one or a few migration events needed to account for the current distribution (e.g. fruit bats, Giannini and Simmons 2003; squirrels, Mercer and Roth 2003; barbets, Moyle 2004; bulbuls, Moyle and Marks 2006; woodpeckers, Fuchs et al. 2007; *Macaranga* trees, Kulju et al. 2007; stingless bees, Rasmussen and Cameron 2007). In effect, the five major tropical forest regions represent five, more or less independent, evolutionary responses to the humid tropical environment (Primack and Corlett 2005).

Have the separate evolutionary histories of the major rainforest regions resulted in functionally different ecosystems? Unfortunately, comparisons among regions are almost always confounded by differences in the modern physical environment, so this is not an easy question to answer. There are many examples of at least superficially convergent evolution between unrelated organisms in different regions, but also many examples of real differences; for example, in pollination (Corlett 2004), seed dispersal (Corlett and Primack 2006), litter decomposition (Davies et al. 2003), and the structure and composition of rainforest understoreys (LaFrankie et al. 2006). These and other differences may be large enough to require differences in conservation strategies between regions (Primack and Corlett 2005; Corlett and Primack 2006). Whether or not this is the case, there is enough evidence for non-convergence to warn against the uncritical extrapolation of the results of ecological studies in one region to the others.

3.3 The transition between TEA and the Australian Region

More has been written about the biogeographical transition between tropical Asia and Australia–New Guinea than any other such boundary. Much of this literature has been aimed at drawing the 'best' line to mark the boundary (Fig. 3.3), but an understanding of the complex geological history (see Chapter 1) shows why this is a fundamentally pointless task. There are two major continental source areas, terminating at the edges of the Sunda and Sahul shelves, respectively, and numerous islands between, varying in size, age, origin, and isolation. It has become standard to refer to most of these islands—Sulawesi, Nusa Tenggara, Maluku, and East Timor—as Wallacea, after the man who first studied this area in detail, Alfred Russel Wallace, the 'father of biogeography'. The Philippines are usually excluded from the definition of Wallacea, although a good case could be made for including them (Cox 2001; van Welzen et al. 2005).

If dispersal across the sea was the only way that terrestrial plants and animals could have reached these islands, then we would expect to see patterns of distribution that largely reflect the relative sea-crossing ability of different groups of organisms. On the whole this is the case (but see Sulawesi, in 3.9.10). In contrast, there are few, if any, consistent environmental differences across the transition. Thus the differences between TEA and the Australian Region today are basically a reflection of the geologically recent and on-going collision between the Australian and Laurasian plates, combined with the still limited ability of the two biotas to cross the declining water gaps. In another 40 million years or so, the water gaps will have shrunk to zero and the two biogeographic regions will merge in a 'Great Australasian Interchange'.

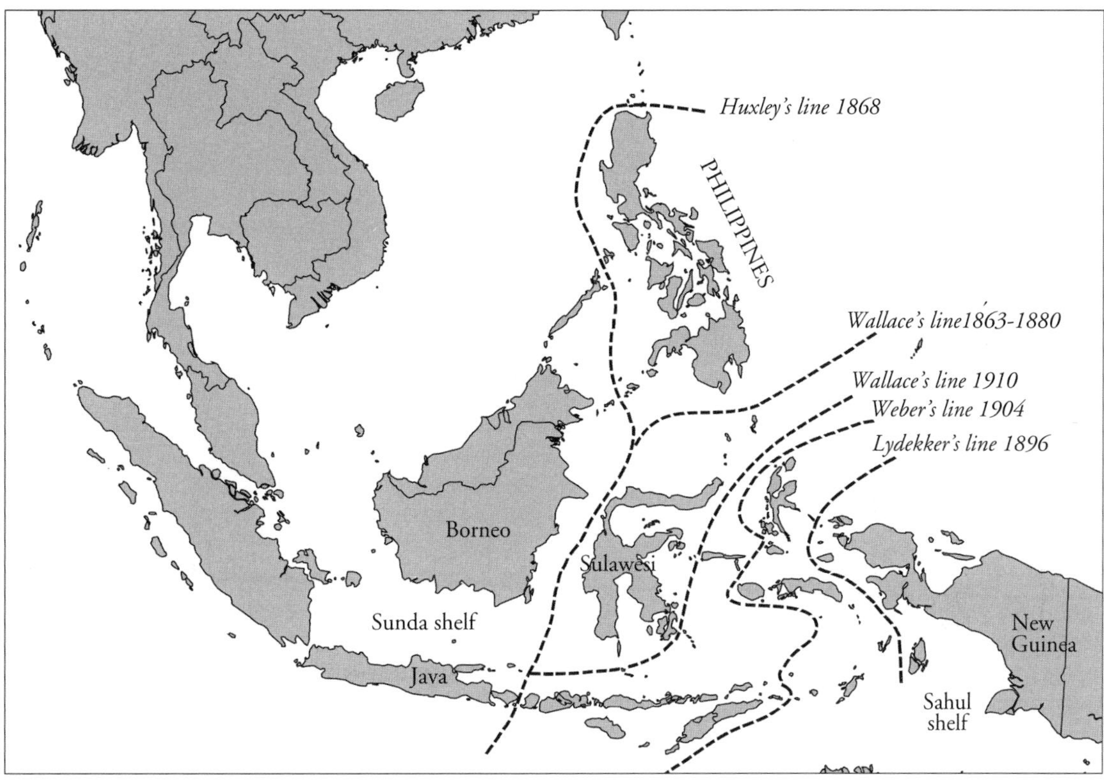

Figure 3.3 Different lines suggested by biogeographers to mark the boundary between the Oriental and Australian faunal regions. Modified from George (1981).

3.4 The transition to the Palaearctic Region

For readers of Chinese, there is almost as much written on the northern boundary of the Asian tropics, as there is in English on the eastern boundary. In this case, the geological history is largely irrelevant, except for the time-scale it provides for migration and evolution. The gradual transition from fully tropical biotas to fully temperate ones, which today takes place between roughly 18°N and 30°N, has probably existed since the origin of the Asian monsoon system in the early Miocene, 20 million years ago (see 1.5), although the position of this transition has shifted north and south in response to global climate-change. In striking contrast to the transition with the Australian Region, therefore, it seems unlikely that dispersal limitation has made a significant contribution to current distributions.

The borders between the tropical and non-tropical regions of the world are of a rather different nature from those between tropical regions. In some parts of the world (e.g. North America and Africa) there are deserts or mountains that act as barriers to migration, but even where this is not the case, as in East Asia and the east coasts of Australia and Brazil, the decline in temperature restricts the spread of many groups of organisms outside the tropics. Minimum temperatures appear to be the key factor, although lack of summer warmth or the length of the growing season may be important for some species. Air temperatures below zero are lethal to the leaves of most tropical plants because of the devastating impact of ice-crystal formation inside plant cells (Guy 2003), and half of flowering plant families have no temperate representatives (Donoghue 2008). In practice, however, the northern limits for tropical plant species vary considerably, probably because of the sublethal effects of low temperatures above zero, as well as the ability of some plants to recover from frost exposure, as long as it is not prolonged enough to kill woody and underground tissues.

Few tropical tree species are able to cross the frost line into the subtropics, although 'speciation spillover'—non-tropical species from tropical genera (Fine 2001)—forms an important element of the subtropical flora in TEA. Moreover, the continuous connection between tropical and temperate vegetation in TEA, since at least the early Miocene, appears to have made this region a major source of cold-tolerant lineages with tropical ancestry (Donoghue 2008). It is usually assumed that non-tropical plants and lineages are prevented from crossing the other way by the costs of frost tolerance, which make them less competitive when the investment is not needed, but there is no direct evidence for this and it is possible that intolerance of persistent high temperatures also plays a role.

The border is even more diffuse for tropical animals, which are often able to avoid brief periods of extreme cold behaviourally or by thermoregulation. Yunnan (21–29°N) straddles the Tropic of Cancer and has a very steep north–south gradient in plant distributions, but little change in the mammal community (Xie et al. 2004). The northern limits for some tropical taxa are probably set more by a seasonal gap in food supply—especially for animals that specialize on fruit, nectar, or large insects—than by climate directly. The fact that billions of birds of hundreds of species migrate between the tropical and temperate region each year, illustrates the porosity of the border, as well as the key role of food supply in controlling animal distributions.

The plant–animal dichotomy is highlighted by the latitudes chosen by botanists and vertebrate zoologists, respectively, for the transition between the tropical Oriental and non-tropical Palaearctic regions. Botanists have generally placed the boundary near the Tropic of Cancer (23°27′N) (e.g. Zhu et al. 2007a), while vertebrate zoologists have put it around 30–35°N (e.g. Xie et al. 2004). Note, however, that the northern boundary of the tropics is confounded with rising altitude in the north-west of TEA, and tropical floras extend north to 27–28° at low altitudes in deep valleys in Myanmar and north-east India (Kingdon-Ward 1945; Proctor et al. 1998; Zhu et al. 2008). In the north-east of TEA, by contrast, there is plenty of land at low altitude, but this was almost entirely deforested several centuries ago. At 29°47′N in Hubei, tropical plant genera slightly outnumber temperate ones in one of the few patches of forest remaining at low altitude (350–790 m) (Lai et al. 2006).

The distributions of invertebrates have received less study than those of vertebrates in the region, but for most groups for which there is information, tropical taxa dominate well north of the Tropic of Cancer (e.g. Fellowes 2006; Meng et al. 2008). The fauna of Hong Kong (22°17′N), 130 km south of the Tropic of Cancer, is almost entirely tropical, although it also lacks some taxa (e.g. stingless bees) that are ubiquitous further south (Dudgeon and Corlett 2004).

3.5 Biogeographical elements in the biota of TEA

In addition to the demarcation of biogeographical regions that can be recognized by their distinct biotas, biogeography has traditionally also tried to identify groups of species with similar spatial distributions—biogeographical 'elements' or 'patterns'. Like biogeographical regionalization, this task has often been viewed as valuable in itself, but it could potentially also contribute to an understanding of how the modern biota was assembled over time and how it will respond to future climate changes. Traditional biogeographical elements were necessarily based largely on the present-day distributions of organisms, while the availability of dated molecular phylogenies has shown that similar modern distributions can arise from very different histories. A modern pantropical distribution for a genus or family, for example, could reflect either an origin on Gondwana before it started to break up, dispersal via high-latitude land bridges in the early Eocene, more recent trans-oceanic dispersal, or various combinations of these three (Primack and Corlett 2005). And if taxa with similar modern distributions do not share similar histories, this may make it less likely that they will respond in the same way to future climate change. However, the increasing availability of dated phylogenies makes it potentially possible to identify biogeographical elements that reflect similarities in space *and* time.

The identification of biogeographical elements within TEA is most advanced in southern China, where this approach has been used for many groups of organisms, including vascular plants (e.g. Qian et al. 2003) and bryophytes (e.g. Zhang and Corlett 2003). Qian et al. (2003), for example, recognize 14 geographical elements in the flora of East Asia, north of the Vietnam–China border, with pantropical, Eastern Asian, and Tropical Asian genera dominant in the tropics and subtropics. There have been no such comprehensive analyses for South-East Asia or TEA as a whole, although there has been a long history of speculation on the origins of montane and apparently Gondwanic elements in the region's biota.

3.6 How many species are there in TEA?

Only a million or so species of living organisms have been described, out of a total of, perhaps, 5 to 10 million species on Earth, so assigning a share to TEA is not possible at present. However, it is possible to estimate, with varying degrees of accuracy, the proportion of species in several well-studied terrestrial taxonomic groups that occur in TEA, and then use these estimates as a rough guide to the overall proportions for both known and unknown species. TEA supports around 800 mammal species, which is 15% of the global total of 5400 (Wilson and Reeder 2005), around 23% (2300 of 10,000) of all bird species, and 16% (932 of 5915) of all amphibian species (Global Amphibian Assessment). The region also supports around 20% of the known species of both butterflies (3500 of 17,500; Robbins and Opler 1997) and tiger beetles (Cicindelidae) (*c.*400 of 2028; Pearson and Cassola 1992), and perhaps as many as 25% of dragonflies (Odonata) (*c.*1450 of 5680; Kalkman et al. 2008). Recent estimates of the global number of angiosperm species range from 220,000 to 420,000, with the uncertainty largely reflecting the proportion of synonyms among published names (Scotland and Wortley 2003). TEA's share of this total is probably around 60,000–70,000 species, i.e. 15–25% of the total (Turner 2001; Bramwell 2002).

Overall, therefore, it appears that TEA supports 15–25% of the known taxa in well-studied groups of terrestrial organisms. If these taxa are representative—and we have no way of knowing if this is true—these percentages can be used as an estimate of the proportion of the total terrestrial biodiversity at the species level that occurs in the region. For comparison, TEA has less than 4% of the Earth's total land area—a reflection of the well-known, if incompletely understood, polar–tropical gradient in species diversity.

3.7 Patterns of diversity within TEA

On a global scale, patterns of vascular plant diversity on the continents are predicted well by the present-day environment, with potential evapotranspiration (PET, a measure of the annual energy input) and the number of wet days per year, along with measures of habitat and topographical heterogeneity, as the best (positive) predictors (Kreft and Jetz 2007). When PET is greater than 500 mm per year—as it is in the tropics, except on the highest mountains—then any further increase in energy input has little effect, and the number of wet days becomes the best single predictor of plant diversity. On a smaller spatial scale, dry-season length (the number of months with <100 mm rainfall) was the best (negative) predictor of tree species diversity within small (<1 ha) forest plots at sites with >2000 mm of rainfall in the Western Ghats of India (Davidar et al. 2005). In the Amazon region, dry-season length is a much better predictor of the *maximum* plot diversity within an area than the average, since there are low diversity plots in all areas, reflecting edaphic or historical factors (Ter Steege et al. 2003).

Too few data have been published in TEA in a comparable form to do a formal analysis. The largest dataset is for the numbers of trees >10 cm diameter at breast height (=1.3 m) in a single hectare (i.e. 100 × 100 m) (Table 3.1). A hectare is too small to average out the local variation in tree diversity in lowland rainforests, as shown by the wide range of values (172–290) for adjacent hectares at Lambir, but some clear patterns emerge. Tropical lowland rainforests on the Sunda Shelf support 100–290 large-tree species in one hectare. Single lowland plots from the Philippines and Sulawesi fit within this

Table 3.1 Numbers of tree species >10 cm in diameter in 1 hectare of forest at sites in Tropical East Asia. Sites arranged in latitudinal order. There was considerable methodological variation between studies (in the shape and contiguity of the sample area, and the treatment of unknown species and large lianas), some of which may have affected the reported numbers of tree species. Individual sources should be consulted for details. Rainfall estimates are usually from the nearest town with long-term records and may not accurately reflect the rainfall at the site. A month is considered dry if the mean rainfall is <100 mm.

Location	Latitude	Altitude	Annual Rainfall	Dry months	Tree Species per Hectare	Forest Type	Source
Gede-Pangrango, Java	6°40′S	1450–1500	3380	1	59	montane	Meijer (1959)
Gede-Pangrango, Java	6°40′S	1600	3380	1	57	montane	Yamada (1975)
Saparua Island, Maluku	3°35′S	<500	3400	?	58	lowland rainforest	Kaya et al. (2002)
Berau, East Kalimantan	2°00′S	100	2500	0–1	160–201	lowland rainforest	Sist and Saridan (1998)
Batang Ule, Sumatra	1°37′S	150	>3000	0	250	lowland rainforest	Rennolls and Laumonier (1999)
Sebulu, E. Kalimantan	1°30′S	70	2300	0–1	276	lowland rainforest	Sukardjo et al. (1990)
Wanariset Sangai, Kalimant.	1°29′S	100	>3000	0	179–259	lowland rainforest	Wilkie et al. (2004)
Lore Lindu, Sulawesi	1°05′S	1100–1200	2000	?	148	submontane	Kessler et al. (2005)
Pasirmayang, Sumatra	1°05′S	100	2500–3000	0–1	230	lowland rainforest	Laumonier (1997)
Gunung Pasir, Kalimantan	1°03′S	20–40	2347	0	29	heath forest	Riswan (1987b)
Wanariset, Kalimantan	0°59′S	50	2350	0	172	lowland rainforest	Kartawinata (1981)
Toraut, N. Sulawesi	0°30′S	150	>2100	0	109	lowland rainforest	Whitmore and Sidiyasa (1986)
Lempake, Kalimantan	0°20′S	40–80	1935	0–1	165	lowland rainforest	Riswan (1987a)
Bukit Timah, Singapore	1°21′N	70–125	2473	0	113	lowland rainforest	Lum et al. (2004)
Pasoh Forest, Pen. Malaysia	2°59′N	70–90	1788	0	206	lowland rainforest	Manokaran et al. (2004)
Malinau, Kalimantan	3°20′N	200	3730	0	207	lowland rainforest	Wunder et al. (2008)
Ketambe, Sumatra	3°41′N	350	3229	0	100	lowland rainforest	Kartawinata (1990)
Gunung Mulu, Sarawak	4°02′N	50	5000	0	223	lowland rainforest	Proctor et al. (1983)
Gunung Mulu, Sarawak	4°03′N	200–250	5000	0	214	lowland rainforest	Proctor et al. (1983)
Gunung Mulu, Sarawak	4°08′N	300	5000	0	73	forest over limestone	Proctor et al. (1983)
Gunung Mulu, Sarawak	4°09′N	170	5000	0	123	heath forest	Proctor et al. (1983)
Lambir Hills, Sarawak	4°11′N	110–150	2734	0	172–290	lowland rainforest	Lee et al. (2002)
Kuala Belalong, Brunei,	4°30′N	55–80	5080	0	197	lowland rainforest	Small et al. (2004)
Temburong, Brunei	4°32′N	250	4000	0	231	lowland rainforest	Poulsen et al. (1996)
Badas, Brunei	4°34′N	10–15	3000	0	77	heath forest	Davies and Becker (1996)
Bukit Sawat, Brunei	4°35′N	10–20	3000	0	121	heath forest	Davies and Becker (1996)
Ladan Hill, Brunei	4°37′N	40–70	3000	0	194	lowland rainforest	Davies and Becker (1996)
Andulau, Brunei	4°39′N	40–50	3000	0	256	lowland rainforest	Davies and Becker (1996)
Danum Valley, Sabah	4°58′N	210–260	2825	0	128	lowland rainforest	Newbery et al. (1992) 1996
Sepilok, Sabah	5°10′N	10–40	3000	0–1	160	lowland rainforest	Nicholson (1965)
Sepilok, Sabah	5°10′N	45–80	3000	0–1	140	lowland rainforest	Nilus (2004) sandstone ridge
Sepilok, Sabah	5°10′N	30–45	3000	0–1	126	lowland rainforest	Nilus (2004) alluvial
Sepilok, Sabah	5°10′N	45–120	3000	0–1	79	heath forest	Nilus (2004)

Kinabalu, Sabah	6°05′N	700	2509	0	148	lowland rainforest	Aiba and Kitayama (1999)
Khao Chong, Thailand	7°58′N	50–300	2700	2–3	100–165	lowland rainforest	Bunyavejchewin (unpublished)
Mt Kitanglad, Mindanao	8°00′N	2065–2360	2500–3500	0	43	montane	Pipoly and Madulid (1998)
Little Andaman, India	10°40′N	<70	3000–3500	3–4	84	lowland rainforest	Rasingam & Parathasarathy (2009)
Little Andaman, India	10°40′N	<70	3000–3500	3–4	83	semi-evergreen	Rasingam & Parathasarathy (2009)
Little Andaman, India	10°40′N	<70	3000–3500	3–4	58	deciduous forest	Rasingam & Parathasarathy (2009)
Little Andaman, India	10°40′N	<10	3000–3500	3–4	43	littoral forest	Rasingam & Parathasarathy (2009)
North Negros, Philippines	10°41′N	1000	4650	0	92	submontane	Hamann et al. (1999)
Cat Tien, Vietnam	11°25′N	<150	2450	4–5	57	semi-deciduous	Blanc et al. (2000)
Cat Tien, Vietnam	11°25′N	<150	2450	4–5	91	seasonal evergreen	Blanc et al. (2000)
Khao Yai, Thailand	14°10′N	780	2360	5	63	seasonal evergreen	Kitamura et al. (2005)
Sakaerat, Thailand	14°30′N	460–540	1240	6	56, 81	seasonal evergreen	Bunyavejchewin (1999)
Mae-Klong, Thailand	14°35′N	100–900?	>1650	6	50	mixed deciduous	Marod et al. (1999)
Huai Kha Khaeng, Thailand	15°40′N	550–640	1475	6	65	seasonal evergreen	Bunyavejchewin et al. (2004)
Palanan, Luzon	17°02′N	85–140	5000	0	100	lowland rainforest	Co et al. (2004)
Doi Inthanon, Thailand	18°31′N	1700	1908	6	67	montane	Kanzaki et al. (2004)
Jianfengling, Hainan	18°40′N	820–830	3500	4?	153	montane	Li et al. (1998)
Wuzhishan, Hainan	18°49′N	850–1100	2430	3–4	217	montane	An et al. (1999)
Diaoluoshan, Hainan	18°50′N	900–1000	2570	3–4	190	montane	Wang et al. (1999)
Bawangling, Hainan	18°55′N	1200	>2000	6	84	montane	
Mengla, Yunnan	21°37′N	700–870	1500	6	124	tropical seasonal	Lan et al. (2008)
Nanjenshan, Taiwan	22°03′N	300–340	3582	0	61	evergreen broadleav.	Sun and Hsieh (2004)
Dinghushan, Guangdong	23°10′N	230–470	1985	4	92	evergreen broadleav.	Ye et al. (2008)
Tianjingshan, Guangdong	24°40′N	480–550	2800	7	73	evergreen broadleav.	Huang et al. (in prep.)
Fushan, Taiwan	24°45′N	600–733	4271	0	43	evergreen broadleav.	Su et al. (2007)
Jinggangshan, Jiangxi	26°32′N	500–750	1800	7	63	evergreen broadleav.	Cao et al. (in prep.)
University Forest, Okinawa	26°49′N	250–330	2456	?	32–42	evergreen broadleav.	Enoki et al. (2003)
Gutianshan, Zhejiang	29°15′N	450–715	1964	?	45	evergreen broadleav.	Zhu et al. (2008)

range, as do the data from Khao Chong, in Peninsular Thailand, and—surprisingly—plots from 700 to 1200 m in tropical China. Plots at higher altitude (>1200 m), higher latitude (north of the tropics), on extreme soil types, or with a dry season of >3 months, support <100 species. One apparent anomaly—the high diversity at Mengla, Yunnan, where a forest with a 6-month dry season is as diverse as many on the Sunda Shelf—can be explained by the persistent fog during the dry season (see 2.2.4).

Data from the more numerous smaller plots are consistent with these general patterns of tree diversity. Soil has a large influence on species diversity at the local scale, with tree diversities often considerably lower on extreme soil types, such as peat (Cannon and Leighton 2004), spodosols/podzols (Riswan 1987b), sand, and soils derived from limestone (Proctor et al.1983; Zhu 2008a) and ultrabasic rocks (Aiba and Kitayama 1999). In Borneo, tree-species diversity is also lower on nutrient-rich alluvial soils than on soils derived from granite or sedimentary rocks, possibly because faster growth rates lead to more rapid competitive exclusion (Cannon and Leighton 2004; Paoli et al. 2006).

The diversity of insects is much greater than that of vascular plants. Novotny et al. (2006) show that tropical and temperate tree species support similar numbers of insect species per 100 m^2 of foliage, suggesting that the latitudinal gradient in leaf-feeding insects can be explained largely by the gradient in plant diversity. However, Dyer et al. (2007) show that larval diets of tropical forest Lepidoptera are more specialized than those of temperate species—i.e. tropical species feed, on average, on fewer plant species, genera, and families—suggesting that increased dietary specialization is an additional reason for high tropical-insect diversity. Other possible contributions to tropical herbivore megadiversity could come from a greater number of herbivore species per plant species and/or a higher turnover of herbivore species among sites (i.e. beta diversity), but Lewinsohn and Roslin (2008) conclude that the currently available data are simply not good enough to allow for quantitative comparisons.

On a regional scale, Beck and Kitching (2007) show that the diversity of hawkmoths (Sphingidae) reaches a peak in northern South-East Asia, rather than in the equatorial region. They suggest that the most likely explanation for this pattern is a combination of the 'normal' pattern of decreasing richness with increasing latitude and a 'peninsula effect', as a result of the larger land area available in the north of the region.

In contrast to plants and insects, the distributions of vertebrates often bear a stronger imprint of history, so the present-day physical environment alone does not accurately predict patterns of diversity. History is expected to be most significant for poorly-dispersed taxa, such as frogs, in which most species are unable to disperse fast enough to track changing climates, and least significant for well-dispersed groups like birds. Patterns of bird diversity are of particular interest because of the high quality of the data (in terms of completeness, accuracy, and fine grain), which is unmatched for TEA for any other group of organisms. Ding et al. (2006) show that primary productivity—as estimated by satellite measures of NDVI (normalized difference vegetation index)—is the best overall predictor of bird species richness in 100 × 100 km quadrats on the mainland, but that even large islands always have fewer birds species than adjacent mainland areas (e.g. Borneo and Taiwan have fewer species per quadrat than the Malay Peninsula and mainland southern China, respectively). Bird diversity was maximum at around 25°N, where Myanmar borders India and China, rather than near the equator. This area also had the highest non-island NDVI. The larger land area available at this latitude, than on the relatively narrow peninsula further south, may be an additional factor, as may the steep and rugged topography, which provides a great diversity of habitats. It would be very interesting to know if other groups of animals have peak diversity in the same region, but this area is one of the least-known in TEA.

Correlation does not mean causation. While it is certainly plausible that highly productive forests provide more potential niches for birds, floristically diverse forests provide more potential niches for herbivorous insects, and continuously warm and wet conditions promote coexistence in plant species (see Chapter 4), the mechanisms that generate these species must have a historical component. A possible general historical explanation for the availability of more species in the warmer, wetter parts of the region is that the older lineages of plants and animals were originally adapted to the warm, wet climates of the early Tertiary and that many of them have been excluded from cooler and/or drier areas as the global climate has deteriorated since the Eocene (Hawkins et al. 2007; Fig. 1.5). The correlations we observe today, therefore, result not just from present-day environments, but also from the continuity of these environments through time while species accumulate.

This historical explanation would also predict a similar decline in species diversity with increasing altitude on tropical and subtropical mountains, as declining temperatures exclude warmth-dependent clades. This expectation is reinforced by the steep decline in the land area available with altitude, as well as an increase in isolation. Land above 3000 m, for example, forms a few, tiny, scattered islands in TEA, except in the Himalayan foothills in the extreme north-west of the region. Ant-species richness declines exponentially with altitude, reaching zero above 2300 m (Brühl et al. 1999) and the species richness of oribatid mites declines with altitude on Mt Kinabalu (Hasegawa et al. 2006), but observations for many other taxa do not match these expectations. Instead, a mid-elevation peak in diversity has been reported for a wide range of organisms, including vascular plants (Teejuntuk et al. 2003; Grytnes and Beaman 2006; Liu et al. 2007; Wang et al. 2007d), ferns and orchids (Kitayama 1992), small mammals (Heaney 2001; Nor 2001), beetles (Stork and Brendell 1990), and moths (Beck and Chey 2008).

Some of these results may reflect methodological problems, including deforestation or severe disturbance in the lowlands, the interpolation of distributions between altitudinal records (so that species richness near the top and bottom of the transect includes only observed species, while species richness at mid-elevations includes observed and interpolated species), small sample sizes, and collecting biases, but others appear to be real. The best general explanation for these results is that they reflect an increase in moisture availability (or a decrease in drought frequency) from the lowlands up to 1000–2000 m on many mountains, particularly in the north of the region (e.g. Liu et al. 2007), but there may also be an historical component, since pollen records suggest an expansion of montane forest environments during the glacial periods that have occupied the majority of the last 2.5 million years (see Chapter 1). It is also important to note that the most species-rich known sites for trees, birds, and vertebrates are all in the lowlands, <600 m above sea-level.

3.8 Subdividing TEA

TEA has been divided in many different ways by biogeographers, but most of the differences between authors involve the area north of the Tropic of Cancer. Within the tropics, four main blocks or subregions are widely recognized, although the names and precise boundaries differ: Wallacea, the Philippines, Sundaland, and Indochina (or Indo-Burma). Wallacea extends further east than this book, and the Indochinese subregion is often extended west of the India–Myanmar border to incorporate parts of north-east India and Bangladesh, but the other subregions are wholly within TEA. Wallacea is defined by the absence of most of the Sundaland vertebrate fauna, rather than any shared taxa, and the same is largely true for the Philippines, although there are also some endemic vertebrate taxa shared by most major islands (see 3.9.9). The Sundaland (or Sundaic or Malayan) subregion is defined to include the islands on the Sunda Shelf (principally Borneo, Sumatra, and Java) and the southern part of the Thai–Malay Peninsula. This subregion was largely covered in tropical rainforest until recently. Characteristic elements of the fauna and flora include the rodent genera *Iomys*, *Sundamys*, and *Sundasciurus*, the *Presbytis* leaf-monkeys, and the many similar co-existing tree species in the subgenera *Rubroshorea* (red meranti) and *Richetia* (yellow meranti) of the genus *Shorea*.

The boundary between Sundaland and the Indochinese subregion (or Indo-Burma) is placed between approximately 7° and 13°N by various authors (Fig. 3.4). Botanists have long favoured the Kangar–Pattani line, near the border between Malaysia and Thailand, while zoologists working on a variety of taxa have preferred a line 400–500 km further north, near the Isthmus of Kra (e.g. Hughes et al. 2003; Gorog et al. 2004; de Bruyn et al. 2005). There are plausible ecological reasons for both lines, with the southern one marking the transition from

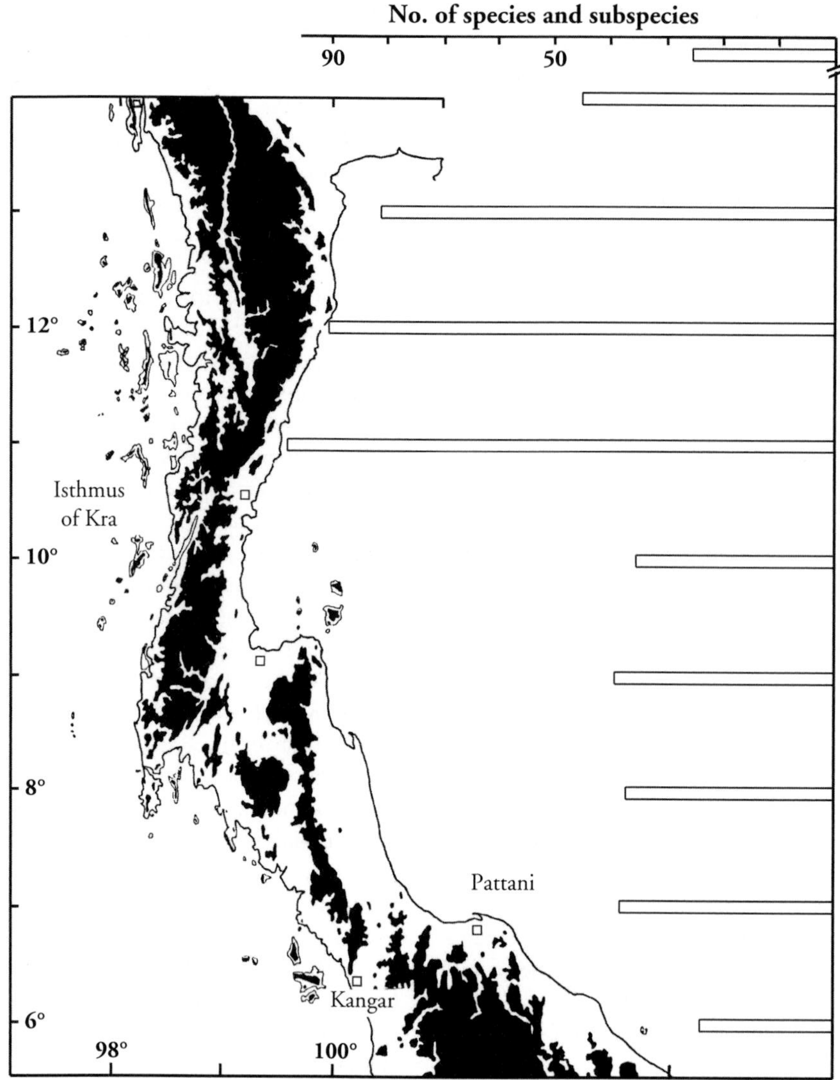

Figure 3.4 Map of the narrowest part of the Malay–Thai Peninsula, showing the Kangar–Pattani line (*c*.7°N), which marks the transition between aseasonal tropical rainforests to the south and seasonal forests to the north, and the Isthmus of Kra, 450 km further north, which zoologists have taken as the boundary between the Sundaland and Indochinese subregions. Also shown are bars indicating the numbers of resident forest bird species and subspecies with distributional limits at that latitude. The maximum is just north of the Isthmus of Kra. From Hughes et al. (2003).

aseasonal tropical rainforests to seasonal forests, and the northern one a transition between seasonal evergreen and mixed deciduous forests, although there are forest outliers in both directions. An eco-physiological explanation for the Kangar–Pattani line is supported by a study that showed higher inherent-drought tolerance in tree species that were distributed both sides of the line, than those that were confined to the aseasonal forests south of it (Baltzer et al. 2008).

Many zoologists, however, have considered vegetation change an insufficient explanation and attributed the sharpness of the transition near the Isthmus of Kra to marine sea-ways, which cut the peninsula in the middle Miocene and, possibly, the early Pleistocene (Woodruff 2003), i.e. they have preferred historical to ecological explanations. Whether sea-levels were ever high enough is still uncertain (see 1.4), as is the extent to which fairly narrow (<50 km) and short-lived (<1 million years) sea-ways would leave this strong an imprint on modern distributions. Moreover, the fossil evidence shows the border between the subregions migrating north and south in response to Pleistocene climate change (Chaimanee 2007), suggesting that there is nothing special about the present location. Characteristic elements of the flora and fauna of the Indochinese subregion include mammalian genera that inhabit relatively open forests, such as *Axis* (deer), *Canis* (jackals and wolves), and *Lepus* (hares).

Finer divisions of the region are possible, but only for subsets of the total biota—vascular plants or vertebrates, for example—or by combining ecological and biogeographical information. This latter approach was used in the ecoregional analysis undertaken by the World Wildlife Fund (Olson et al. 2001; Wikramanayake et al. 2002). Ecoregions are small areas with characteristic species and ecosystems, delineated using maps of the original vegetation. A total of 63 ecoregions are recognized within the boundaries of TEA: examples include the Bornean lowland rain forests, Luzon montane rain forest, Sumatran peat swamp forests, and Central Indochina dry forests. Xie et al. (2004) combined information on physical geography with distributional data for plants and mammals to produce a system of biogeographical divisions for China.

3.9 Island biogeography

TEA is a region with numerous islands. A key distinction can be made between 'continental islands', which were connected to the Asian mainland by dry land at the last glacial maximum (LGM), when water levels were 120 m lower than today (see 1.4), and 'oceanic islands', which have never been connected. The biotas of continental islands are controlled largely by extinction after isolation, since an island of a given size can support fewer species than a similar area on the mainland. For example, a random 500 km^2 area of forest on the Malay Peninsula would, until recently, have supported elephants, rhinoceroses, and tigers, but none of these could maintain a viable population on a forested island of the same size.

The biotas of oceanic islands, by contrast, are limited by factors that affect dispersal and establishment. The marine barrier that terrestrial organisms must cross to reach such an island acts as a filter, resulting in floras and faunas that are 'unbalanced' or 'disharmonic' in comparison with those of the mainland. Terrestrial organisms reach islands by either active flight (birds, bats, some insects), passive wind transport (spores, tiny wind-dispersed seeds, tiny flying insects, 'ballooning' spiders), active swimming (elephants, pythons, monitor lizards), passive floating (coastal plants, some invertebrates), 'rafting' on driftwood, floating trees, or larger 'floating islands' (reptiles, small mammals, invertebrates), or hitching a lift on or in such organisms (vertebrate-dispersed plants, some invertebrates). The variety of practical dispersal mechanisms declines as the width of the barrier increases, so the biotas of oceanic island, unlike those of continental islands, vary with distance from the mainland. If dispersal is frequent enough to lead to successful colonization, but rare enough to limit gene flow from the mainland, island populations can differentiate genetically, resulting, on older islands, in the evolution of island-endemic races, subspecies, species and—eventually—higher taxa.

There are also islands in TEA that do not fit clearly in either group, including several that appear to have been connected at some time in the past, but not at the LGM. In these cases, the arguments tend to become circular: an island has one or more species that, it is believed, could not have crossed a water gap, so there must have been a dry-land connection, which in turn is evidence that these species cannot cross water gaps... and so on. It is possible, however, that the same distribution patterns could be produced by a permanent, but narrow, water gap—perhaps a few kilometres wide—across which, given time, almost any bird could fly, any large vertebrate could swim, even salt-sensitive frogs could raft, and any seed or invertebrate could be blown or carried. The diversity of island biogeographies in TEA will be illustrated here by some of the best-studied islands and island groups.

3.9.1 Continental islands on the Sunda Shelf

The main factor determining the number of species found on the continental islands of the Sunda Shelf is island area. For all mammals (Heaney 1984) and for primates only (Harcourt 1999), the log–log relationship between species number and island area is highly significant and explains >90% of the variation in species number. Comparison of the species–area relationship for islands (slope = 0.235) and for four areas on the Malay Peninsula (slope = 0.104) shows that substantial extinction has taken place since isolation, e.g. a 100 km^2 island is estimated to support 15 species of mammals, compared with 65 in the same area on the mainland (Heaney 1984) (Fig. 3.5). The faunas on smaller islands tend to lose the larger species in each guild, and the largest species (orangutans, tigers, leopards, tapirs, rhinoceroses, and elephants) are only on the largest islands. Carnivore-richness decreases more rapidly than other groups on smaller islands and they are absent from smallest ones, presumably because carnivores necessarily live at lower densities than their prey.

Neither the distance from the mainland nor the water depth in between have a significant impact on the diversity of all mammals, or just primates, on continental islands (Heaney 1984; Harcourt 1999).

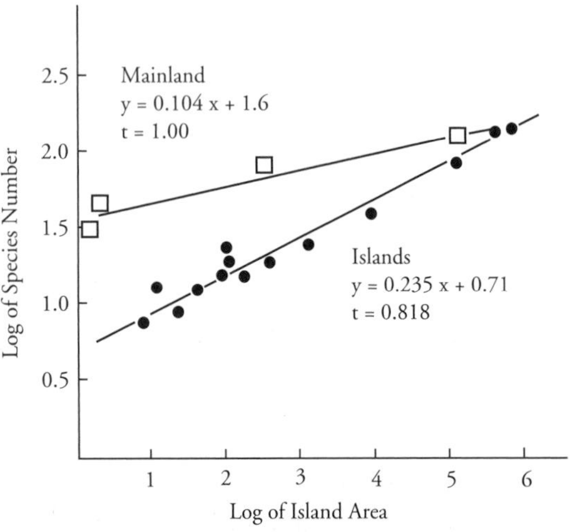

Figure 3.5 Species–area relationships for islands on the Sunda Shelf (R^2 = 0.95) and four areas on the Malay Peninsula (R^2 = 0.94) show that substantial extinction has taken place since the islands were formed by the flooding of the shelf at the end of the last glacial period. From Heaney (1984).

Water depth, however, can influence which species are present. Islands isolated by relatively deep water will have lost their terrestrial connection when the climate was considerably cooler and drier than today, so they may lack species that can only move through closed forest. Meijaard (2003) used the modern mammalian faunas of 215 South-East Asian islands to deduce whether or not they were forested at the time they became isolated. He found that few forest-dependent species had reached the islands in the Java Sea, off the east coast of eastern Borneo, in the Sunda Strait, off the west coast of the Malay-Thai Peninsula, and in the Gulf of Thailand, suggesting that they did not support closed forest when they became isolated. Conversely, most islands west of Sumatra, north-west of Borneo, in the Malacca Straits, and around Palawan, apparently did. Other evidence from Palawan, however, suggests that there were also extensive grasslands or sparsely wooded savanna at the glacial maximum (Bird et al. 2007). Islands, such as those in the Riau and Lingga Archipelagos, which are isolated today by shallow seas, also support many forest-dependent species, but this says nothing about last-glacial vegetation, since these islands remained connected until the early Holocene, when the climate was similar to that of today.

As discussed in Section 1.5, an increasing amount of molecular genetic evidence suggests that free movement across the Sunda Shelf between the major islands has not been possible for rainforest organisms during most of the Pleistocene, and possibly not since the early Pliocene. Borneo appears to have been particularly isolated, which may partly explain why its flora contains a significantly higher proportion of endemic species than other parts of Sundaland (Van Welzen et al. 2005). Java has few endemics, but its flora is more similar to that of Wallacea than the rest of the Sunda Shelf, presumably as a result of the dry climate of most of the island.

3.9.2 Hainan and Taiwan

Hainan and Taiwan are two similar-sized continental islands (34,000 and 36,000 km^2, respectively), with broad connections to the Chinese mainland at glacial low sea levels (Fig. 3.6). As expected, their faunas are essentially continental, but with considerable differentiation at the subspecies level or, occasionally, species level. It appears that neither island is large enough to support a big cat, with the 15–20-kg clouded leopard (*Neofelis nebulosa*) the biggest species on both, although both also support the bigger, but largely herbivorous, Asiatic black bear (*Ursus thibetanus*, <180 kg), and a variety of smaller carnivores.

3.9.3 Ryukyu (Nansei) Islands

The Ryukyu (Nansei) Islands form a 1500 km-long archipelago between Taiwan and the southern Japanese island of Kyushu (Fig. 3.6). At present, the islands are separated into three groups by deep-water gaps (>1000 m) between the southern and central islands, and the central and northern islands, respectively. Pleistocene low sea-levels would have joined together the islands within each group and also, apparently, the southern islands to Taiwan and the northern islands to the main islands of Japan. The region is tectonically active, so the antiquity of the deep-water gaps is unclear. Animal distributions, living and fossil, have been used to argue for a landbridge as far as Amami Island, in the central group, in the early Pleistocene (Oshiro and Nohara 2000), but this may be underestimating the ability of terrestrial vertebrates to cross relatively narrow stretches of open sea. The northern islands (including Yakushima), share species with the main islands of Japan that do not occur further south, including the Japanese macaque (*Macaca fuscata*), Japanese weasel (*Mustela itatsi*), and Sika deer (*Cervus nippon*), while the central islands (including Okinawa and Amami Island) have endemic genera of murid rodents and the Amami rabbit (*Pentalagus furnessi*). The gap between the central and northern groups of islands, the Tokara Strait, is often considered the northern limit of the Oriental Region.

3.9.4 Ogasawara (Bonin) Islands

The definition of TEA used in this book does not include any large areas of deep water, so there are no really remote oceanic islands. The most remote islands that could be considered part of the region, since TEA appears to be the source of most of their biota, are the subtropical (27°N) Ogasawara or Bonin Islands, in the extreme north-east, 1300 km east of the Ryukyus. These small (<24 km^2) and widely scattered islands are of Tertiary volcanic origin and >1000 km from the nearest large land mass (Shimizu 2003). They have been inhabited by people for less than 200 years and, until recently, were covered in subtropical evergreen broad-leaved forest. Their isolation is reflected in the absence of any native freshwater fish, frogs, snakes, and non-flying mammals, the occurrence of one endemic lizard species, two endemic bats (one now extinct), and five endemic birds (three now extinct), and the very

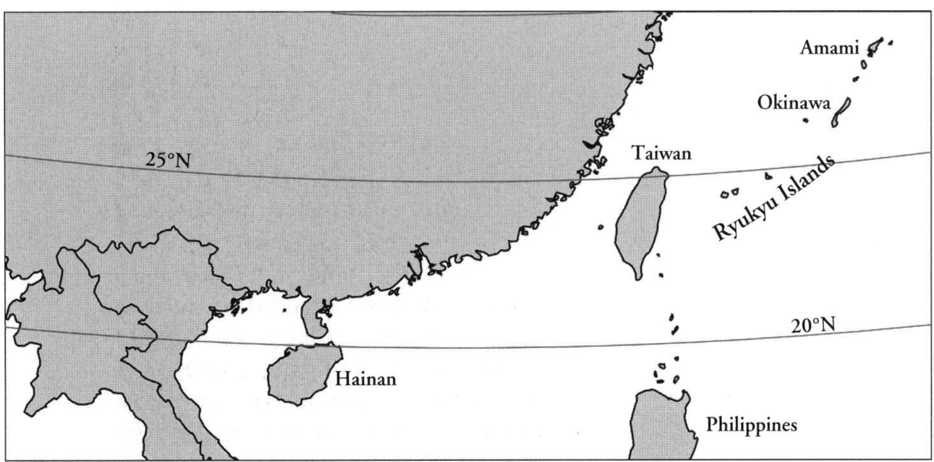

Figure 3.6 Map showing the locations of Hainan, Taiwan, and the Ryukyu (Nansei) Islands.

high rates of endemism in the vascular plant flora (43%). The tree flora is impoverished compared with mainland forests at similar latitude and the poorly dispersed family Fagaceae is absent. The invertebrate fauna also has a high rate of endemism, lacks some poorly dispersed taxa (e.g. *Apis* bees), and includes several examples of adaptive radiation within the archipelago (e.g. landsnails in the genera *Hirasea* and *Mandarina*). In recent decades, the islands have suffered greatly from invasive introduced species (see 7.4.10).

3.9.5 Palau Islands

The oceanic Palau Islands (7°N) (Fig. 3.11) are somewhat closer to TEA than the Ogasawara Islands, being only 845 km east of Mindanao, but they are also a similar distance from both the Moluccas (Maluku) and New Guinea, so their biogeography is more complex. The islands are of mixed volcanic and coral origin, with a total area of 500 km^2, 365 km^2 of which comprises the largest island, Babeldaob. They have had human inhabitants for at least 3000 years and have probably suffered prehistoric faunal extinctions, like many other islands in the Pacific, but there is no fossil evidence for this (Steadman 2006). The islands support a single, endemic, frog species, with close relatives in New Guinea, four native species of snakes, and a diverse fauna of skinks and geckoes, some of which are endemic (Crombie and Pregill 1999). The historical mammalian fauna consisted of two fruit bats, one now extinct, and one insectivorous bat. The 31 species of resident landbirds include two endemic genera, each with one species, and seven additional endemic species (Steadman 2006). The flora is relatively rich for a Pacific island and includes a number of endemics (Donnegan et al. 2007). As with the Ogasawara Islands, many exotic species are now established on Palau, including rats, mice, several species of birds, a toad, several reptiles, and many invertebrates and plants.

3.9.6 Krakatau

The early stages of the colonization process on a newly formed island have been extensively studied on the Krakatau Islands, between Java and Sumatra (Fig. 3.7). In 1883, the 800-m high volcanic island of Krakatau exploded in an eruption that killed around 40,000 people, mostly from the resulting tsunamis. Although the surviving islands are on the Sunda Shelf, the eruption is thought to have sterilized them, so the current biota has all arrived by dispersal over at least 16 km of sea—the distance to the nearest island that retained at least part of its biota—and probably over the 40 km from Sumatra or Java (Thornton et al. 2002). As expected, there are no frogs and no large-seeded primary forest trees. The only mammals are bats and rats, the latter probably introduced by people. The reptiles include pythons (*Python reticulatus*) and monitor lizards (*Varanus salvator*), which are excellent swimmers, and other snakes, geckos, and skinks that could have survived on floating trees and other driftwood, although human-assisted dispersal is impossible to rule out. The bird list includes 55 recorded species, not all of which have resident populations, but the weakly flying babblers and pheasants are absent. The invertebrate fauna is also dominated by species that can be dispersed long distances in the air.

3.9.7 Andaman and Nicobar Islands

The best-studied of the larger oceanic islands are those that form the 700-km archipelago stretching between south-western Myanmar and north-western Sumatra (Fig. 3.7). From north to south, this archipelago consists of the continental Preparis Island (part of Myanmar), followed by the oceanic Coco Islands (Myanmar), Andaman Islands (India), and Nicobar Islands (India). At glacial low sea-levels, the Andamans and Cocos formed a single large island, which was still separated by 70 km from Myanmar. Deep channels also separated the Andamans from the Nicobars at this time, divided the Nicobars into three, and separated them from Sumatra. There are also two additional oceanic islands in this area, Narcondam and Barren Islands (both Indian), 150 km east of the Andamans. Barren Island is an active volcano, while Narcondam is volcanic, but apparently dormant. The total land area of the oceanic islands in the archipelago is approximately 6400 km^2 and the natural vegetation is evergreen or semi-evergreen rainforest, with smaller areas of deciduous forest.

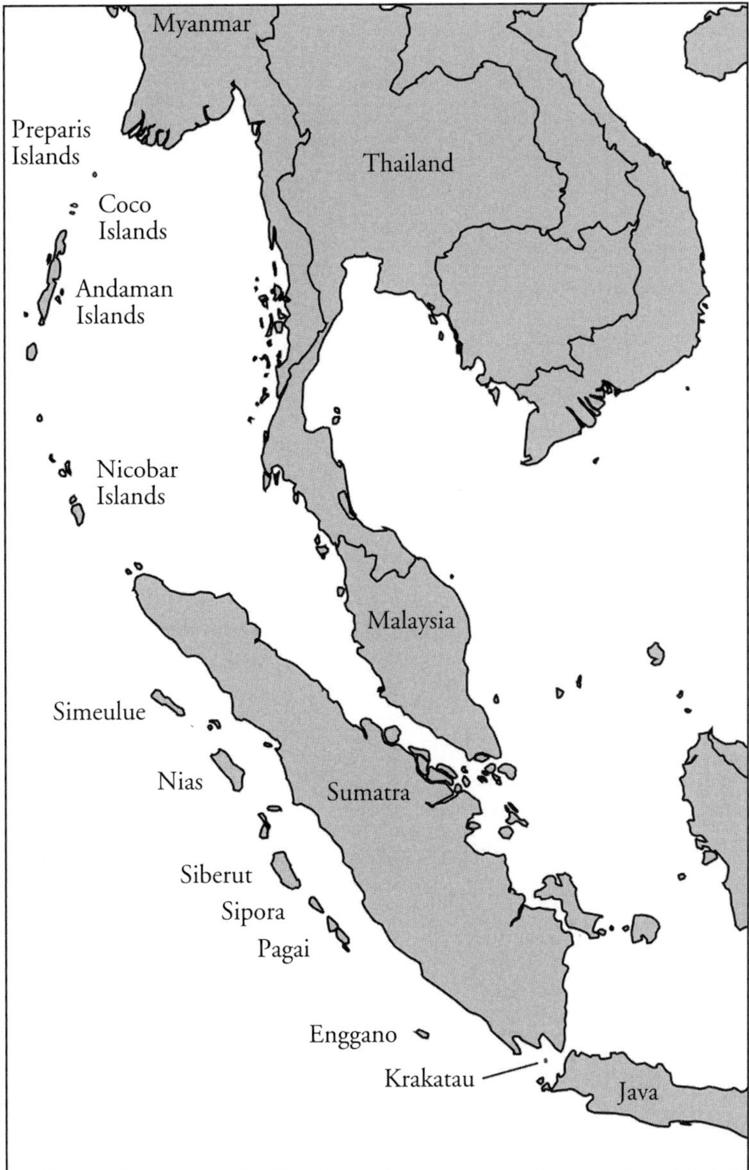

Figure 3.7 Map showing the locations of the major islands in the west of Tropical East Asia.

The oceanic nature of the islands is reflected in their vertebrate faunas, despite the relatively narrow water-gaps. The mammalian fauna is impoverished, with no native primates, ungulates, carnivores, or other large species, but includes endemic species of tree shrew (*Tupaia nicobarica*), rat (*Rattus palmarum*), shrew (two *Crocidura* species), and fruit bat (*Pteropus faunulus*) (Wilson and Reeder 2005). The wild-pig populations have also been considered native, but are more likely to be prehistoric human introductions. The bird fauna lacks barbets, trogons, pheasants, and babblers (vs. 39 species of babblers in south-west Myanmar) and includes only 3 species of bulbuls (vs. 17 in south-west Myanmar), but is relatively rich in the strong-flying pigeons (8 species) and hawks (9 species) (Ripley and Beehler 1989). There are no hornbills in the Andamans and Nicobars, but a single endemic species lives on isolated Narcondam Island. Altogether there are 17 endemic bird species in the Andamans and/or Nicobars, plus a number of endemic subspecies and races. The amphibian and reptile faunas also include a high proportion of endemics (Das 1999).

It is fairly easy to imagine the ancestors of the native vertebrate fauna crossing 70 km of open water on floating rafts of vegetation or, in the case of the birds and bats, blown by a storm. The absent or under-represented groups consist mostly of species too large for natural rafts (primates, ungulates, carnivores) or too weak flyers (babblers) to survive even a wind-assisted passage. In contrast to the vertebrate fauna, the vascular plant flora is a fairly rich (Table 3.1), although the poorly dispersed family Dipterocarpaceae is represented by only 7 species in the Andamans and none in the Nicobars (Hajra et al. 1999).

3.9.8 Islands off the west coast of Sumatra

The chain of near-equatorial islands off the west coast of Sumatra (Fig. 3.7) is an excellent illustration of the problems that arise when the main evidence for past terrestrial connections is the biota itself. These islands sit above the Sunda megathrust (see 1.6) and are subject to rapid subsidence and uplift during earthquakes (Stone 2006), as well as longer term changes, so present water depths are unlikely to be a reliable guide to the past extent of dry land. Several of the larger islands are separated from Sumatra by water that is too deep to have allowed a connection at the last glacial maximum, even with considerable tectonic activity, but earlier landbridges—late Pliocene to mid Pleistocene—have been invoked to explain the surprisingly diverse vertebrate faunas on some of them.

Simeulue (1738 km^2) and Enggano (443 km^2) probably never had land connections, as shown by their impoverished mammalian faunas and the presence of endemic species of birds and reptiles. Simeulue also has an endemic subspecies of long-tailed macaque, but what was previously thought to be an endemic subspecies of pig turns out to be a feral population of the Sulawesi warty pig (*Sus celebensis*) (Wilson and Reeder 2005). Enggano has two poorly known species of *Rattus* rat that appear to be endemic. Nias (4771 km^2) is also surrounded by deep water, but it supports a diverse subset of the Sumatran mammalian fauna, including ungulates and carnivores, which suggests a land connection in the not-too-distant past. The presence of several forest-dependent mammal species is also evidence that closed-canopy forest persisted on these islands during glacial times (Meijaard 2003).

The most interesting and puzzling faunas are on the Mentawai Islands, particularly the large islands of Siberut (4030 km^2), Sipura (845 km^2), and the closely adjacent islands of North and South Pagai (combined area 1675 km^2). These islands may have been connected via Siberut to the continental Batu Islands and thus Sumatra during Pleistocene low sea-levels, but probably not at the last glacial maximum (Inger and Voris 2001). In addition to ten apparently endemic rodent species (Wilson and Reeder 2005), these islands support an amazing four or five species of endemic primates: a gibbon, a leaf-monkey, a snub-nosed monkey, and one or two macaques, depending on whether the Siberut species is separated from the one present on the other three large islands (Ziegler et al. 2007).

3.9.9 The Philippines

The geological history of the Philippines (Fig. 3.8) is both complex and controversial (see 1.3). Of all the numerous islands, only Palawan, Mindoro, and some associated small islands originated as parts of mainland Asia and there is no evidence to suggest that they have acted as rafts, bringing mainland plants and animals into the archipelago. The other islands are of very varied age (>30 Ma to recent) and origin, but many are now less isolated than they were when they first appeared above sea-level. At Pleistocene low sea-levels, most of the land area of the Philippines was in six major islands: Greater Palawan, Greater Sulu, Greater Mindanao, Greater Luzon, Greater Negros-Panay, and Mindoro. These Pleistocene islands were separated by narrow but deep straits. Present-day patterns of similarity and dissimilarity between islands reflect this history more than present-day geography (Heaney et al. 1998).

Figure 3.8 Map of the Philippines. Only Palawan may have had dry-land connections with Borneo in the past.

Only Palawan may have ever been connected to Borneo and it is not clear if, or when, this last happened. Narrow submarine canyons in the straits between them apparently exceed 200 m in depth (Heaney 1991), so it would require either a greater drop in sea-level than is currently believed to have occurred or tectonic changes to completely eliminate the water barrier. Palawan supports a subset of the Bornean fauna, including many species not found elsewhere in the Philippines, but it also has several endemic species and subspecies, which argue either for a long period of separation or a narrow but enduring water-gap (Heaney et al. 1998; Inger and Voris 2001). Evidence for a considerably drier climate at the last glacial maximum and a long history of human occupation further complicate interpretations of Palawan's modern biota (Bird et al. 2007; Lewis et al. 2008).

Mindanao is the next richest island, in terms of its vertebrate fauna, but, like the rest of the Philippines, the missing groups (including pheasants, gibbons, pangolins, porcupines, dogs, bears, mongooses, mustelids, and mouse deer) suggest that there has never been a land connection with Borneo. Mindanao has 15 endemic bird species and 13 endemic mammals, including endemic genera of tree shrews and insectivores. Luzon, the northernmost large island, is only 25 km from Mindanao, yet it lacks colugos, tree shrews, and tarsiers, and supports an endemic radiation of weird and wonderful murid rodents, including giant arboreal leaf-eaters and long-snouted earthworm specialists (Fig. 3.9) (Heaney et al. 1998)! The large-mammal fauna of Luzon consists of a macaque (*Macaca fascicularis*), a pig (*Sus philippensis*), a deer (*Cervus marianus*), and two civets (*Paradoxurus hermaphroditus, Viverra tangalunga*), currently all considered native, although macaques, pigs, deer, and civets have been moved to other islands by people. New species of bats and rodents are being described from the Philippines every year, with most of them endemic to single islands.

The fossil fauna of the Philippines includes elephant-like stegodonts, on at least Luzon and Mindanao, and both elephants and rhinoceroses on Luzon (Bautista 1991). There are also fossils of extinct *Bubalus* species—smaller relatives of the water buffalo—on Luzon and Cebu. Together with the living tamaraw, *Bubalus mindorensis*, on Mindanao, these fossils suggest that this genus may have once been present throughout the Philippines (Croft et al. 2006). Although at first sight the presence of these large mammals might seem to argue for a terrestrial connection in the past, the narrow gaps between the Pleistocene islands were well within the swimming range of modern elephants, and presumably also stegodonts, rhinoceroses, and water buffaloes. Large carnivores have never been present, which helps explain both the large sizes of some of the living and fossil rats, and the relatively small sizes of the extinct elephants, stegodonts, and *Bubalus* species.

Figure 3.9 The Northern Luzon shrew-rat, *Rhynchomys soricoides*, is an example of the weird and wonderful radiation of endemic murid rats in the Philippines. These shrew-rats are confined to montane forest and show extreme morphological specialization for a diet of earthworms and other soft-bodied invertebrates (Balete et al. 2007). Photograph courtesy of Lawrence Heaney, © the Field Museum.

3.9.10 Sulawesi

The complex geological history of Sulawesi (Fig. 3.10) was mentioned in Chapter 1, and the relationship between this and the assembly of its current biota is still controversial. Plant diversity is comparable to the large islands of the Sunda Shelf (Roos et al. 2004), which is consistent with the idea that part of the ancestral flora was stranded there by the opening of the Makassar Straits in the late Eocene (see 1.3). The fauna, in contrast, is relatively impoverished and largely consistent with later, overwater dispersal, from Borneo. The Makassar Straits were still deep at the last glacial maximum, but only about 45 km wide at the narrowest point. A possible exception to this overwater origin of the fauna is the phylogenetically isolated, pig-like babirusa (*Babyrousa babyrussa*), which is perhaps old enough to have crossed the last land connection from Borneo (van den Bergh et al. 2001). The ancestors of the endemic frog fauna may also have arrived overland from Borneo, while rafting is more plausible for the relatively salt-resistant snakes (Inger and Voris 2001). Two marsupials, the bear cuscus (*Ailurops ursinus*) and Sulawesi dwarf cuscus (*Strigocuscus celebensis*), may have arrived on a micro-continental fragment rifted from the margins of Australia–New Guinea (Moss and Wilson 1988). They are apparently sister taxa, despite their great morphological differences, and diverged from their nearest living relatives around 25 million years ago, which is consistent with this story (Ruedas and Morales 2005).

Sulawesi has no pheasants, barbets, trogons, or bulbuls, and only one babbler, two woodpeckers,

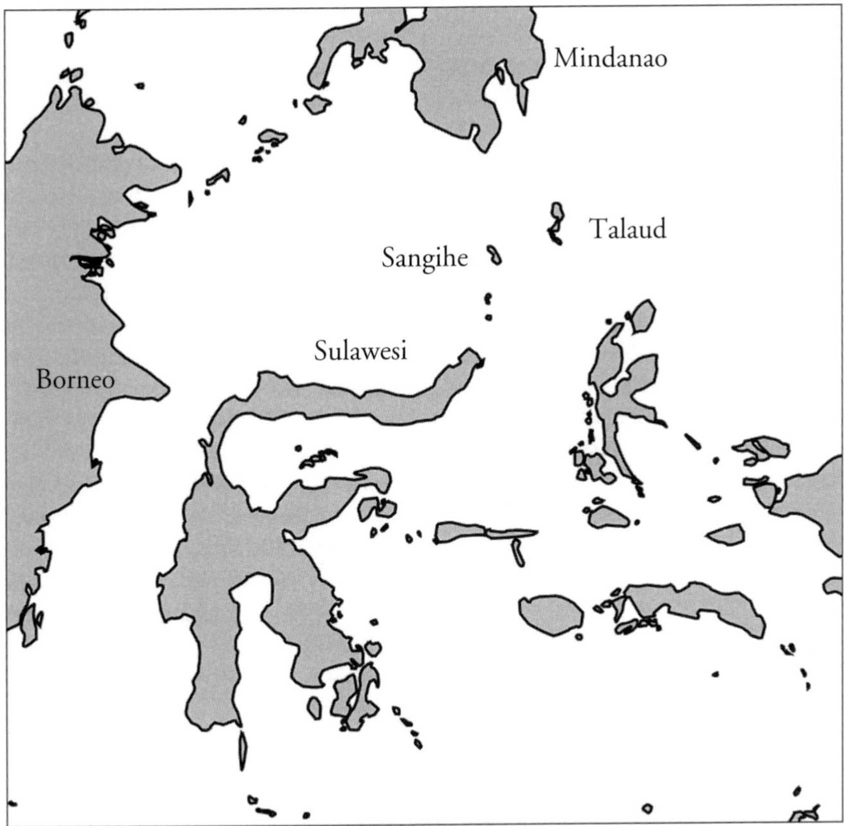

Figure 3.10 Map of Sulawesi, also showing the small oceanic islands in the Sangihe and Talaud groups.

and two hornbills. There are also three species of honeyeaters (Meliphagidae), an Australian bird family that does not reach Borneo. The mammalian fauna includes many endemic rodents, a rich bat fauna, tarsiers (Fig. 5.13), an endemic radiation of macaques, an endemic civet genus (but no other carnivores), a pig, and two species of anoa (*Bubalus*)—relatives of the mainland water buffalo and the living and extinct *Bubalus* species of the Philippines. The bear cuscus appears to occupy the arboreal leaf-eater niche occupied by colobine monkeys on Borneo. The Pleistocene fossil fauna includes a number of extinct genera, including stegodonts and elephants (van den Bergh 1999).

3.9.11 Sangihe and Talaud Islands

These two small groups of oceanic islands form a chain between the northern tip of Sulawesi and the southern tip of Mindanao (Fig. 3.10). The Sangihe Islands are dominated by active volcanoes, while the Talaud Islands are non-volcanic and low-lying. The largest islands are Sangihe Island (700 km^2) in the Sangihe group and Karakelang (976 km^2) in the Talaud group. The archipelago supports only 31 native mammal species, including five endemics: a fruit bat and two rats in the Talaud group, and a tarsier and squirrel in the Sangihe group (Riley 2002a). The bird fauna is also impoverished, but there are ten endemic species: six in the Sangihe group, three in the Talaud group, and one shared (Riley 2002b, 2003). This vertebrate fauna appears to have reached the island mostly from Sulawesi, with dispersal to Sangihe aided by island stepping-stones during glacial low sea-levels (van den Bergh et al. 2001). Sangihe Island had stegodonts in the Pleistocene, illustrating once again the swimming ability of this group.

3.9.12 Nusa Tenggara

Nusa Tenggara (literally, 'south-east islands') or the Lesser Sunda Islands, consists of a group of medium and small oceanic islands that extends over 1900 km east from Java (Monk et al. 1997) (Fig. 3.11). Geologically, they consist of an inner volcanic arc, including Lombok (4740 km^2), Sumbawa (15,415 km^2), Komodo (340 km^2), Flores (17,150 km^2), Alor (2125 km^2), and Wetar (3600 km^2), and an outer group of more varied and complex origins, including Sumba (11,057 km^2), Timor (31,280 km^2), and the Tanimbar Islands (5060 km^2). Most of Nusa Tenggara has strongly seasonal rainfall and some areas are among the driest in South-East Asia.

The western islands have largely Asian biotas, but Timor is only 480 km from Australia and supports many taxa of Australian origin. The flora of Nusa Tenggara is deficient in poorly dispersed families, with the Fagaceae absent and only a single species of dipterocarp, *Dipterocarpus retusus*, confined to the western islands. The Pleistocene mammal fauna showed the same pattern of giant rats, diverse bats, and dwarfed stegodonts as the oceanic islands of the Philippines, although late-Pleistocene extinctions and many recent mammalian introductions complicate the modern picture. Deliberate introductions to Flores started with the Sulawesi pig (*Sus celebensis*) at least 7000 years ago, followed by the Eurasian pig (*S. scrofa*), macaques (*Macaca fascicularis*), porcupines (*Hystrix javanica*), and civets (*Paradoxurus hermaphroditus*), with additional species in colonial and modern times (van den Berg et al. 2008b). Similarly, the avifauna lacks barbets, trogons, and native bulbuls, while there is a single, endemic species of hornbill, on Sumba, a single, widespread woodpecker, and a single

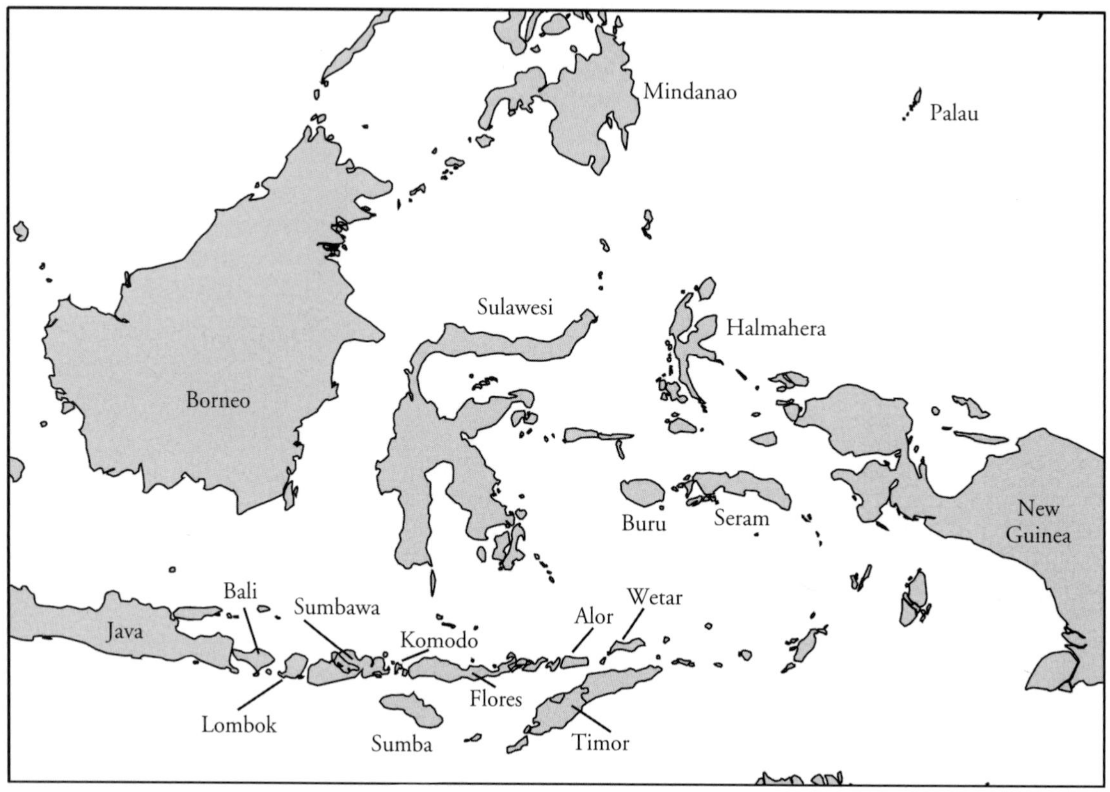

Figure 3.11 Map of the Wallacean region showing the islands of Nusa Tenggara (including Lombok, Sumbawa, Komodo, Flores, Alor, Wetar, Sumba, and Timor) and Maluku (Halmahera, Buru, Seram), plus the Palau Islands.

babbler, which reaches as far as Timor. Giant tortoises, which once lived on at least Flores and Timor, are now extinct and the Komodo dragon (*Varanus komodoensis*)—the world's largest living lizard—is now confined to Komodo, western Flores, and a couple of nearby islands.

3.9.13 Maluku

The Maluku Islands or Moluccas are an archipelago of small islands of complex and varied geological origin lying between Sulawesi and New Guinea (Fig. 3.11) (Monk et al. 1997). The total land area is 78,000 km^2 and the largest islands are Halmahera (18,000 km^2), Seram (17,000 km^2), and Buru (9,500 km^2). The islands are mostly mountainous and most were covered until recently in rainforest. These islands are not part of TEA as defined for this book, but they support many tropical Asian taxa and are mentioned here briefly because of their biogeographical interest. The flora is largely Asian, but relatively impoverished. Lowland dipterocarp forest on Seram, for example, supports a single dipterocarp species, *Shorea selanica*, which can represent 75% of the total basal area (Edwards et al. 1993) and there are only two Fagaceae in Maluku. Asian influence on the fauna declines rapidly as one moves east. The non-bat mammalian fauna consists of several marsupials, including some only known from Holocene fossils, a variety of endemic murids, and recently introduced shrews, rodents, pigs, deer, and civets (Flannery 1995). The bird fauna includes parrots, honeyeaters, and megapodes of New Guinea origin.

CHAPTER 4

The ecology of plants: from seed to seed

4.1 Introduction

There are fundamental biological differences between terrestrial plants and animals that justify the treatment of their ecologies in separate chapters. Plants can make their own food from water, carbon dioxide, mineral nutrients, and sunlight, while all terrestrial animals depend on food chains that are ultimately based on plants. Plants are immobile as adults and relatively long-lived, while animals are mostly mobile and relatively short-lived. Although much of modern ecology consists of the interactions between plants and animals, these interactions are never symmetrical and it usually makes more sense to consider the plant and animal perspectives separately. Even in relationships where both partners benefit—mutualisms—these benefits are not a result of inter-species collaboration, but a consequence of each species pursuing its separate and very different interests.

4.2 Study sites

Most information on the plant ecology of TEA comes from a small number of well-studied sites (Table 4.1). Many of these form part of the pan-tropical network of standardized Forest Dynamics Plots coordinated by the Center for Tropical Forest Science of the Smithsonian Tropical Research Institute (CTFS; Losos and Leigh 2004). Eleven plots in East Asia are part of this network, stretching from Bukit Timah Nature Reserve (1°N) in Singapore, just north of the equator, to Fushan (25°N) in subtropical northern Taiwan. Similar plots on the Chinese mainland, forming part of the Chinese Forest Biodiversity Monitoring Network, extend the range north to Gutian (29°N) and Tiantong (30°N), in Zhejiang Province. In each plot, every tree at least 1-cm diameter is marked, mapped, measured, and identified, with the whole process being repeated every few years, giving information on growth, mortality, and recruitment. Most published information so far has come from the large (50–52 ha) equatorial lowland rainforest plots, at Pasoh and Lambir, although an increasing amount is now coming from the equally large, seasonal, dry, evergreen forest plot at Huai Kha Khaeng in Thailand. The other Asian plots are smaller (2–25 ha) and most were established more recently.

Other key sites for research in South-East Asia include: on the island of Borneo, Gunung Palung National Park (1°S) in Indonesia; the Kuala Belalong Field Studies Centre (4°N) in Brunei; and the Danum Valley Conservation Area (5°N) and Mount Kinabalu National Park (6°N) in Sabah; in Thailand, Khao Yai National Park (14°N) and the Mae Klong Watershed Research Station (14°N); and in subtropical Japan, the forests of northern Okinawa Island (27°N) and the Yakushima Island Biosphere Reserve (30°N).

Research at all the sites mentioned above focuses on relatively undisturbed primary forest or comparisons between primary forest and a variety of disturbance types. Much of the information on the ecology of human-dominated landscapes in the region comes from Singapore (1°N), which has been deforested over the last 200 years and still has some remnants of primary lowland rainforest (Corlett 1992), and Hong Kong (22°N), which was

Table 4.1 Key ecological study sites in Tropical East Asia, arranged in latitudinal order.

Location	Latitude	Major Vegetation Types
Gunung Gede-Pangrango National Park, Java	7°S	montane forests
Gunung Palung National Park, Kalimantan	1°S	various lowland forest types
Barito Ulu research area	0°N	various lowland forest types
Bukit Timah Nature Reserve, Singapore	1°N	disturbed fragment of lowland rainforest
Republic of Singapore	1°N	various primary and secondary vegetation types
Pasoh Forest Reserve, Peninsula Malaysia	3°N	lowland rainforest
Ketambe, Gunung Leuser National Park, Sumatra	4°N	lowland rainforest
Gunung Mulu National Park, Sarawak	4°N	various lowland forest types, limestone caves
Lambir Hills National Park, Sarawak	4°N	lowland rainforest
Kuala Belalong, Ulu Temburong Nat. Park, Brunei	4°N	lowland rainforest
Danum Valley Conservation Area, Sabah	5°N	lowland rainforest, logged forests
Sepilok Forest Reserve, Sabah	5°N	various forest types
Mount Kinabalu, Sabah	6°N	elevational and edaphic range of forest types
Khao Chong,Thailand	8°N	lowland rainforest
North Negros Forest Reserve, Philippines	11°N	montane and submontane forests
Cat Tien National Park, Vietnam	11°N	various seasonal forest types
Khao Yai National Park, Thailand	14°N	seasonal evergreen and deciduous forests
Sakaerat Environmental Research Station, Thailand	15°N	seasonal evergreen and deciduous forests
Mae-Klong Watershed Research Station, Thailand	15°N	mixed deciduous forest
Huai Kha Khaeng Wildlife Sanctuary, Thailand	16°N	seasonal evergreen and deciduous forests
Palanan Wilderness Area, Luzon, Philippines	17°N	lowland rainforest
Doi Inthanon National Park, Thailand	19°N	various lowland and montane forest type
Jianfengling National Nature Reserve, Hainan	19°N	lowland and montane forest
Bawangling National Nature Reserve, Hainan	19°N	lowland and montane forest
Xishuangbanna National Nature Res., Yunnan	21–22°N	various forest types
Nanjenshan Nature Reserve, Taiwan	22°N	evergreen broadleaved forest
Hong Kong Special Administrative Region	22°N	various secondary vegetation types
Dinghushan National Nature Reserve, Guangdong	23°N	evergreen broadleaved forest
Ailao Shan National Nature Reserve, Yunnan	25°N	subtropical montane forest
Fushan, Taiwan	25°N	evergreen broadleaved forest
Northern Okinawa Island, Japan	27°N	evergreen broadleaved forest
Gutianshan National Nature Reserve, Zhejiang	29°N	evergreen broadleaved forest
Yakushima Island Biosphere Reserve, Japan	30°N	broadleaved and coniferous forests

Sites with CTFS forest dynamic plots (Losos & Leigh 2004) are shown in bold.

deforested much earlier and has no primary forest (Hau et al. 2005).

4.3 Trees from seed to seed

Most vegetation in TEA is dominated by angiosperm trees. These have complex, multi-stage lifecycles involving many, more-or-less independent, processes (Fig. 4.1). Adding to this complexity is the fact that many of these processes involve interactions with other organisms, including animals, microbes, and other plants. Most research in the region has focused on only one or two of these processes, so it is not possible to compare entire lifecycles across plant species. Instead, the major stages and associated processes are first considered separately, before the implications of these processes for the maintenance of plant species-diversity are discussed. This stage-by-stage approach is not ideal, however, since it understates the importance of interactions

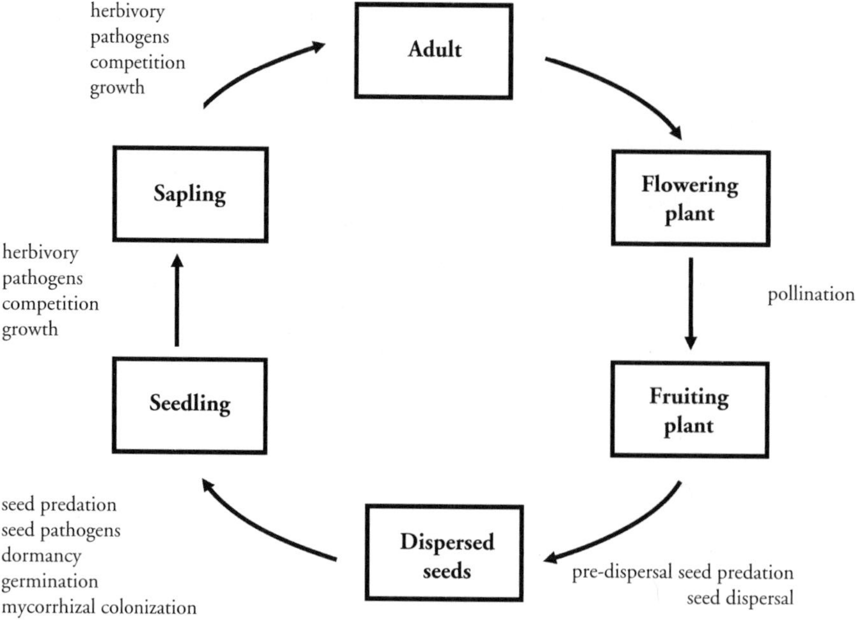

Figure 4.1 The life-cycle of an angiosperm tree, showing the major processes involved.

between stages: for example, high mortality by seed predation may be compensated by a lowered mortality at the seedling or sapling stage, as a result of decreased density. The focus throughout is on forests, because these are the best-studied systems, but non-forest systems are mentioned where data are available. Growth forms other than trees are considered more briefly in the next section (4.4).

4.3.1 Pollination

(See also flower visitors in 5.2.7.)

Pollination is the transfer of pollen grains from the anther to the stigma. In most plants there is no seed production without pollination, although a significant minority of species retains some capacity to set seeds by self-pollination, without any external agent. Outcrossing and long-distance pollen dispersal appear to be the norm in tropical forest trees, suggesting that there is strong selection against inbreeding (Ward et al. 2005; Scofield and Schultz 2006; Naito et al. 2008b). There is evidence for widespread pollen-limitation of seed production in nature, with many species setting more seed after hand-supplementation of pollen, than under natural conditions (Knight et al. 2005). Pollen limitation may not only limit seed quantity, but also offspring quality, by reducing pollen competition and thus selectivity among gametes (Colling et al. 2004).

Globally, an estimated 20% of all angiosperms and most gymnosperms are pollinated by wind, but the proportion is much lower in tropical lowland forests. This is presumably because the low windspeeds below the canopy reduce the effectiveness of aerial pollen transport, while the wide spacing of conspecific plants in species-rich vegetation favours the directed movement of pollen by animal vectors. Wind pollination is expected to be more common in open sites and in the relatively low-diversity forests at higher altitudes and latitudes, and on extreme soil types. There have been few studies of wind pollination in TEA, but counts of pollen grains suspended in the air, mostly in urban areas, and pollen preserved in sediments (see 1.5), combined with information on floral morphology and pollen characteristics, support these predictions. The grasses and sedges of open habitats are wind-pollinated, as are the conifers

Figure 4.2 The wind-pollinated flowers of *Quercus* (*Cyclobalanopsis*) *edithiae* in Hong Kong. Photograph © the Hong Kong Herbarium.

and oaks (*Quercus*, including *Cyclobalanopsis*; Fig. 4.2), which dominate some low-diversity forests in the north and above 1000 m, and *Casuarina equisetifolia*, which forms single-species stands on some tropical shores. However, there is insufficient information for many other species with small, inconspicuous flowers that grow in habitats where wind would appear to be a suitable pollen vector (Corlett 2004).

There is a large regional literature on animal pollination in TEA (reviewed in Corlett 2004), but most of it simply reports visual observations of flower visitors and their behaviours in the flower. Contact with both anthers and stigmas is taken as evidence of pollination and experimental manipulations—such as the exclusion of possible pollinators—are rare. Most studies are also very short in comparison with the life-span of the plant and cover only a tiny fraction of the plant's geographical range. Thus the statement that 'plant X is pollinated by animal Y' should often be read as 'in one population of plant X, in a particular year, the flowers were visited by animal Y in a manner that could potentially result in pollen transfer, if the visitor had previously visited a conspecific plant with compatible pollen'.

Bees and wasps

The most important animal vectors of pollen in TEA are bees. Social bees dominate in lowland rainforests, while non-social bees may be more important in other forest types (Kato et al. 2008). Unlike most other pollen vectors, for which pollen has no value, bees gather pollen to feed their larvae. There is no data for the tropics, but European solitary bees need the entire pollen content of 7–1100 flowers to rear a single larva (Müller et al. 2006). Natural selection on the bees will, therefore, tend to maximize their pollen-collection efficiency, which is likely to reduce their pollination efficiency. Pollination success depends on the bee failing to completely groom pollen from the part of its body that will contact the stigma. Another potential problem with bees as pollinators is that they are 'central-place foragers', flying to and from a nest. Bees visiting a large pollen or nectar source, such as a flowering tree, may, therefore, simply fly back and forth between tree and nest, without transferring pollen to other trees of the same species.

Despite these conflicts between bee behaviour and pollination, the sheer abundance of bees in the tropics and the efficiency with which social bees, in particular, discover and exploit new resources, ensure their dominant position. It seems that plants must either use bees or, as many do, exclude them from their flowers. Bilabiate (two-lipped) flowers, which deposit pollen on the back of the bee, from where it is not easily groomed, have evolved multiple times in bee-pollinated angiosperms (Westerkamp and Classen-Bockoff 2007).

Figure 4.3 *Trigona collina* stingless bees at the entrance to their nest. Photograph courtesy of Sara Leonhardt.

Lowland rainforests in TEA can support more than 20 coexisting species of stingless bees (Apidae, Meliponini, with most species traditionally placed in the genus *Trigona*; Fig. 4.3), as well as up to four species of honeybee (Apidae, Apini, *Apis*), typically one large, one medium, and one small species. Both stingless bees and honeybees are highly social and form long-lived colonies with numerous workers. *Trigona* and *Apis* have overlapping size ranges, but most *Trigona* species are relatively small, and the largest social bee is the 18-mm giant honeybee, *Apis dorsata*. The dance language that honeybees use to communicate the locations of floral resources to their nest mates has been intensively studied (e.g. Riley et al. 2005), but while some stingless bees can communicate the 3-dimensional locations of food resources, the mechanisms of communication and recruitment are not well understood (Nieh 2004). Both genera are generalist foragers on a wide range of plant species and flower types, but they are mechanically excluded from some flower types, where access to nectar is restricted by tightly closed petals or a very long corolla tube, or where nectar is absent and the pollen reward is enclosed in porose anthers that must be buzzed at the correct frequency.

Social bee colonies store food and can thus survive through periods of low resource-availability, but some species also move their nest sites. This is taken to the extreme in *Apis dorsata*, which forms colonies of up to 100,000 workers that occupy an unprotected comb, usually attached to the exposed branches of a large tree (Fig. 4.4). In this species, most colonies appear to be migratory, apparently tracking the availability of sources of nectar and pollen sufficiently large to support their huge colonies. In seasonal climates these migrations are annual, but in the aseasonal lowland rainforest at Lambir, Sarawak, the colonies appear only during the mass-flowering episodes that occur at irregular, multi-year intervals (Itioka et al. 2001). Between these episodes, the colonies apparently make use of the more reliable floral resources in other forest types. *Apis dorsata* is also unusual in that it will forage before sunrise and after sunset, and several nocturnal-flowering canopy tree species at Lambir appear to depend on this behaviour (Momose et al. 1998).

Stingless bees are less diverse and abundant in disturbed, open habitats and, although up to three *Apis* species may persist in deforested areas, the cavity-nesting *A. cerana* (Fig. 5.8) is usually most abundant. Stingless bees are also absent from most montane and subtropical forests in TEA, where *A. cerana* is often the only highly social bee present. This species extends north in East Asia to 46°N

Figure 4.4 Nests of the giant honeybee, *Apis dorsata*, in a *Koompassia* tree at Lambir Hills National Park during the 1996 general flowering episode. Picture taken by the late Professor Tamiji Inoue and used courtesy of Eiko Inoue.

and will continue foraging at temperatures as low as 10–12°C. This combination of tolerance to disturbance and to low temperatures makes *A. cerana* the single most important pollinator in Hong Kong (22°N) (Corlett 2001).

Many other bee species occur in TEA and are important pollinators for flower types that *Apis* and *Trigona* avoid and in habitats where they are less abundant. Bumblebees (Apidae, Bombini, *Bombus*) are large, social bees that form small, usually annual, colonies. They occur throughout the northern part of TEA and in mountains as far south as the Philippines, Java, and Sumatra, but not Borneo. Carpenter bees (Apidae, Xylocopinae, *Xylocopa*) are large, non-social bees that visit mostly large, showy flowers and appear to be the sole pollinators of some species with porose anthers. In TEA, *Xylocopa* and *Bombus* species have largely complementary distributions in space and time, with *Xylocopa* most abundant in lowland, tropical habitats and, where the two genera co-occur in the north of the region, active only in the summer.

Unlike *Apis* and *Trigona*, *Bombus*, and *Xylocopa* are able to force their way into flowers with tightly closed petals, an ability they share with another group of large bees, the Megachilidae (Sakagami et al. 1990). This family includes what may be the world's largest bee, *Megachile* (*Chalicodoma*) *pluto*, in the northern Moluccas, with females reaching almost 4 cm in length. Some bees can access nectar protected by long corolla tubes or similar structures by means of their long tongues. Species of *Amegilla* and related genera (Apidae, Anthophorini) are important pollinators of such flowers, particularly herbs, in both the forest understorey and open habitats. There are also some long-tongued species in the diverse and widespread family Halictidae. In contrast, the short-tongued bee families Colletidae and Andrenidae are significant only in the northern part of the region, where they are seasonally abundant.

Not all bees are generalist foragers. The genus *Ctenoplectra* (Apidae) consists of short-tongued bees found throughout TEA that feed their larvae on a mixture of pollen and oil. This is apparently derived entirely from oil-producing members of the Cucurbitaceae, in the genera *Thladiantha* and *Momordica*, which they pollinate (Schaefer and Renner 2008).

Wasps (the term applied to all Hymenoptera that are not bees or ants) are much less important as pollinators than bees. The larger species are basically carnivorous, but many adults visit flowers for nectar and may sometimes be significant as pollinators. The conspicuous exception to the general unimportance of wasps in pollination is the well-studied relationship between the figs (*Ficus*, Moraceae) and their tiny, short-lived, wasp pollinators (Agaonidae, subfamily Agaoinae). With a few exceptions, each of the 300 or so species of figs in TEA has a single species of pollinator, and *vice versa*, and neither can reproduce without the other (Harrison 2005). The diversity and abundance of figs in TEA demonstrates the success of this relationship.

Other insects

In tropical rainforests in TEA, the most important pollinators after the bees are beetles (Coleoptera) and flies (Diptera). Although several authors have described a distinct beetle-pollination 'syndrome' of flower characters, the relationships between beetles and flowers are diverse and complex. Flower-visiting beetles in TEA range from highly adapted flower-specialists, through more-or-less generalist herbivores, to species attracted by the deceitful mimicry of their foods or the sites used for egg-laying. Most beetle-pollinated flowers have a strong odour and many are pale in colour and open in the evening or at night, but there are exceptions to all these. Most records from TEA refer to the plant family Annonaceae, which appears to be predominantly beetle-pollinated, or species in the Araceae, Dipterocarpaceae, or Arecacae (palms).

In many species of Annonaceae, the inner petals form a chamber by pressing tightly over the compact stamens and carpels. Beetles, apparently attracted by the odour, enter the chamber during the female stage and leave during the male stage, carrying pollen on their bodies. In most cases, the reward for the beetles is not clear. In the Araceae, a chamber is typically formed by the spathe that surrounds the inflorescence of numerous small flowers. In many species this chamber acts as a trap that retains the beetles during the female stage. In the dipterocarps there is no pollination chamber and the beetle pollinators of 20 species at Lambir apparently switch their diet from young leaves to petals and other floral parts during general flowering episodes (Kishimoto-Yamada and Itioka 2008a). Beetles are also often associated with palm inflorescences in TEA, but there have been no detailed studies of their role in pollination.

The importance of flies as pollinators in TEA has probably been underestimated, although members of at least 25 dipteran families have been recorded visiting flowers in the region (Corlett 2004). As with the beetles, flower-visiting flies range from highly adapted flower specialists, through more-or-less generalist feeders on liquids, to species normally associated with decaying organic matter that are attracted by deceit. Pollination by deceit seems to be particularly widespread in the orchids (e.g. Bänziger 1996a), but also occurs in several other plant families. The most dramatic examples of flowers imitating fly food—in this case rotting meat—are in the plant family Rafflesiaceae, where the huge flowers attract flies in the families Calliphoridae (blow flies) or Sarcophagidae (flesh flies) (Bänziger 1991, 1996b) (Fig. 4.5). The flowers of at least some Rafflesiaceae are endothermic, maintaining tissue temperatures several degrees above air temperature (Patiño et al. 2000).

Endothermy is also found in the trap-flowers of *Alocasia odora* (Yafuso 1993) and other fly-pollinated members of the family Araceae. Many species in the fly genus *Colocasiomyia* (Drosophilidae) have evolved intimate mutualisms with members of the Araceae, in which the flies depend on the inflorescences and infructescences as breeding sites, while the plants depend on the flies for pollination (Sultana et al. 2006; Takenaka et al. 2006). In many cases, two *Colocasiomyia* species are associated with one host plant, with one species using the lower (female) part of the inflorescence for oviposition and larval development, while the other uses the upper (male) part.

Butterflies and moths (Lepidoptera) are conspicuous flower visitors in TEA, particularly in tropical deciduous forests, where they appear to be more important than beetles and flies (Kato et al. 2008). Most of the well-studied examples of dependence on lepidopterans involve long-tongued hawkmoths (Sphingidae) and/or butterflies (mostly Papilionidae) visiting flowers in which the nectar is at the bottom of a long corolla tube—pale, scented, crepuscular or nocturnal flowers are visited mainly by hawkmoths and brightly coloured, scentless, diurnal flowers by butterflies—but a great variety of flower types are visited and apparently pollinated.

Figure 4.5 The giant flowers of *Rafflesia kerrii* are pollinated by blow flies (Calliphoridae). Photograph courtesy of Hans Bänziger.

Recently, a very different example of moth pollination was discovered in some species of the large genus *Glochidion* and related genera in the Phyllanthaceae. Here, the tiny flowers are pollinated at night by female *Epicephala* moths, attracted by species-specific floral odours, whose larvae consume some of the seeds produced by the flowers they pollinate (Kawakita and Kato 2006; Okamoto et al. 2007). As with the better-known relationship between figs and their pollinating wasps, the moths appear to be at least locally species-specific, and each partner is dependent on the other for successful reproduction.

Other insect orders appear to play a relatively minor role in pollination, although the importance of the ubiquitous flower-feeding thrips may have been underestimated. Thrips have been credited with pollinating six co-occurring species of *Shorea* (Dipterocarpaceae) at Pasoh, Malaysia (Appanah 1993), although some of the same species are pollinated by beetles at Lambir. Thrips are weak fliers and the thrips-pollinated *Shorea acuminata* at Pasoh showed a higher selfing rate (38%) and shorter mating distances (<100 m) than has been found in bee-pollinated dipterocarps, such as *Neobalanocrapus heimii* at the same site (Naito et al. 2008b). Cockroaches (Blattodea), bugs (Hemiptera), and other insect orders may also have a role in pollinating some plant species.

Vertebrates

Birds are considerably less important as pollinators in TEA, than they are in the Neotropics and probably also less important than in Africa, New Guinea, or Australia. Nectar-feeding birds in Asia are also less diverse and specialized than their Neotropical counterparts (Fleming and Muchhala 2008). More than 50 bird species in 18 families have been recorded taking nectar from flowers in TEA, but most of these reports are from a small number of widely cultivated ornamental trees, with large flowers and abundant, easily accessible nectar (particularly *Bombax* and *Erythrina*). Except on the eastern margins of the region, in Wallacea, where nectarivorous honeyeaters (Meliphagidae) and parrots of Australian affinities (see 3.9) are widespread, almost all well-documented cases of pollination of wild plants by birds in TEA involve the sunbirds (Nectariniidae), flowerpeckers (Dicaeidae), or white-eyes (Zosteropidae).

The sunbirds (Fig. 4.6) are the most specialized flower-birds in TEA, with both morphological and physiological adaptations for exploiting nectar that parallel those in the Neotropical hummingbirds (Primack and Corlett 2005). Most have more or less elongated bills (up to 55 mm long in some spiderhunters (*Arachnothera*)) and long, narrow,

Figure 4.6 Fork-tailed sunbird, *Aethopyga christinae*, visiting *Erythrina* flowers. Photograph courtesy of Henry T.H. Lui.

tubular tongues that can be protruded beyond the tip of the bill. They are recorded as visiting a wide range of flowers, but most known or suspected cases of effective pollination involve either mistletoes (Loranthaceae) or gingers (Zingiberaceae). Most of these bird-pollinated plants have tubular or brush-shaped, scentless flowers, that are often red in colour. The most extreme adaptations to bird pollination occur in some species pollinated by spiderhunters that have floral tubes >30 mm in length.

Flowerpeckers and white-eyes are shorter-billed than most sunbirds and most species of both appear to be more frugivorous than nectarivorous. Flowerpeckers are reported as the major pollinators only of mistletoes, particularly species with explosive dehiscence (Davidar 1985). White-eyes are the commonest flower-visiting birds in the northern part of TEA and on some mountains in the tropics, and may be important pollinators for some large flowers.

As with birds, bats appear to be much less important as pollinators in TEA than they are in either the Neotropics or Australia. Old World nectar-feeding bats are also much less specialized than their Neotropical counterparts (Fleming and Muchhala 2008). Around a dozen species of pteropodid bats (Pteropodidae) have been recorded visiting flowers in TEA and it is likely that other species also do so. These bats span a huge size range (15–1500 g) and include both nectar specialists, with long tongues, narrow muzzles, and reduced teeth (*Eonycteris* and *Macroglossus*), and predominantly frugivorous species (*Cynopterus*, *Pteropus*, *Rousettus*).

Flowers adapted for bat pollination tend to be large—or large aggregations of small flowers, as in *Parkia*—and robust, with pale or drab colouration, opening at night and with a strong nocturnal odour. The plants offer unobstructed access to the flowers for flying bats. The reward is usually copious amounts of nectar, but several Asian species of Sapotaceae offer a detachable, sugar-rich, fleshy corolla, and *Freycinetia* (Pandanaceae) provides edible bracts. Most bat-pollinated species are in the families Bignoniaceae, Bombacaceae, Fabaceae, Lecythidaceae, Musaceae, Lythraceae, Myrtaceae, Pandanaceae, and Sapotaceae.

Non-flying mammals have only a very minor role in pollination in TEA, although both treeshrews and squirrels have been implicated in specific cases (Corlett 2004). A bizarre new example was discovered recently, involving the widespread rainforest understory palm, *Eugessona tristis*, and seven species of small (<1 kg) mammals, including treeshrews, the slow loris, a squirrel, and rats (Wiens et al. 2008). The large inflorescences produce copious supplies of nectar over a long period. This nectar is fermented by yeasts into an alcoholic solution that can approach the concentration of beer and gives the inflorescences a 'strong alcoholic smell reminiscent of a brewery'! The pollinators show no signs of intoxication and the authors speculate that the regular consumption of alcohol may have an unknown benefit.

4.3.2 Seed dispersal

(See also frugivory in 5.2.8.)

Pollen must be moved to a conspecific stigma, but the target for seed dispersal is less clearly defined. Dispersal *away* from the parent and siblings reduces competition and the impact of species-specific herbivores and pathogens. The advantages of dispersal *towards* a suitable microsite for germination and growth into an adult plant are obvious, although which particular site is most suitable is often impossible to predict in advance. In any case, while pollinators can be 'paid on delivery' at the recipient flower, seed-dispersal agents must be 'paid in advance' at the fruiting plant, making it almost impossible for the plant to direct seed-dispersal agents to particular targets.

A number of plant species in TEA, including some members of the families Acanthaceae, Euphorbiaceae, Fabacae, and Urticaceae, the genus *Impatiens*, and some species of *Viola*, eject their seeds mechanically, by a variety of mechanisms, for distances ranging from a few centimetres to several metres. The role of these mechanisms in seed dispersal has not been studied in TEA, although limited data from other parts of the world suggests that this initial mechanical dispersal phase is often followed by a second phase involving ants or, for larger seeds, possibly rodents (Vander Wall and Longland 2004).

The impracticality of directing seeds to specific microsites might be expected to favour the use of abiotic dispersal agents such as wind, but most seeds are much larger than pollen, so wind-dispersal in forests is restricted to tiny-seeded plants like orchids or to the upper forest canopy, where winds are strongest. Most members of the tree family Dipterocarpaceae are wind-dispersed (Fig. 4.7), as are many canopy legumes and large woody lianas, but only in open, herbaceous vegetation is wind-dispersal dominant. Dispersal by water is not limited by seed size, but is effective only in areas subject to regular flooding: mangroves, seasonally inundated swamp forests, and riverine vegetation.

Figure 4.7 The two-winged fruits of a *Dipterocarpus* species (Dipterocarpaceae).

The regional literature on seed dispersal (reviewed in Corlett 1998) consists largely of observations of fruit consumption by animals, and in most cases there is little direct evidence for the effectiveness of frugivores in seed dispersal. Any animal that consumes fruits and deposits intact seeds away from the parent crown may contribute to seed dispersal, but frugivores vary hugely in both the quantity of seeds they remove and the quality of the sites where they are deposited. If all fruits are eventually removed, then a low-quality disperser could have a negative influence on the plant by removing seeds that would have been more effectively dispersed by an alternative vector. On the other hand, frugivores, unlike abiotic dispersal agents, can be satiated by abundant fruit supplies, leaving a variable proportion of the fruit crop undispersed (Hampe 2008). If a low-quality disperser feeds only on fruits that would otherwise remain beneath the parent plant—as terrestrial herbivores do—then any contribution they make to seed dispersal is positive for the plant.

Despite the many post-dispersal sources of mortality discussed below, differences in the effectiveness of different seed-dispersal mechanisms persist into the spatial pattern of the adult plants. At Pasoh, the average size of the spatial clusters of conspecific trees increases from species with wingless, dry fruits (dispersed, if at all, by scatter-hoarding rodents), to wind-dispersed species, to fleshy fruited animal-dispersed species; within this latter group, cluster size increases with fruit size, presumably as a consequence of larger frugivores with larger home-ranges eating larger fruits (Seidler and Plotkin 2006). Mean cluster-size ranged from around 50 m in wind-dispersed trees to more than 150 m in the species with the largest, animal-dispersed fruits (>5 cm diameter).

In the following account, observations from throughout TEA have usually been combined, because there is insufficient data to consider different vegetation types separately. Despite the lack of quantitative data, however, it is evident from comparing plant species lists that there is a decline in the numbers of tree species producing large (>*c*.2.5-cm diameter) fleshy fruits with both altitude and latitude. Moreover, this decline, which is gradual within tropical lowland and lower montane forests, accelerates above *c*.1500 m and north of about 23–24°N. The simplest explanation is that this reflects a similar decline in the diversity and abundance of large frugivorous birds and mammals, but Moles et al. (2007), from a global dataset, describe a similar sudden drop in seed size at the margins of the tropics, suggesting that the smaller fruits could also be a consequence of selection on the size of seeds.

Birds

Birds are the most important seed-dispersal agents in TEA, in terms of both the quantity and diversity of seeds dispersed. Seed-dispersing birds in TEA range in size from 5-g flowerpeckers (Dicaeidae) to 3000-g hornbills (Bucerotidae) (Fig. 4.8). Birds have no teeth and typically swallow fruits whole, so the largest fruit that can be removed is determined by the maximum gape width, which is, in turn, approximately related to body size. Large birds can swallow small fruits, but not *vice versa*, so the number of potential avian frugivores increases as fruit size decreases. But even small birds may peck seed-containing pieces from large, soft, many-seeded fruits, such as some species of figs (*Ficus*).

Figure 4.8 A wrinkled hornbill, *Aceros corrugatus*, eating a fig. Photograph © Tim Laman.

The most important avian dispersal agents for large seeds in large (>25 mm diameter) fruits are the hornbills (Bucerotidae) and the fruit pigeons (a clade of Columbidae with thin-walled gizzards and short, wide guts). Some hornbills and fruit pigeons are also notable for their large daily flight distances (Leighton and Leighton 1983, Kinnaird et al. 1996; Symes and Marsden 2007). Fruits of intermediate size (*c*. 15–25-mm diameter) are available to a wider range of birds, including the larger species of pheasants (Phasianidae), cuckoos (Cuculidae), thrushes (Turdidae), starlings (Sturnidae), barbets (Megalaimidae), and the crows and their relatives (Corvidae). Even in intact forest communities, however, most fleshy fruits are <15-mm diameter and small-fruited plants dominate in secondary forests and shrublands. These fruits are available to all but the tiniest frugivores, the flowerpeckers and white-eyes, which have gape limits <9 mm. The most important avian dispersal agents for small fruits are the bulbuls (Pycnonotidae; Fig. 4.9), which are abundant in almost all vegetation types in TEA, and, in forests, probably also the babblers (Timaliidae). The babblers are not as highly frugivorous as the bulbuls and most will not cross open areas, but they are diverse and abundant, particularly in the north of the region, and all species for which there is dietary information eat at least some fruit.

Fruits eaten by birds are either 'unprotected'—that is, they have only a thin outer peel—or they are dehiscent, so the birds eat only the unprotected inner part (usually the seed plus a fleshy aril). Most bird-fruits in TEA appear black or red to human eyes, but a minority appear orange, yellow, blue, white, green, or brown. Birds are tetrachromatic, with an additional visual pigment that is sensitive in the ultraviolet, but, although bird plumage colours make use of UV signals (Stevens and Cuthill 2007), it appears—from admittedly small sample sizes—that few fruits in TEA reflect strongly in the UV (Corlett, unpublished data). The nutritional reward for fruit consumption is most often sugars, particularly glucose and fructose (Ko et al. 1998), but some bird-fruits are high in lipids (<70% of the dry weight of the fruit pulp) and some birds appear to prefer these.

Bats

The fruit bats (Pteropodidae) in TEA appear to be considerably less important as seed-dispersal agents than those in the Neotropics, which belong to the unrelated bat family Phyllostomidae. The difference is particularly striking in the early stages of woody plant succession, where bat-dispersed genera dominate in the Neotropics (e.g. *Cecropia*, *Muntingia*,

Figure 4.9 A Chinese bulbul, *Pycnonotus chinensis*, swallowing a fruit of *Melia azedarach* in Hong Kong. Photograph courtesy of Samson So.

Piper, Solanum, Vismia), while bird-dispersed species (such as *Macaranga*) dominate in TEA. Bats are excellent dispersal agents for small-seeded plants, because the seeds are swallowed and scattered widely when defecated in flight, but most pteropodids do not swallow seeds >2–3 mm in diameter, instead depositing them under temporary 'feeding roosts' near the fruiting plant. The resulting clumping of seeds from the same source reduces one of the major advantages of seed dispersal. Large-seeded bat-dispersed plants must either have mechanisms to tolerate this clumping or rely on a small proportion of seeds being dropped elsewhere.

Fruit bats in TEA range in size from 15 to 1500 g and the larger species are capable of carrying away fruits that are as big as, or bigger than, the largest that can be swallowed by frugivorous birds. Unlike birds, bats have teeth and while hanging upside down from one foot can use the other foot to manipulate a fruit, allowing for complex processing. Bat-fruits often have a protective peel, which is first removed. Large seeds are then dropped before the rest of the fruit is taken into the mouth and crushed between the tongue and hard palate (the bony roof of the mouth). Only the juice is swallowed, while the remainder of the pulp is discarded as a fibrous wad. In relatively fibrous fruits, such as figs, many small seeds may end up in the wad, but a variable proportion is swallowed with the juice. Fruit bats lack colour vision and most specialized bat-fruits are—like bat-flowers—pale or drab in colour, with a nocturnal odour, and held away from the foliage and other obstructions. Odour appears to play a dominant role in the location of ripe fruits (Hodgkison et al. 2007; Borges et al. 2008). Most parts of TEA have some fruits that are eaten only by bats, but there is also a considerable amount of dietary overlap with frugivorous birds, particularly with regard to small-fruited figs. This overlap increases towards the northern limits of TEA, which are also the northern limits of pteropodid distributions (Funakoshi et al. 1993, Nakamoto et al. 2007).

Primates

The seed-dispersal literature for TEA suggests that, in intact forest communities, the primates are second only to the birds in their importance as dispersal vectors. There may be some bias in this assessment, since primates have received a disproportionate amount of research attention, but primates form a large proportion of the canopy animal biomass and many species consume a lot of fruit. The tarsiers feed almost entirely on arthropods, lorises feed largely on plant exudates and arthropods, and most colobine monkeys (Cercopithecidae, Colobinae: leaf monkeys, langurs, snub-nosed monkeys) destroy most seeds they eat, but the orangutans, gibbons, and macaques are highly frugivorous.

The orangutans (Hominidae: two *Pongo* species) are now confined to shrinking areas in Borneo and Sumatra, but were widespread in mainland TEA in the late Pleistocene, only a hundred or so tree-generations ago (Corlett 2007a). They are the largest arboreal frugivores (40–100 kg) and eat fruits of many different types, including the biggest species available. Seeds may be spat out, broken, or swallowed and defecated intact. There has been no detailed study of their role in seed dispersal, but orangutans can clearly move large numbers of seeds relatively long distances.

Gibbons (Hylobatidae: 13 or more species in four genera; Fig. 4.10) occupied forests throughout mainland TEA and on the large continental-shelf islands of Borneo, Sumatra, Java, and Hainan within historical times, but have been eliminated from most of China over the last 1000 years, and from much of the rest of the region over the last 100 years (Corlett 2007a). Gibbons may well be the most effective of all mammalian seed-dispersal agents, consuming large quantities of fruit, swallowing most seeds, and then defecating them, intact, over their large home-ranges. A study in Borneo found that >90% of seeds were dispersed >100 m from the parent plant (McConkey and Chivers 2007). Gibbons prefer thin-skinned, pulpy, yellow-orange fruits in most studies, but also eat large quantities of some fruit species that do not match these criteria (e.g. McConkey et al. 2002).

Macaques (Cercopithecidae, Cercopithecinae: *c.*20 *Macaca* species; Fig. 4.11) are the most widespread non-human primates in TEA, occurring naturally on islands with no other primates (e.g. Sulawesi, Taiwan, Yakushima) and as human introductions

Figure 4.10 A pair of white-handed gibbons at Khao Yai National Park, Thailand. Photograph courtesy of Billy C. H. Hau.

Figure 4.11 A rhesus macaque, *Macaca mulatta*, with cheek pouches full of food. Photograph courtesy of Laura Wong.

on others. Although they have been recently eliminated from many parts of their former range, they are usually the last primates to succumb to habitat loss and persecution, and are sometimes the only large mammals to survive in human-dominated landscapes, as in Hong Kong and Singapore. Tropical macaques are highly frugivorous and fruit is preferred, when seasonally available, by macaques living in seasonal climates in the north of the region.

In areas where macaques coexist with gibbons, there is considerable overlap in their fruit diets, but macaques process fruits very differently from gibbons, so the consequences for seed dispersal can be very different. The most important differences reflect the smaller size-threshold for swallowing seeds in macaques. Fruit is masticated thoroughly, damaging some seeds, indigestible parts are then spat out, and only small seeds (mostly <4 mm-diameter) are swallowed, unless the tight adherence of the fruit flesh to the seed makes separation in the mouth difficult (Corlett and Lucas 1990; Yumoto et al. 1998; Otani and Shibata, 2000; Otani 2004). The consequences for larger seeded plants are not entirely negative, however, since macaques have cheek pouches in which they store fruits (Fig. 4.11), which are then processed as the animal moves between trees. A proportion of the larger seeds are, therefore, dropped or spat out up to 100 m from the parent plant. Thus macaques disperse many seeds, but whether or not a macaque visit is of overall benefit to a fruiting plant will depend on whether more effective alternative dispersal agents are available.

Carnivores

Fruit is easy to find, ingest, and digest, so it forms part of the diet of many animals that have specialized adaptations to other foods. Among the order Carnivora, only the hypercarnivorous cats and linsangs appear to totally avoid fruits. In TEA, both species of bear (Ursidae), most species of civets (Viverridae), and several species in the families Canidae (*Canis aureus*, *Nyctereutes procyonoides*, and *Vulpes* spp.) and Mustelidae (e.g. *Melogale moschata* (Fig. 4.12); Zhou et al. 2008a; *Martes flavigula*; Zhou et al. 2008b) can live for at least part of the year on fruit-dominated diets. Carnivores swallow their food without much chewing, so most seeds are passed through undamaged, and most species have large home-ranges, so seed-dispersal distances can be large.

The civets are the most consistently frugivorous, particularly members of the subfamily Paradoxurinae, the palm civets (e.g. *Paguma larvata*; Zhou et al. 2008c; Fig. 5.18), which are nocturnal and largely arboreal. Although they have been understudied in comparison with primates, fruit bats, and birds, civets are probably important seed-dispersal

Figure 4.12 The Chinese ferret-badger, *Melogale moschata*, is largely carnivorous, but fruits dominate its diet from September to December in subtropical China and it can disperse seeds over long distances (Zhou et al. 2008a). Photograph courtesy of Paul Crow, KFBG.

agents in forests. Moreover, several species of civets (e.g. *Paguma larvata, Paradoxurus hermaphroditus,* and *Viverricula indica*) thrive in degraded, human-dominated landscapes, if they are not directly persecuted, and may then be the only dispersal agents for large fruits with large seeds. The characteristics of fruits consumed by civets have not been studied in detail, but most reports suggest a preference for relatively large, thin-skinned, sugar-rich fruits. Although most civets apparently swallow most seeds in the fruits they eat, there are reports of larger seeds being dropped from the mouth (e.g. Tsang and Corlett 2005) and a single observation of the small-toothed palm civet (*Arctogalidia trivirgata*) squeezing the fluid out of the juicy fruits of *Ficus hispida* and then discarding the residue, including most of the seeds, from the mouth (Duckworth and Nettelback 2007).

Terrestrial herbivores

Many ripe, fleshy fruits reach the ground uneaten, either as a result of being actively shed by the tree (e.g. *Choerospondias axillaris*) or because they are dropped or knocked down by animals foraging in the canopy. The fate of such fruits has received relatively little attention in TEA, but it is clear that many are eaten by large terrestrial herbivores, including probably all species of pigs, deer, cattle, tapirs, elephants, and rhinoceroses. The indehiscent pods produced by some dry-forest legumes may be consumed by the same animals. Unlike carnivores, terrestrial herbivores have teeth and jaws adapted to breaking up tough plant materials, but seeds may escape by being too small, too hard, or too distasteful. Intact small seeds have been reported in the faeces of all these animals and the large seeds of wild mangos (*Mangifera*) have been found germinating in rhinoceros and elephant dung (Fig. 4.13). Asian elephants have been reported to prefer large, yellow, sweet-smelling fruits with large, hard seeds (Kitamura et al. 2007). Some studies suggest that they consume less fruit and fewer species than their African relatives, but fruits from 29 species were recorded by mahouts in Myanmar as eaten by work elephants (Campos-Arceiz et al. 2008a) and potential seed-dispersal distances are very large (<6 km; Campos-Arceiz et al. 2008b). Deer (*Tragulus*, *Muntiacus*, and *Rusa*) also disperse some large, hard seeds by regurgitation (Chen et al. 2001; Prasad et al. 2006; Chanthorn and Brockelman 2008).

Figure 4.13 Seeds of a wild mango, *Mangifera* sp., germinating from elephant dung at Khao Yai National Park, Thailand.

All terrestrial herbivores may also disperse seeds that adhere to the outside of the animal (Mouissie et al. 2005). This mechanism has not been investigated in TEA, but casual observations suggest that adaptations for external dispersal (hooks, bristles, etc.) are common only in plants of disturbed habitats, where they are probably dispersed mainly by people and their livestock.

Rodents

Rodents may disperse very small seeds that escape destruction in the mouth, are then swallowed, and appear intact in their faeces. They are more significant, however, as dispersal agents for large, dry seeds without a fleshy reward, which they store for future consumption (Forget et al. 2005). Rodents may also be involved in the secondary dispersal of large seeds from fleshy fruits after the flesh has been removed by the primary dispersal agent. If the seeds removed by a rodent are then stored in scattered caches of one or a few seeds buried in the soil—scatter-hoarding—this can be a very effective means of dispersal, even if most cached seeds are eventually retrieved and eaten. Many rodents, however, store seeds in their nests, which are usually in locations from which germination and establishment are unlikely, even if a seed is forgotten. Most direct evidence for scatter-hoarding in TEA comes from the north of the region, where rodents in the genera *Apodemus* and *Leopoldamys* disperse large, hard seeds (mostly Fagaceae) tens of metres (e.g. Cheng et al. 2005; Abe et al. 2006; Xiao et al. 2006). However, dry fruits that appear to be adapted for dispersal in this way occur throughout TEA, including all the Fagaceae and many other plants with large, wingless, dry fruits, including some dipterocarps. *Apodemus* is absent from the tropics, but *Leopoldamys*, *Maxomys*, and at least two species of ground squirrels are known to scatter-hoard seeds in tropical forests (Yasuda et al. 2000; Kitamura et al. 2004, 2008) and other rodents are suspected to, although the consequences for seed dispersal have not yet been investigated in any detail.

Ants

Ants are ubiquitous in TEA and can often be seen carrying seeds, but whether this represents seed dispersal or seed predation (see below) is generally unknown. Seeds specialized for ant dispersal usually have a lipid-rich elaiosome that acts as an attractant, a handle for carrying the seed, and a food reward. Elaiosomes have evolved multiple times within the angiosperms (Dunn et al. 2007). The seed itself is discarded in or near the nest. In TEA, several species of rainforest herb in the genus *Globba* (Zingiberaceae) have been shown to be dispersed by ants in this way (Pfeiffer et al. 2004; Zhou et al. 2007; Fig. 4.14). Other, unstudied, plants have similar structures. Seed-dispersal distances are short (<10 m and often <1 m). In general, larger ants take larger seeds and disperse them further (Pfeiffer et al. 2004, 2006). Ants also remove small seeds without elaiosomes from the ground and from vertebrate droppings. In these cases, they are probably acting as seed predators, but this has rarely been tested.

People

In additional to the dispersal of seeds and fruits with adaptations for external attachment mentioned above, people disperse many small seeds without specialized adaptations in and on their clothing, in

Figure 4.14 Workers of *Oecophylla smaragdina* carrying a seed of *Globba franciscii*. Photograph courtesy of Martin Pfeiffer.

mud on their shoes, attached in various ways to their vehicles (Von der Lippe and Kowarik 2008), and as contaminants in transported grain, soil, and horticultural stock (Hodkinson and Thompson 1997).

4.3.3 Gene flow by pollen and seeds

Gene flow within and between plant populations can occur by the movement of pollen or the movement of seeds. It is often assumed that pollen-mediated gene flow is much greater than seed-mediated gene flow, as is the case with most temperate plants that have been studied. However, there is no reason to assume this is true in the tropics, where both processes usually make use of animal vectors. There is little data from TEA itself, but mean gene-dispersal estimates from genetic data for tropical trees elsewhere are typically in the range 20–1000 m, depending on plant species, site, and the methods used (e.g. Hardy et al. 2006). Unfortunately the tropical dataset is biased towards species with large, wind- or rodent-dispersed seeds, where dispersal distances are likely to be short, but the little data for insect-pollinated trees with fleshy, vertebrate-dispersed fruits suggests that, in such species, gene flow in pollen and seeds is of similar magnitude (Hardesty et al. 2006; García et al. 2007).

This is consistent with estimates of pollen- and seed-movement distances, which tend to fall between 100 and 1000 m for vectors as diverse as bees (for pollen), birds (for pollen and seeds), and primates (for seeds). There are exceptions in both directions, but these exceptions account for a minority of plant species. Thrips (pollen) and rodents (seeds) move genes <100 m, while large bees, large bats, and fig wasps (pollen) and large birds, civets, and elephants (seeds) can move genes >1000 m. Wind can move pollen and tiny, dust-like seeds >1000 m, but large seeds <100 m. Pollen-mediated gene flow is probably relatively more important in the north of the region, where wind becomes increasingly important in pollination and wind and rodents in seed dispersal.

4.3.4 Seed predation and seed pathogens

(See also granivory in 5.2.9.)

Seed predation (i.e. the consumption of seeds) can be divided into pre-dispersal seed-predation, which takes place while the seeds are still on the parent plant, and post-dispersal predation, which takes place on the ground and typically involves different groups of organisms. Pre-dispersal seed-predation can greatly reduce the numbers of seeds available for dispersal and may also reduce the ability of the plant to attract dispersal agents. Post-dispersal predation can not only reduce the numbers of seeds available for germination, but may also change their spatial pattern, if predation rates vary between habitats and microhabitats, or with distance from the parent plant (see below). Moreover, seeds vary widely in their size, nutritional value, and defences, so even generalist seed-predators are selective, potentially influencing the species composition of the resulting plant community.

Despite its obvious importance, there have been relatively few studies of seed predation in TEA. Most information on pre-dispersal predation comes from studies of the diets of known seed predators, particularly colobine monkeys, and can rarely be used to assess the quantitative impact on the plants. Most Asian colobine monkeys are largely folivorous, but fruits and/or seeds form a significant part of the diet of many species (Kirkpatrick 2007). Most fleshy fruits eaten by these monkeys are taken before they are ripe and the seeds are probably destroyed, although this has rarely been checked. Predation is most obvious when they feed on dry fruits, such as those of dipterocarps. Other pre-dispersal predators of dipterocarps include other primates, squirrels, long-tailed parakeets (*Psittacula longicauda*), beetles, and moth larvae (Curran and Leighton 2000; Nakagawa et al. 2005; Sun et al. 2007). In three mast-fruiting events at Pasoh, 13–56% of all dipterocarp fruits were destroyed by vertebrate and insect pre-dispersal seed-predators (Sun et al. 2007). Beetles (Curculionidae) and moths were the major pre-dispersal seed-predators of three co-occurring species of oaks (*Quercus* and *Cyclobalanopsis*) in evergreen broad-leaved forest in south-west China (Xiao et al. 2007b).

Post-dispersal seed-predation has received somewhat more attention in TEA. In most studies the major seed predators were rats (Muridae), but this result may be biased by both the experimenters' use of seeds that are too large for removal by most ants (>20 mg) and the fact that most studies were in

Figure 4.15 A bearded pig, *Sus barbatus*, in Borneo. Photograph courtesy of Siew Te Wong.

landscapes that have lost many of their larger vertebrates. In a near-intact lowland rainforest, vertebrate community in Borneo, the major post-dispersal predators of the large seeds of dipterocarps were nomadic bearded pigs (*Sus barbatus*; Fig. 4.15), which were attracted by the temporarily superabundant food supply (Curran and Leighton 2000). Other recorded post-dispersal seed-predators in TEA include wild boar (*Sus scrofa*), pig-tailed macaques (*Macaca nemestrina*), ground squirrels, porcupines, pheasants, the emerald dove (*Chalcophaps indica*), ants, and a variety of other insects. The seed-predation rates, recorded in experimental studies over 1–4 weeks, range from 0 to 100% for different sites, species, and experimental designs, and losses of >50% are common (e.g. Blate et al. 1998; Zhang et al. 2005). McConkey (2005) showed that the presence of gibbon faeces doubled the rate at which seeds were removed. The influence of seed characters on removal rates has not been consistent across studies, although very hard or toxic seeds are usually avoided.

Seed pathogens have received even less attention than seed predation, both in TEA and elsewhere (Gallery et al. 2007; Pringle et al. 2007). Yet 'fungal exclusion' experiments using fungicides, usually result in a significant reduction in seed mortality and one that is selective by species and microhabitat (e.g. Leishman et al. 2000; O'Hanlon-Manners and Kotanen 2006). There is also evidence that those seed species that do persist in the soil for long periods, have chemical defences against pathogens (Veldman et al. 2007). This suggests that further research, particularly in the hot, wet, fungus-promoting, rainforest understorey, is needed.

Seed predation is of particular theoretical interest in the lowland rainforests of Sundaland, because escape from seed predation is one of the more plausible hypotheses to explain the evolution of supra-annual, synchronized mast-fruiting in lowland dipterocarp forests (see 2.6.2). Populations of seed predators are reduced in the long periods of low fruit-production, so during a mast-fruiting episode they are rapidly satiated and the majority of seeds escape predation (Janzen 1974). In keeping with these predictions, there was a time lag of several months between increasing fruit production at Lambir and increases in small-mammal populations (Nakagawa et al. 2007). Synchronization across large areas is needed to prevent highly mobile seed predators, such as wild pigs, from simply tracking the crops across the landscape (Curran and Leighton 2000).

In its original formulation, the predator-satiation hypothesis ignored pre-dispersal predation, but this provides an additional selective force for tight synchronization by reducing the crop sizes on individual plants and thus reducing the ability of any individual plant to satiate seed predators by itself (Sun et al. 2007). One major problem with the predator-satiation hypothesis is that the evidence for it largely comes from dipterocarps; but numerous non-dipterocarp species also join most mast-fruiting events, including some species with tiny seeds, which are of no interest to vertebrate seed-predators, and many with fleshy fruits, which must, at the same time, avoid satiating their dispersal agents.

Both predators and pathogens may influence the pattern of recruitment after seed dispersal (Nathan and Casagrandi 2004). The number of dispersed seeds almost always declines with distance from the parent tree, but recruitment does not necessarily show the same pattern. In a classic paper in 1970, Daniel Janzen suggested that, in general, we would expect few seeds to survive close to adults, despite the large number that are deposited there, because of pathogens and predators attracted to the vicinity of the adult plant by the high density of seeds. We would also expect low recruitment at large distances from the adult, because few seeds are dispersed that far. He, therefore, predicted a recruitment peak at intermediate distances. Joseph Connell proposed a similar mechanism involving seedling herbivores, so this is now known as the Janzen–Connell hypothesis.

Despite a number of well-documented examples, there is no evidence that Janzen–Connell effects are widespread at the seed stage (Hyatt et al. 2003). Indeed, the recruitment pattern predicted by Janzen–Connell is only one possibility. An alternative, proposed by Hubbell (1980), is that survival increases with distance from the parent, as proposed by Janzen–Connell, but less rapidly than dispersal decreases, so there is a monotonic decline in recruitment with distance. A third alternative, attributed to McCanny (1985), is that survival is *higher* nearer the adult, as a result of predator satiation, so that recruitment falls off very rapidly with distance. Nathan and Casagrandi (2004) show that the Janzen–Connell pattern is expected when host-

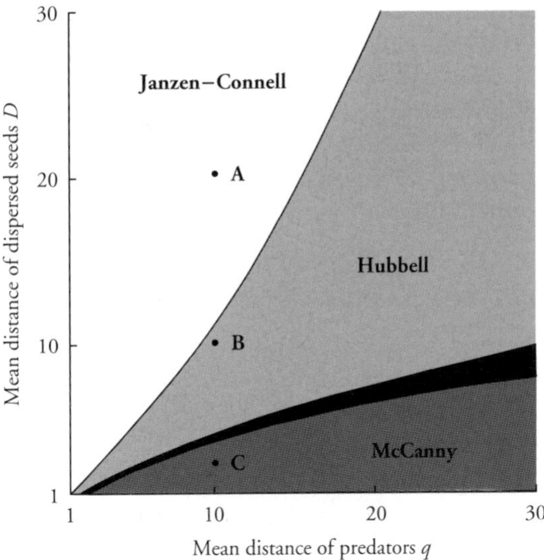

Figure 4.16 If the number of seeds declines with distance from the parent tree, while the probability of seed or seedling survival increases because of escape from pests and pathogens, we may observe a recruitment peak at intermediate distances: the Janzen–Connell pattern. This is expected when host-specific predators or pathogens are less mobile than the seeds. If dispersal and predation distances are similar, then there will be a monotonic decline in recruitment with distance—the Hubbell pattern. If mortality is caused by mobile generalist predators, then survival may be higher near the parent tree, as a result of predator satiation, and recruitment will fall off very rapidly with distance—the McCanny pattern. From Nathan and Casagrandi (2004).

specific invertebrate seed-predators or pathogens are less mobile than the seeds, so they do not locate seeds that are dispersed far away (Fig. 4.16). The Hubbell pattern is expected when dispersal and predation distances are similar, and the McCanny pattern is most likely with mobile, generalist, vertebrate predators that are likely to find even widely dispersed seeds, but are satiated by the abundant seeds in the vicinity of the parent.

Takeuchi and Nakashizuka (2007) experimentally tested the influence of seed distance and density on the fates of two wingless dipterocarp species at Lambir. These seeds are very poorly dispersed, while the major predators were mobile generalist rodents, so it is not surprising that the results fitted the McCanny pattern, with mortality increasing with distance from the tree. If, as the available evidence suggests, mobile generalist vertebrates are the major causes of seed mortality in TEA, then Janzen–Connell, distance-dependent survival is unlikely, even in well-dispersed plant species. However, the role of invertebrate seed-predators and pathogens, which are likely to be both more host-specific and less mobile, has received little attention in TEA, so more work is needed to resolve this question.

4.3.5 Germination and seedling establishment

The stage between a seed's escape from predators and pathogens and its appearance above ground as a seedling has been under-researched in TEA, and much of what we know comes from experiments in the laboratory or greenhouse, aimed at understanding the potential for seed storage and germination in connection with *ex situ* conservation and reforestation.

Many tropical forest seeds are desiccation-sensitive ('recalcitrant' or 'intermediate') and cannot be dried for long-term storage (Tweddle et al. 2003; Daws et al. 2006). These seeds are generally large (>0.5 g) and have a thin covering (seed coat and endocarp), they are metabolically active when shed, and they germinate rapidly. From the rather few data available, it appears that the majority of non-pioneer, woody plants in tropical, evergreen forests have desiccation-sensitive seeds, and that this proportion declines as the risk of drying increases, in more open and/or more seasonal sites. However, all eight non-pioneer tree species investigated in tropical seasonal rainforest in SW China were desiccation-sensitive (Yu et al. 2008a). The pioneers of treefall gaps and open habitats have desiccation-tolerant ('orthodox') seeds, as do most small-seeded herbs and the majority of plants in drier climates, except for species that are dispersed at the wettest time of the year. Globally, an estimated 92% of all angiosperm species have desiccation-tolerant seeds. Presumably, desiccation sensitivity is not, in itself, of any benefit to the plant, but is instead a side-effect of producing seeds that are ready for immediate germination and growth in an environment where seeds are unlikely to be subjected to water stress. Note also that species with desiccation-sensitive seeds vary greatly in their *resistance* to desiccation, (i.e. their ability to slow water loss) (Yu et al. 2008a).

Dormancy is to some extent independent of desiccation sensitivity, although most desiccation-sensitive seeds are not dormant. Dormant seeds, by the generally accepted definition, do not germinate under conditions that would normally be favourable for germination, so a dormant, desiccation-sensitive seed makes little ecological sense. Dormancy can be morphological (e.g. the undeveloped embryos of most *Ilex* species; Tsang and Corlett 2005), physical (e.g. a water-impermeable seed or fruit coat), physiological, or a combination of these (Baskin and Baskin 2004). There are rather few data from TEA, but this suggests a trend from seeds that are mostly non-dormant when ripe in tropical lowland rainforest, to seeds that are mostly dormant in dry tropical forests (Ng 1978; Kanzaki et al. 1997; Khurana and Singh 2001; Marod et al. 2002). The seeds of two-thirds of the lowland rainforest tree species tested by Ng (1978), under standard nursery conditions, germinated within 12 weeks, while the remaining third included a wide range of germination strategies. In a study of 45 non-dipterocarp tree species at Pasoh, Kanzaki et al. (1997) found that, while most primary forest species had short-lived seeds, 16% had a mean life-span exceeding one year when buried. In the strongly seasonal climate of northern Thailand, seeds dispersed in the late dry or early wet seasons tend to germinate quickly, while those dispersed at other times of the year are slower (Blakesley et al. 2002).

Rapid germination is an obvious advantage, if the environment is suitable for immediate growth, since it will reduce the risk of seed loss to predators or pathogens. Even if growth beyond the seedling stage is not immediately possible—for example, in deep shade—rapid germination will be advantageous if seedlings are at less risk than seeds. There seems to be little direct evidence for this, but the fact that so many rainforest tree species opt for rapid germination and so few for a persistent seed bank suggests that seedlings are less vulnerable. Tropical rainforests in TEA do have substantial seed banks—up to 5415 seeds per m^2, although the higher figures are for disturbed areas—but these are dominated by a few species of small trees and shrubs (Metcalfe and Turner 1998; Jankowska-Błaszczuk and Grubb 2006; Tang et al. 2006). These include both light-demanding pioneers, which need the high red/far-red ratio of a canopy gap to break seed dormancy (Turner 2001), and tiny-seeded, shade-tolerant shrubs, which need litter-free microsites for establishment (Metcalfe et al. 1998). Seed banks apparently incorporate a wider cross-section of the flora in drier forest types, but there is little available information (Marod et al. 2002).

Seeds arrive at a site without their mycorrhizal symbionts, which are acquired during germination from the soil. Spatial and temporal variation in the fungal community could, therefore, influence seedling establishment and subsequent growth (Lovelock et al. 2003; Theimer and Gehring 2007). It has also been suggested that one advantage of the ectomycorrhizal (ECM) association in the Dipterocarpaceae (Fig. 4.17) and Fagaceae—as opposed to the arbuscular mycorrhizal associations of most other tropical trees—is that it allows seedlings to link immediately to a common ECM network, thus saving on start-up costs and potentially allowing the movement of nutrients and photosynthate from established plants. There is evidence for at least some of these benefits in low diversity ECM-dominated forests elsewhere (e.g. McGuire 2007), but not yet from the species-rich forests of TEA.

There is less information on the fate of seeds in more open vegetation types, but a priori it seems likely that establishment from seed will be more difficult on surfaces subject to drying and extreme temperatures, than in the relative uniformity of the forest understorey. The failure of most attempts at reforestation by simply broadcasting seeds directly on the soil surface at open sites—mimicking natural seed dispersal—illustrates these problems (e.g. Engel and Parrotta 2001; Woods and Elliott 2004; Doust et al. 2006). Seed burial greatly increases the chance of establishment in the open, but must be a rare event in nature. In dry forests, seedling emergence tends to be patchy in both space and time (Marod et al. 2002; personal observations). Ground fires kill most seeds on the soil surface (Elliott et al. 1994), but may also encourage the germination of fire-adapted seeds and those species represented in the soil seed-bank. The removal of litter may also be important for species that require exposed mineral soil to establish.

4.3.6 The seedling stage

The initial sizes of seedlings are determined by the sizes of the seeds from which they grow, and may,

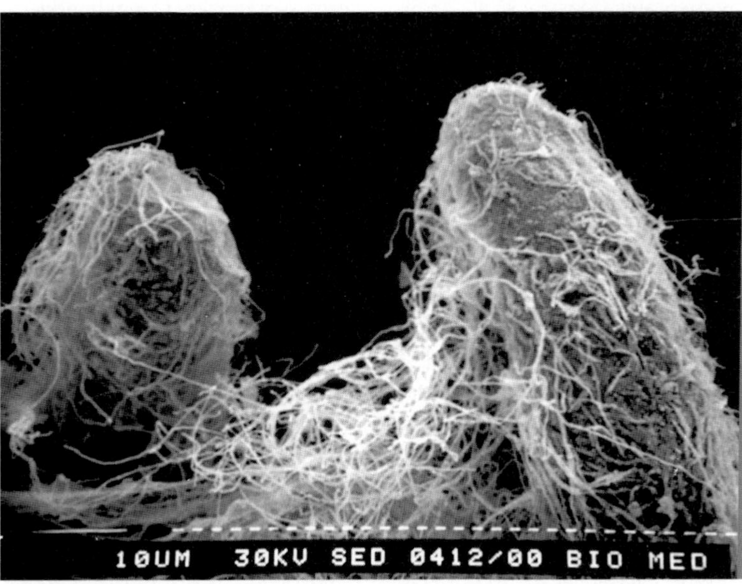

Figure 4.17 Scanning electron micrograph of two root tips of *Hopea nervosa* surrounded by fungal hyphae of an unidentified ectomycorrhizal *Inocybe* species (Agaricales, Basidiomycota). Photograph courtesy of Francis Brearley.

therefore, vary over several orders of magnitude (e.g Tsujino and Yumoto 2004). Large seeds and seedlings tend to have a survival advantage, particularly in deep shade, which partly compensates for the smaller numbers of such seeds produced per year and their generally less efficient dispersal (Moles and Westoby 2004; Baraloto et al. 2005). Seedling mortality is particularly high in the first year or two—typically 100% for shade-intolerant species that germinate in deep shade, but usually >20%, even for species that germinate in their optimum environment. There is a strong functional trade-off between growth and survival at the seedling stage, with species spread along a gradient from slow growth and high survival, to fast growth and low survival (e.g. Kurokawa et al. 2004; Gilbert et al. 2006; Aiba and Nakashizuka 2007a). Rapid growth characterizes light-demanding species, while high survival is found in shade-tolerant species. Once established, a shade-tolerant tree seedling may spend years, or even decades, in the understorey, waiting for a canopy opening to provide enough light for growth up into the canopy.

Few studies have identified the proximate causes of seedling mortality: many just disappear between censuses and others 'dry out' for no obvious reason. Fungal pathogens are a likely suspect (Bell et al. 2006), but herbivorous invertebrates (Massey et al. 2005), vertebrate browsers, and falling debris are other candidates, along with drought and, in drier, more open forests, fire (Saha and Howe 2003; Marod et al. 2004). Lowland tropical forests show great interannual variation in seedling densities, driven largely by variations in recruitment (Metz et al. 2008). Seedling densities are typically low (mean $8.9\,m^{-2}$ at Pasoh), so competition between seedlings is much less important as a cause of mortality than the asymmetric competition between seedlings and canopy trees (Paine et al. 2008).

Light availability is the most important abiotic factor for seedling survival in closed-canopy forests, but it is not the only one, and it is relatively less important in drier, more open forests (Markesteijn et al. 2007). Conversely, the importance of drought tolerance in controlling survival is most obvious in areas with a dry season (Marod et al. 2002, 2004; Engelbrecht et al. 2007), but irregular droughts of varying severity occur even in the wettest areas and species that specialize on more drought-prone soils or topography show adaptations to surviving these extremes (Yamada et al. 2005a). Poorter and Markesteijn (2008) identified three functional groups of seedlings in seasonal forests in Bolivia: drought avoiders, with deciduous leaves and taproots; drought resisters, with tough tissues (i.e. high dry matter content); and light-demanding moist-forest pioneers, with extensive root systems. There is evidence for a trade-off between tolerance for shade and tolerance for drought, reflected in the differential allocation of biomass to leaves and roots. Seedlings in dry forests also vary greatly in their ability to survive ground fires (Marod et al. 2004).

In contrast to the mortality due to abiotic factors, which is controlled by spatial patterns in the environment, the mortality due to pathogens and insect herbivores is at least partly controlled by the local density of conspecific seedlings, as well as the abundance of the species on larger scales (Webb and Peart 1999; Bell et al. 2006). Indeed, it is not just conspecific neighbours that are a problem: seedling survival is reduced by being surrounded by seedlings that are closely related, presumably because close relatives share some pathogens and herbivores (Webb et al. 2006). Unlike the seed predators discussed above, there is no evidence for seedlings satiating herbivores, so Janzen–Connell effects—with maximum recruitment at intermediate distances from the parent—are more likely for seedling survival, than for seed survival. Indeed, a global meta-analysis of the effect of distance on propagule survival found that, while there was no general effect of distance on seed survival, there was a significant positive impact on seedling survival (Hyatt et al. 2003). Note, however, that the suggested advantages of seedling access to a common ectomycorrhizal network (see 4.3.5) could potentially *increase* seedling survival and growth nearer to parent trees, reinforcing the pattern resulting from limited seed-dispersal and, in extreme cases, leading to local dominance by ectomycorrhizal trees (McGuire 2007).

4.3.7 Sapling to adult

The same patterns—a strong growth–survival trade-off (e.g. Russo et al. 2008) and, at least partly, density-dependent mortality—continue into the

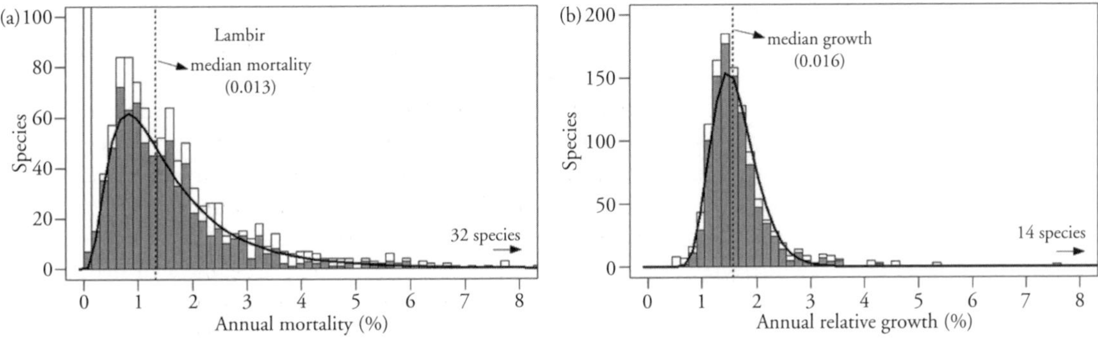

Figure 4.18 (a) The annual mortalities and (b) relative growth rates of saplings at Lambir Hills National Park are surprisingly similar for most species, suggesting that the growth–survival trade-off plays a minor role in species coexistence at this site (Condit et al. 2006).

sapling stage and beyond. Mortality declines with size, while growth rates increase in closed-canopy forests, as a result of the steep, vertical gradient in light availability (King et al. 2006a). In the aseasonal lowland rainforests at both Pasoh and Lambir, most species had annual sapling mortalities around 1% (Fig. 4.18). In the drier, more open forest at Huai Kha Khaeng, the modal mortality was similar, but there was also a long 'tail' of species with higher mortalities, which probably reflects the intense annual dry season and the occurrence of ground fires (Condit et al. 2006).

Interspecific differences in sapling survival in closed-canopy forests largely reflect differences in shade tolerance. This, in turn, is correlated with wood density (King et al. 2006b), growth rate, carbohydrate storage (Poorter and Kitajima 2007), resistance to herbivores and pathogens, and the ability to recover from stem breakage (Curran et al. 2008). Wood density is highly correlated with the total cost of supporting a given-sized crown at a given height and, at Pasoh, both the mortality and growth rate of pole-sized trees (8–20 cm dbh) decline with increasing density (King et al. 2006b). In contrast, large differences in crown architecture between sapling species do not seem to significantly affect the costs of height growth (Aiba and Nakashizuka 2007b). Sapling growth-rates also vary within a species, largely due to variations in the size of the crown and its degree of exposure to light (King et al. 2005). Edaphic factors (water and nutrients) also continue to influence growth and survival, so that many species become increasingly confined to their preferred soil and topography (Russo et al. 2005). The mechanisms behind this 'ecological sorting' are not clear, but may involve a trade-off between low mortality on nutrient-poor or drought-prone soils and fast growth on better soils (Russo et al. 2008).

Mortality in many species continues to be density-dependent beyond the seedling stage. Peters (2003) found evidence for density-dependent mortality for most tree species studied at Pasoh, and in all three size-classes investigated: 1–5 cm dbh, 5–10 cm, and >10 cm. In most cases, this mortality was consistent with the spread of natural enemies (pathogens or insect herbivores) from adults to their nearby offspring and among high densities of juveniles. Older plants may be less likely to be killed directly by pathogens and insect herbivores than seedlings, but the cumulative impact on their carbon and nutrient balance may be fatal in the long term. In Amazonia, the mortality of trees >10 cm in diameter is best predicted by low wood-density (reflecting the species-level growth–survival trade-off) and slow relative growth rate (reflecting low individual tree vigour) (Chao et al. 2008). Sudden death from pests or pathogens is also possible in adult trees, however, as shown by the dramatic recent die-off of an abundant canopy tree, *Mesua nargassarium*, in the lowland rainforest at Sinharaja, Sri Lanka, possibly caused by a fungal pathogen (Chave et al. 2008), and the virtual elimination of two dominant *Cryptocarya* species from a subtropical lowland forest plot at Dinghushan, Guangdong, by a geometrid moth identified as *Thalassodes quadraria* (Huang 2000).

The potential final size of the adult trees of a species influences the growth characteristics of its juveniles. Among trees of similar diameter at Pasoh, growth rates increase with potential final height (King et al. 2006a). The explanation for this is not clear, but taller species may simply commit everything to height growth, since most do not reproduce until near their final size. Mortality increases as a tree reaches its maximum potential size, suggesting the possibility of a finite, size-dependent, life-span (King et al. 2006a).

Overall, the characteristics of trees in lowland rainforests appear to vary principally along two major axes: one related to adult tree size and one related to shade tolerance (Turner 2001; Nascimento et al. 2005; King et al. 2006c). For communication purposes, it is often useful to dissect this continuum and assign tree species to one of three or four guilds: pioneer (small, light-demanding, low wood-density, high growth-rate, high mortality); sub-canopy (or understorey; small, shade-tolerant, moderate to high wood-density, lower mortality); and canopy and emergent (tall, variable wood-density and growth-rate, low mortality). Turner (2001) recognizes an additional group of large pioneers (Fig. 4.19). This classification ignores the possibility of ontogenetic changes in plant strategy,

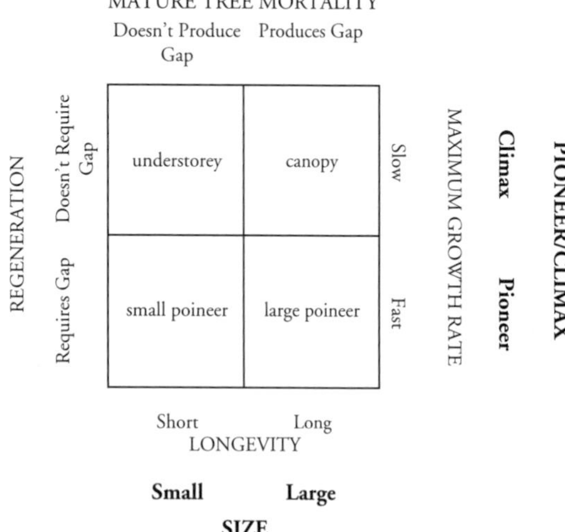

Figure 4.19 The four main guilds of trees in tropical lowland rainforest. Redrawn from Turner (2001).

although these appear to be rare (Gilbert et al. 2006). It also ignores the range of other abiotic and biotic factors to which tree species have been shown to adapt. Other axes of variation, particularly moisture requirements and fire tolerance, appear to be equally or more important in drier and more open forest types, although less information is available on these (Marod et al. 2002, 2004). Where cyclones (typhoons) are an important agent of disturbance, a trade-off between resistance to damage, which is positively related to wood density, and resilience, which is negatively related, may be another important axis (Curran et al. 2008).

The generally low mortalities of trees beyond the seedling stage imply long life-spans, with expected maximum longevities around 200–500 years for most species but <1000 years for some (Kurokawa et al. 2003; Nascimento et al. 2005). These time-scales are much longer than those for which we have ecologically relevant data in most of TEA. Catastrophic events, such as major cyclones and other wind storms, extreme droughts, and unusual fires, which are rare on a human time-scale, may have influenced the structure and species composition of the forests we see today. At Danum Valley, Sabah, the forest structure suggests that it is still recovering from a major drought in 1878 (Bischoff et al. 2005). At Huai Kha Khaeng, Thailand, evidence from the few species with annual tree-rings and from other sources, suggest that the present structure and floristics are a legacy of a catastrophic disturbance—most likely a cyclone or other wind event—around 150 years ago (Baker et al. 2005). Probably all forests in TEA are subject to rare, community-wide, catastrophic disturbances (Whitmore and Burslem 1998; see also 2.2.6, 2.2.7, 2.2.8) and it is likely that current ecological studies, which focus on time-scales of years or decades, are missing a key element in the ecology of tropical forest plants.

4.3.8 Sprouts and resprouting

The ability to resprout after damage is found in most woody plants across a wide range of vegetation types (Vesk and Westoby 2004). The fire-ecology literature recognizes a sprouter/non-sprouter dichotomy after major fires, but most plants can sprout after the less extreme damage caused by falling vegetation, wind, herbivory, trampling, cutting of firewood, or light ground-fires. Sprouting can occur from various places, depending on the species and the type of damage. Most saplings broken by pigs, during nest construction at Pasoh, resprouted from buds on the stem (Ickes et al. 2003) and breakage scars from a variety of causes are frequent on sapling stems in all types of forest. More severe damage, as from fires, leads to sprouting from protected buds at the base of the stem or on the root crown.

A smaller proportion of species is also able to produce sprouts from the underground root system. Root sprouts may be mistaken for seedlings, unless they are dug out, although they tend to be more vigorous and have more adult foliage. In some species, root sprouts are not just a way of replacing a damaged stem without going through the hazardous stages from seed dispersal to seedling establishment, but also a means of clonal spread away from the parent plant. The near-annual burning, to which many deciduous forests in tropical Asia are now subject, favours such species over those with alternative strategies (Saha and Howe 2003). In Brazil, some dry-forest species can still sprout from remnant roots left in the soil for many years after deforestation (Vieira et al. 2006), but there are no comparable data from TEA.

Vigorous sprouting is also characteristic of many montane and subtropical evergreen broad-leaved trees. Sprouts dominate succession after cultivation for hill rice in montane forest in northern Thailand, since tree stumps are retained during forest clearance and only a single crop is grown, while seeds are more important after the more prolonged cultivation of opium poppies (Fukushima et al. 2008). In the broad-leaved evergreen forest at Fushan, Taiwan, 83% of new sprouts survived more than a year, while fewer than 15% of seedlings did (Su et al. 2007). Sprouts can also dominate succession after timber harvest in these forests (Li et al. 1999; Aiba et al. 2001; Nanami et al. 2004). As a result, clear-cut stands may regenerate directly into a forest very similar to that which occupied the site before cutting, although a minority of species shows little or no sprouting capability (Wu et al. 2008).

Resprouting and clonal spread have not yet been incorporated into most models of tropical forest dynamics, perhaps because of the research bias towards undisturbed lowland rainforests. Even in undisturbed rainforests, however, differences in the ability of species to resprout after damage could have a significant effect on species composition, and differential survival after logging, anthropogenic fires, or unnaturally high (or low) densities of large herbivores could influence the process of recovery from human impacts. Although most species show some ability to resprout, there is considerable variation between species in the percentage that do so successfully. This is probably related to the more general growth–survival trade-off mentioned above, with resprouting more likely to be successful in species with adequate carbohydrate storage (Poorter and Kitajima 2007). In tropical rainforests in north Queensland, the highest resprouting ability after a major cyclone was shown by the species with low wood-densities that sustained the greatest mechanical damage (Curran et al. 2008).

4.4 Other life forms

4.4.1 Lianas

Lianas—woody climbers that use other plants for support—are a relatively neglected component of the forests of TEA. This is partly because they are difficult to identify in the field, but also because their high canopy-to-stem ratio means that their importance to the forest is not obvious from the ground. By the time they reach 2-cm diameter, most lianas have already reached the canopy, while trees of the same diameter are still stuck in the understorey (Kurzel et al. 2006). This neglect is unfortunate because, apart from their direct contribution to woody plant diversity (*c.*10–30%; Zhu 2008b), the biomass of the forest, and the food supply for animals, lianas can limit tree growth and fecundity, increase tree mortality, and suppress regeneration in gaps. Lianas cause mechanical damage to their hosts and compete both above and below ground. In addition, their relatively large allocation to nutrient-rich foliage over woody tissues means that lianas make a disproportionate contribution to litterfall and nutrient cycles (Cai and Bongers 2007; Kusumoto and Enoki 2008). The legumes are usually the first or second most diverse family of lianas in tropical Asian forests (Zhu 2008b), but their role in nitrogen-fixation has not been studied.

The diverse phylogenetic origin of lianas is apparent in the variety of their climbing mechanisms and stem anatomies (Putz and Mooney 1991), so one should probably not expect to be able to make generalizations about the ecology of lianas any more than one can for tropical trees. In general, liana abundance appears to be greatest in tropical, dry forests and declines both with increasing rainfall and with declining temperature. The success of lianas in seasonally dry forests has been attributed to a dry-season growth advantage that, in turn, reflects access to ground water through exceptionally deep, and wide, root systems (Schnitzer 2005; but see also van der Heijden and Phillips 2008). The decline in liana abundance is also very rapid north of the Tropic of Cancer, probably because thin, uninsulated stems and large vessel-elements make them very vulnerable to freezing conditions. The same problem presumably limits their abundance with increasing altitude, although they can be very common in some tropical lower montane forests.

At the landscape level, liana abundance generally correlates with disturbance (Schnitzer and Bongers 2002). Lianas are favoured by treefall gaps, in part because their flexible stem anatomy allows adult lianas to survive a treefall and then grow laterally into the gap. Larger scale disturbances, such as cyclones and logging, also increase the biomass and diversity of lianas, and the cutting of lianas before logging is widely recommended as a means of reducing their adverse impacts, both during and after tree harvest.

Rattans—spiny, climbing, palms—are diverse and abundant in the region's tropical lowland forests, with up to 30 species coexisting in some areas. Although rattans are the most important forest product after timber in much of the region, very little is known about their ecology. Growth rates reported from natural populations and cultivation trials range from 0.2 to 4.0 m per year (Bøgh 1996).

4.4.2 Ground herbs

The herbaceous flora of the forest floor is another neglected component of closed-canopy forest in TEA. Their occurrence is usually fairly sparse and patchy, and their taxonomic diversity is considerably lower than that of the woody plants, but their ecological diversity is high. Cover, density, and species-richness, increase with altitude between 200 and 850 m in Borneo (Poulson and Pendry 1995). Tropical forest herbs are perennial and, while many of the herbs of seasonal forests die back to underground storage organs every year, this does not happen in aseasonal rainforests, where many dicot herbs have some secondary thickening at the base and there is no sharp distinction between herbs and shrubs. In addition to pteridophytes, prominent rainforest herb families include the Araceae, Cyperaceae, Marantaceae, Orchidaceae, and Zingiberaceae among the monocots, and Acanthaceae, Begoniaceae, Gesneriaceae, Melastomataceae, and Rubiaceae among the dicots (Poulson and Pendry 1995; Poulsen 1996). Myco-heterotrophs—chlorophyll-free plants that are parasitic on fungi—are common and include members of the families Burmanniaceae (*Gymnosiphon*, *Thismia*), Orchidaceae (e.g. *Aphyllorchis*, *Galeola*), and Triuridaceae (*Sciaphila*). In most cases, the fungal hosts are ectomycorrhizal, so these herbs can be viewed as 'epiparasites' on adjacent trees (Leake 2005).

4.4.3 Epiphytes

Epiphytes have also been under-studied in TEA and very little is known about their ecology. This is unfortunate because, despite their relatively low contribution to forest biomass, epiphytes make a major contribution to plant diversity, provide habitat for numerous animal species, and play an active role in nutrient cycling (e.g. Hsu et al. 2002). At Danum, a single species (or species complex), *Asplenium nidus*, the bird's nest fern, can contribute a tonne of dry mass per hectare and support more than half the invertebrate biomass in the entire forest canopy (Ellwood et al. 2002; Ellwood and Foster 2004; Fig. 4.20). Studies of the same species in tropical north Queensland suggest that the young stages of *A. nidus*, where the water supply is not buffered

Figure 4.20 Giant epiphytic bird's nest fern, *Asplenium nidus*, in crown of *Koompassia excelsa* at Danum Valley, Sabah. Photograph courtesy of Roman Dial.

by accumulated dead organic matter, are vulnerable to dry periods lasting a month or more, and that even large plants can be killed by prolonged droughts (Freiberg and Turton 2007). In general, water supply seems to be the key determinant of epiphyte abundance and diversity, but it is certainly not the only factor. On Mt Kinabalu (*c.*6°N), the species-richness of vascular epiphytes reaches a maximum at 1200–1500 m (Grytnes and Beaman 2006)—below the cloud zone—while the abundance, and probably the diversity, of bryophytes continues to increase above this altitude (Frahm 1990).

Many vascular epiphytes in TEA have associations with ants. The best-studied examples are the ant-house epiphytes (e.g. *Lecanopteris*, *Myrmecodia*), which provide nesting spaces for ant colonies in exchange for nutritional benefits (Beattie 1989); but there are also many examples of ant-gardens, where the ants incorporate seeds of epiphytic plants in the organic walls of their nests, with the plant roots apparently stabilizing the walls (Kaufmann and Maschwitz 2006). A common feature of many of these associations is directed seed-dispersal by the ants—an obvious advantage in an environment where seeds must be deposited in suitable canopy microsites to establish. One major group of epiphytes in TEA, however, the orchids, is rarely associated with ants and apparently depends on the undirected dispersal of huge numbers of dust-like seeds by air movements.

4.4.4 Hemi-epiphytes and stranglers

Hemi-epiphytes are plants that start life as epiphytes, but then send stem-like aerial roots down to the ground. The great majority of hemi-epiphytes in TEA are figs (*Ficus* spp.). A minority of hemi-epiphytic fig species (only 1 of 27 species at Lambir; Harrison et al. 2003) become stranglers, which produce anastomosing aerial roots that eventual enclose and kill their host, leaving a free-standing fig tree (Fig. 4.21). Hemi-epiphytic figs are diverse but rare in forests, although a few species can be abundant on roadside trees. Their low density in forests seems to reflect a combination of dispersal limitation (Laman 1996a), seed predation by ants (Laman 1996b), and the need for microsites with good water-retention for establishment (Laman 1995). Mortality rates at Lambir were very high (4.7% per year), with three-quarters of deaths resulting from the fall of the host trees (Harrison 2006).

Figure 4.21 A strangling fig, *Ficus depressa*, at Lambir Hills National Park. Photograph courtesy of Rhett Harrison.

4.5 The maintenance of species-diversity in tropical forests

The 52-ha (0.52-km^2), lowland rainforest plot at Lambir in Borneo supports as many tree species (1175 species) as the entire temperate forest of the northern hemisphere—Asia, Europe, and North America (1166 species) (Wright 2002). Lambir is extreme, but forests with >100 tree species in a single hectare are widespread in the region (Table 3.1). The antiquity of warm, wet environments in TEA may explain the evolution of a large pool of tropical rainforest tree species, but how can more than a hundred species coexist in a single hectare? Why don't the most competitive species increase

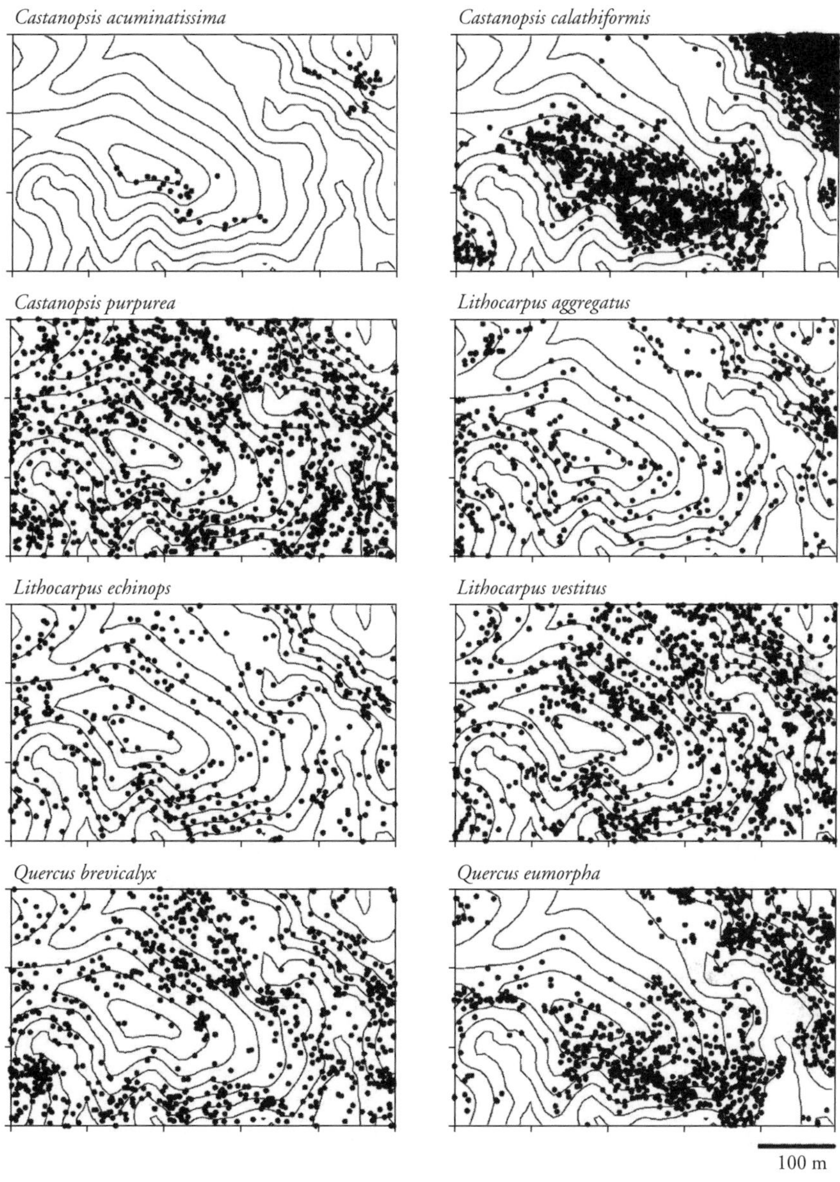

Figure 4.22 Spatial distribution of Fagaceae trees with dbh ≥1 cm in a 15-ha study plot in tropical montane forest at Doi Inthanon National Park in northern Thailand. Contour lines are drawn at 10-m intervals. From Noguchi et al. (2007).

in abundance until the other species are excluded? At least a dozen more or less plausible hypotheses have been put forward to explain plant species-diversity in tropical forests and only the ones with most current support are considered here.

The simplest hypotheses involve classical niche-differentiation, with coexisting species avoiding competitive exclusion by using resources differently. There is plenty of evidence for niche differences in tropical forests associated with small-scale topography (Fig. 4.22), soils, and the availability of specific nutrients (Bunyavejchewin et al. 2003; Enoki 2003; Itoh et al. 2003; Miyamoto et al. 2003; Sri-Ngernyuang et al. 2003; Paoli et al. 2006; Noguchi et al. 2007). Indeed, the complex topography and heterogeneous soils of the Lambir site have undoubtedly contributed to its record diversity (Baillie et al. 2006). The vertical gradient in light intensity, from canopy to forest floor, could be similarly partitioned among species of different adult heights. This gradient is steepest near the forest floor, which could potentially increase the number of understorey species that can coexist (Turner 2001). However, there is no evidence to suggest that the trees at Lambir have divided up the soil, topography, and height spectrum so finely that there are 1175 different niches available, so heterogeneity in soil and topography can only explain part of the observed species-diversity.

The widespread trade-off between growth and survival described earlier could also contribute to diversity, by allowing the coexistence of species differentiated along an axis from fast growth (short life-span) to slow growth (long life-span), with the former superior in recently disturbed sites, such as large tree-fall gaps, where resources are temporarily abundant, and the latter persisting in the understorey, where resources are scarce. It is not clear how many distinct niches could be accommodated along this axis, but there are good examples of the extremes: large-gap specialists, such as most *Macaranga* species, and slow-growing survivors, such as *Eusideroxylon zwageri* (Borneo ironwood). However, in the most species-rich forests, such as Lambir, growth rates and mortalities are far from being spread evenly over the entire range, but instead are surprisingly similar for the majority of species (Fig. 4.18). This suggests that the growth–survival trade-off plays a fairly minor role in species coexistence at these sites.

Negative density-dependence of growth and mortality has been shown for some species at every stage from seeds to mature trees (Peters 2003; Uriarte et al. 2004; Freckleton and Lewis 2006). There is a considerable amount of evidence that the major factor in this pattern is the impact of pests and pathogens—host-specific or shared only by close relatives. This mechanism could potentially play a major role in species coexistence, since multiple, otherwise identical, species could coexist, if they did not share their major natural enemies and the densities of the survivors of each species were sufficiently low to allow the establishment of other species. It could also help explain the decline in plant species-diversity as one moves away from the aseasonal tropical lowlands, if dry and/or cold periods keep pests and pathogens in check and reduce their effectiveness in limiting plant densities. Although the evidence available so far is fairly persuasive, there is now a real need to move beyond purely observational studies of density dependence to field experiments that manipulate plant densities (e.g. Bell et al. 2006).

All the above mechanisms for coexistence depend on differences between species. Many different species of plants can coexist in the same area, if each species limits its own population more than it limits others, because it is confined to a particular soil type, or gap size, or it shares more pests and pathogens with close relatives. A decade ago, however, Steve Hubbell and his collaborators proposed a radical alternative to these mechanisms, one that depends on similarities between species rather than differences (Hubbell 2001, 2006; Volkov et al. 2005; Adler et al. 2007). Neutral theory assumes that all species are ecologically equivalent, with the same per capita birth and death rates, and the same chance of immigration from the regional species pool (or 'metacommunity'). Population dynamics in the local community is driven solely by random variation in births, deaths, and immigration. Even random variation would eventually drive all but one species to extinction, but the process of competitive exclusion is extremely slow, as a result of dispersal and recruitment limitation, i.e. the failure of species to reach all sites where they could grow, plus the failure to recruit on many sites that they do reach. Thus, even very slow rates of speciation in the metacommunity are sufficient to balance species-extinctions.

Despite these extreme assumptions, neutral models predict some fundamental ecological patterns, such as species–area relationships and species–abundance distributions, as well or better than models based on species differences (Volkov et al. 2005; Adler et al. 2007). This, combined with abundant evidence for dispersal and recruitment limitation in the well-studied tropical forest plots, means that neutral theory cannot be ignored. However, it does have serious problems, most obviously that its basic assumption is wrong: species are *not* all the same. Large differences in birth, death, or immigration rates will lead to rapid competitive exclusion, unless strong stabilizing mechanisms are present; even small fitness differences will accelerate local extinction without at least weak stabilization (Chesson 2000). The potential stabilizing mechanisms are the difference-dependent processes considered above. Thus, the very different birth, death, and dispersal rates of a fast-growing pioneer, confined to large gaps, and a slow-growing, shade-tolerant, subcanopy tree do not lead to one excluding the other, because they occupy very different niches.

The various mechanisms proposed for species coexistence, including neutrality, are not mutually exclusive. Neutral processes may be important for the majority of plant species that have adapted for survival and growth in the commonest local micro-environment (e.g. the numerous similar shade-tolerant species in high-diversity lowland rainforests; Condit et al. 2006; Hubbell 2006), while other mechanisms may account for the coexistence of species that have adapted to less common environments, such as large gaps or extreme soils. It may be that neutrality is important in the hyperdiverse lowland rainforests of aseasonal climates, while niche differentiation dominates in the harsher conditions of more seasonal climates. Even if neutral theory is simply wrong—as most ecologists believe—it may still be useful, both as a 'null hypothesis' against which the significance of differences between species can be assessed, and as a stepping stone to a more realistic theory (Leigh 2007).

The need for a mechanistic, process-based understanding of plant community dynamics is becoming increasingly obvious—and increasingly urgent—as intact natural communities are replaced by communities disturbed by logging, air pollution, nitrogen deposition, rising carbon dioxide levels, climate change, and the loss of pollinators and seed-dispersal agents (Purves and Pacala 2008; Chapter 7). The challenge is no longer just to explain why existing communities are the way they are, but to *predict* how they will change in the future. In species-rich, lowland tropical forests, it will probably never be possible to model every species separately, but a better understanding of the key processes may permit simplifications that that will still retain enough biological information to be useful.

4.6 Forest succession

Succession is the unidirectional change in community composition that occurs when space and other resources for establishment are made available by a natural or anthropogenic disturbance. The concept can be applied to microbes, animals, and plants, but only the latter are considered here. In the simplest case, where the soil is left intact but all plants are eliminated from the disturbed site, succession occurs because species differ in their ability to disperse to the site and in their performance after arrival. This typically leads to initial dominance of the site by well-dispersed, short-lived, fast-growing, light-demanding species—pioneers—and their replacement over time by relatively poorly dispersed, longer lived, slower growing, more shade-tolerant species. Large areas of successional vegetation were rare in most landscapes before recent human impacts, but 'pioneer' characteristics were selected for by a wide range of natural disturbances, including extreme winds, landslides, and riverbank erosion.

The situation is more complicated where plant species survive through the disturbance on the site, in vegetative form or as seeds. A relatively mild disturbance of the canopy, such as by a typhoon, may result in little or no successional change, since the dominants simply resprout after damage. The greater the degree of disturbance, the longer the window for establishment of light-requiring pioneers, from dispersed seeds or the soil seed-bank, but in general these do not become dominant unless a large fraction of the trees making up the previous canopy is killed.

At the other extreme, disturbances that impact soils as well as vegetation, such as prolonged cultivation or grazing, or the use of heavy logging machinery, not only eliminate residual plants from the site, including the seed bank, but may also make it unsuitable for fast-growing pioneers. Such sites can become dominated by slow-growing, light-demanding species that occur naturally on permanent, open sites, such as riverbanks, cliffs, and ridge tops (Corlett 1991; Fig. 2.10).

The spatial extent of the disturbance is a key characteristic when most plants must be dispersed into the site. Dispersal limitation is likely to become significant, even for pioneers, at sites more than a few hundred metres from the nearest seed sources (e.g. Weir and Corlett 2007). Many late-successional trees in TEA, such as the wind-dispersed dipterocarps and species with large, fleshy fruits that depend on large-gaped birds or mammals for dispersal, spread very slowly into secondary forests (Turner et al. 1997). The rate of forest recovery following human impact depends not only on the nature and spatial extent of the disturbance, but also on how recovery is measured. In general, forest structure, soil-nutrient stocks, and species-richness recover far more rapidly than species-composition (Chazdon 2003) and there is little evidence that secondary succession will eventually converge on the species-composition of the original, pre-disturbance forest.

4.7 Phylogeny and community assembly

Molecular phylogenies are becoming a powerful tool for investigating the processes that structure species-composition in plant communities (Kraft et al. 2007). This is one of the most rapidly developing areas of modern ecology (e.g. Webb et al. 2008), so only a brief outline of some of the issues can be given here. The species and higher taxa present in a community are not simply a random sample from the regional species-pool. This non-randomness is what ecologists call 'structure' and non-random patterns of evolutionary relatedness are termed phylogenetic community structure. If, as generally seems to be the case, closely related species are ecologically more similar than distantly related ones, then phylogenetic structure can give an indication of the ecological processes involved in the assembly of communities from the regional species-pool.

If environmental filtering limits the range of ecological strategies possible at a site, the result is likely to be 'phylogenetic clustering', with co-occurring species more closely related than expected by chance. This might happen, for example, if regular fires excluded all but the most fire-tolerant species, leading to dominance by a few clades in which this trait was found (Verdu and Pausas 2007). Conversely, competitive exclusion, or the impact of natural enemies, may limit the similarity and thus relatedness of co-occurring species, and thereby generate 'phylogenetic over-dispersion'. Finally, an absence of phylogenetic structure could reflect either a balance between environmental filtering and competitive-exclusion or community-assembly processes that are independent of plant traits, as in the Neutral Model.

In reality, the answers appear to depend strongly on the spatial and phylogenetic scales at which the questions are asked (Swenson et al. 2007). Over-dispersion appears to be more common at small spatial scales ($<25\,m^2$) and near the tips of the phylogenetic tree (i.e. species within genera), while clustering occurs at larger scales ($>100\,m^2$ and/or all flowering plants). This is consistent with what is known about the likely mechanisms responsible for these patterns, with environmental filtering resulting from broad patterns of soil type, topography and fire frequency, while over-dispersion reflects the tendency of closely related species to share herbivores and pathogens.

CHAPTER 5

The ecology of animals: foods and feeding

5.1 Introduction

Animals do a lot more than eat, but an animal's diet has consequences for all other aspects of its biology. It, therefore, makes sense to use diet to structure an account of what we know about animal ecology in TEA. Also, it must be admitted, diet is the most widely studied aspect of animal ecology and, in contrast to the plants considered in the previous chapter, we know little about the reproductive biology of most animal species in TEA.

One major way in which diets differ is in their nutritional quality. The quality of a particular food depends, in part, on the digestive capabilities of the animal that is eating it, but all animals require a minimum amount of energy and protein in order to maintain bodily functions and all animals must pay a price to process foods that are high in indigestible compounds, such as lignin, keratin, and chitin, or contain defensive chemicals, such as phenolics. Some specialists can survive on diets of very low nutritional quality, but they grow slowly and reproduce late in comparison with related species adapted to higher quality foods.

Another important way in which diet influences animal ecology is through the phenology of the food supply. As discussed in section 2.6, some major plant resources, such as young leaves, flowers, fruits, and seeds, are available for only part of the year, or even at several-year intervals (Fig. 2.14, 2.15). Specialist feeders on these resources must match their life-histories to their food supply, with consequences for the food supply of the carnivores that, in turn, feed on them. Generalists can switch resources, taking advantage, for example, of the roughly complementary seasonalities of the fruit and invertebrate supplies at high latitudes in TEA (see 5.6). Other foods are available year-round, including wood, herbivore dung, and the living and dead bodies of vertebrates, so these resources can support extreme specialists, such as termites, dung beetles, and tigers.

5.2 Herbivores

Most of the biomass available to eat in any terrestrial community consists of plant material, so most animals eat plants. In contrast to foods eaten by carnivores, plant foods are extremely variable and generally of low nutritional quality. The major problems are the high fibre content, because each plant cell is enclosed in a tough, digestion-resistant cell wall, the generally low protein content, the presence of chemical defences (phenolics, terpenes, alkaloids, etc.), and also the frequent occurrence of micronutrient deficiencies (e.g. sodium). Food nutritional quality is lowest in wood, which is largely indigestible fibre; intermediate in leaves and fine roots; and highest in underground storage organs, phloem sap, nectar, seeds, and fruits. The great majority of the individuals, species, and biomass of herbivores consists of insects, but vertebrate herbivores are significant in particular communities (e.g. natural grasslands), with particular food types (e.g. seeds and fruits), and in particular times and places.

5.2.1 Leaf-eaters

Leaves are arguably the most vulnerable part of a plant. They must be thin and relatively nutritious, in order to function efficiently in photosynthesis. They cannot be hidden, they must be available most of the

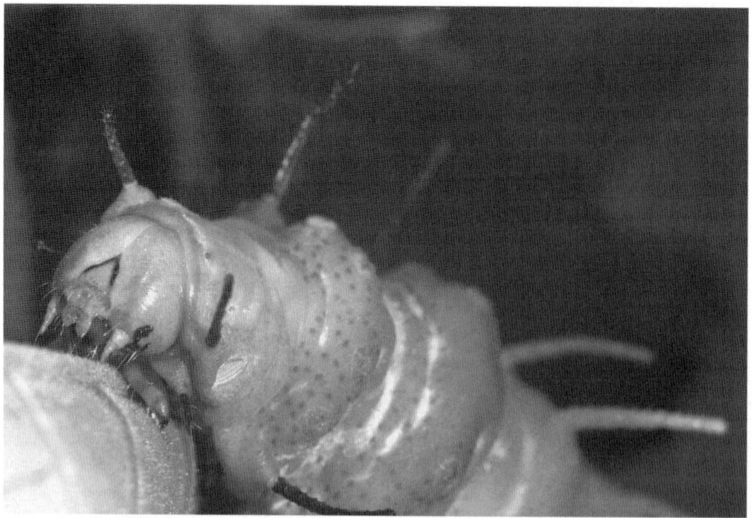

Figure 5.1 The larvae of butterflies and moths (Lepidoptera) are major leaf-eaters in tropical forests. This is a close-up of a feeding larva of an atlas moth, *Attacus atlas*, the largest moth in the world. Photograph courtesy of David Lohman.

Figure 5.2 Adult beetles are also major leaf-eaters in tropical forests. This is *Altica cyanea* (Chrysomelidae). Photograph courtesy of Yiu Vor.

year, and they need to be replaced regularly, which involves going through a soft and particularly nutritious young-leaf stage. Most leaf-eaters (folivores) are insects, particularly larval Lepidoptera (moths and butterflies; Fig. 5.1) and adult Coleoptera (beetles; Fig. 5.2). Other important groups of leaf-eaters in TEA are the Orthoptera (grasshoppers, crickets, katydids) and Phasmatodea (leaf and stick insects). Sap-sucking Hemiptera are considered separately below.

A study in lowland rainforest in New Guinea estimated there were 1600–2600 species of leaf-chewing herbivorous insects on the 152 plant species in 1 hectare of forest (Novotny et al. 2004). Most herbivorous insects specialize at the plant genus or family level, although some appear to feed only on a single species (Novotny and Basset 2005). Most herbivorous insects also specialize *within* the leaf, feeding preferentially on the more nutritious tissues and avoiding the major veins (Choong 1996). Leaf-miners—insects that feed in tunnels or chambers within the leaf—take within-leaf specialization a stage further. The mining habit has arisen many times in the Lepidoptera and Coleoptera, and is also common in one family of Diptera (Agromyzidae, the miner-flies) and some Hymenoptera. Leaf-miners are widespread and diverse in tropical forests, but generally occur at such low densities that their impact on their hosts is probably small.

Leaf-eating vertebrates are too big and long-lived to specialize to the extent that most invertebrates do, but all species that have been studied are selective, feeding heavily on some plant species and avoiding others. Eating small amounts of soft, young leaves from the least chemically defended species probably requires no special adaptations, and this is a common supplement to a fruit-based diet (see 5.2.8). The toughness of older leaves necessitates dental adaptations for efficient processing and the high fibre content, which provides the toughness, also reduces digestibility.

The colobine monkeys (leaf monkeys, langurs, surilis, doucs, snub-nosed monkeys, and the proboscis monkey) are known for their morphological and physiological adaptations for eating leaves, including sharp teeth, enlarged salivary glands, and a ruminant-like, multi-chambered stomach, in which symbiotic bacteria degrade cellulose and defensive chemicals (Kirkpatrick 2007; Fig. 5.3). However, the proportion of leaves in the diets of Asian colobines varies widely between species, with some preferring seeds and (often unripe) fruit. The northernmost populations of *Trachypithecus* langurs appear to be the most highly folivorous of TEA's primates (e.g. Huang et al. 2008). The ratio of young to mature leaves eaten also varies between colobine species, with young leaves preferred by most species when they are available, but mature leaves often dominating seasonally.

Macaques are not specialized folivores and leaves generally make up a relatively small fraction of their diet. However, in the mixed coniferous–broadleaved montane forest on Yakushima Island (30°N), mature leaves constitute 38% of the annual

Figure 5.3 Colobine monkeys, such as these Phayre's leaf monkeys (*Trachypithecus phayrei*), are specialized feeders on leaves and other relatively indigestible plant foods. Photograph courtesy of Andreas Koenig.

diet of the Japanese macaques (*Macaca fuscata*), reflecting the low availability for much of the year of the preferred fruits, flowers, and seeds in this relatively extreme environment (Hanya 2004). The macaques here are very selective feeders, choosing leaves with a relatively high protein/fibre ratio and low condensed tannin content (Hanya et al. 2007). The problems of a leaf-dominated diet in a non-specialist primate were dramatically illustrated by the mass mortality of macaques at lower altitudes on the same island in 1998–99, following exceptionally poor fruit-production in autumn 1998 (Hanya et al. 2004). Leaves are also relatively important in the diets of other northern macaque populations, such as the Tibetan macaques (*M. thibetana*) on Mt Emei (30°N), Sichuan (Zhao 1999). A similar pattern is shown by the gibbons, which are generally the most frugivorous of Asian primates, with leaves most important in the northernmost populations studied (Bartlett 2007).

Terrestrial herbivores that consume substantial amounts of leaves include even-toed ungulates (Artiodactyla: deer and cattle), odd-toed ungulates (Perissodactyla: Malayan tapir, Javan and Sumatran rhinoceroses), and the Asian elephant (Fig. 5.4). Animals that do not have to climb trees can be very large, which permits a low-quality diet, since the basal metabolic rate (i.e. energy expenditure at rest) is proportional to the ¾ power of body mass, while gut capacity remains the same proportion of body mass. Despite their size, however, even elephants are very selective about what parts of which plants they eat (Campos-Arceiz et al. 2008a). At the same time, large herbivores lose access to the majority of the leaf production in a closed-canopy forest, which takes place in the canopy. Rainforest herbivores get around this inherent leaf scarcity by making use of openings in the forest canopy, such as tree-fall gaps, swamps, riverbanks, and man-made clearings, or by diversifying their diet to include fallen fruit or roots. Some forest ungulates may also follow troops of macaques and other primates, in order to feed on fruit and leaves dropped from the canopy (e.g. Majolo and Ventura 2004).

This diversification of diet is only practical for the smaller species, but the biggest terrestrial herbivores, the elephant and rhinoceroses, can improve their access to leaves by pushing down saplings to

Figure 5.4 Asian elephants (*Elephas maximus*) are the largest herbivores in tropical Asia. Photograph courtesy of Siew Te Wong.

bring their leaves and young shoots within reach (Strickland 1967; Schenkel and Schenkel-Hulliger 1969; Sukumar 2003). Such feeding activities have the potential to impact not only individual plants, but also the structure of the vegetation (e.g. Matsubayashi et al. 2006). Cristoffer and Peres (2003) argue that the presence of a forest megafauna in the Asian and Africa tropics, and its absence in the Neotropics, has had a pervasive impact on many aspects of forest ecology, from vegetation structure to the diversity of small herbivores. Unfortunately, the low densities of the surviving populations of very large vertebrates in TEA (not to mention the fairly recent extinction of the previously widespread and abundant stegodonts), makes this theory impossible to test. It is also unclear whether Neotropical forests have always lacked a megafauna or if this is an artefact of the megafaunal extinctions that followed human arrival in the New World.

In more open forest, a grassy understorey can support higher densities of terrestrial herbivores. Wharton (1966) likened the large herds of wild cattle (gaur and banteng) he saw in 'savanna forest' in Cambodia 40 years ago to the ungulate communities of East Africa. Grass has often been considered a poorer food than dicotyledonous 'browse', because of its relatively low protein and high fibre content, but dicots typically have higher contents of lignin and defensive chemicals, so this assumption is probably not justified (Codron et al. 2007). The correlated morphological, physiological, and behavioural differences that have been reported between browsers and grazers probably have other, less simple, explanations (e.g. Clauss et al. 2003; Codron et al. 2007).

Many terrestrial herbivores, and some predominantly arboreal primates, have been recorded eating soil and/or drinking seeping water at specific sites ('licks') (e.g. Moe 1993; Klaus et al. 1998; Matsubayashi et al. 2007; Fig. 5.5). In most cases, the major benefit of this behaviour appears to be the supplementation of a plant-based diet with soil minerals, particularly sodium, which is relatively deficient in plant foods (see 6.4). Access to such mineral supplements may increase the carrying capacity of some areas for vertebrate herbivores. However, other potential benefits, including neutralization of plant toxins and adjustment of gut pH, may be more important at particular sites and with particular animals, and the ecological function of this behaviour is still debated (e.g. Voigt et al. 2008).

Old World fruit bats also consume leaves, apparently as a calcium and protein supplement to their fruit-dominated diets, as well as a back-up food in

Figure 5.5 Camera-trap photo of a sambar deer (*Rusa unicolor*) at a salt lick in Taman Negara National Park, Malaysia. Photograph courtesy of Kae Kawanishi.

times of reduced fruit supply (Nelson et al. 2005; Nakamoto et al. 2007). The bats avoid the need for colobine-like gut specialization or ungulate-like body mass by thorough mastication in the mouth, followed by swallowing only the liquid portion and ejecting a fibrous pellet. Presumably this technique does not work on all leaves, which may explain why fruit bats consume fewer species of leaves than they do of fruits. Fruit bats are known to rely heavily on odour to detect ripe fruits (Hodgkison et al. 2007), but it is not known if odour is involved in leaf selection. Birds, in contrast, rarely eat many leaves and the only birds in TEA for which vegetative plant parts have been reported to make up a significant proportion of the diet are some species of parrots (e.g. Walker 2007).

All leaves have both mechanical and chemical defences, but the amount the plant invests in leaf defence varies hugely between species. In the most heavily defended leaves, such as those of *Eusideroxylon zwageri* (Borneo ironwood), defensive chemicals—in this case condensed tannins and lignins—can make up more than a quarter of the dry weight (Kurokawa et al. 2004). Despite the huge range of apparently defensive chemicals present in leaves (Turner 2001), as well as wide variations in leaf mechanical properties (Lucas et al. 2000; Peeters et al. 2007; Dominy et al. 2008), many studies have suggested that the content of phenolics is the major single determinant of the amount of herbivory (e.g. Eichhorn et al. 2007).

Some tropical plants are also defended against herbivores indirectly by interactions with mutualistic carnivores, particularly ants. In TEA, these interactions range from highly specialized cases, where the plant provides both nesting space ('domatia') and food (nectar or solid food bodies) to resident colonies of specialized ants (Fig. 5.6), to the more numerous species that simply provide food (usually nectar) to attract ants from the surroundings. Studies have shown that both types of interactions provide protection against herbivores, but in the pioneer genus *Macaranga* the defence is much more effective in species with resident colonies of specialized ants (e.g. *M. triloba*) than in those that depend on attracting ants from their surroundings (e.g. *M. tanarius*) (Heil et al. 2001).

Many plants—mostly woody dicots—have specialized pits, pockets, or hair tufts in the axils of veins on their leaves, which are inhabited by mites. These mites are usually either predatory or fungus-feeding species and there is increasing evidence that the 'leaf domatia' can reduce attacks by fungi and herbivores by protecting beneficial mites from their predators (Romero and Benson 2005). Another form of indirect defence—the attraction of carnivorous

Figure 5.6 A young leaf of *Macaranga indistincta* in Sabah being patrolled by *Crematogaster* ants from a colony that lives in the hollow stem of the tree. Photograph courtesy of Brigitte Fiala.

arthropods by the release of volatile organic compounds (VOCs) (Heil 2008)—has not yet been studied under field conditions in TEA. Yet another big unknown is the defensive role of the diverse and ubiquitous endophytic fungi living within apparently healthy leaves of all plants (Herre at al. 2005; Arnold 2008).

Mature leaves are present year-round in all but the driest areas of TEA, but the availability of young leaves varies greatly from month to month (see 2.6.1). In the northern half of TEA, there is usually an annual peak in leaf production between March and May, and a mid-winter low, while in the equatorial region, leafing patterns are much less regular, except when they have been synchronized by major droughts (Ichie et al. 2004; Itioka and Yamauti 2004). In seasonal climates, animals that depend on young leaves adapt by synchronizing their lifecycles with the annual patterns of availability or by diet-switching, but the severe ENSO drought in 1997–98, at normally aseasonal Lambir, Sarawak, caused a drastic local loss of chrysomelid beetle species (Kishimoto-Yamada and Itioka 2008b).

5.2.2 Shoot-, bark-, and wood-feeders

Trunks, branches, and twigs make up the majority of the above-ground biomass in a forest, but their role as food for herbivores has received little attention, except in planted tree crops. Among the most damaging herbivores in plantations are those—mostly moth larvae—that bore into young shoots and often kill them. Older branches and the trunk are attacked by a variety of beetles (particularly in the family Cerambycidae), as well as moth larvae. The diversity and abundance of wood-borers increases in dying and dead trees (see 5.3), with relatively few species able to attack healthy plants.

Bark has been reported in the diets of many mammals in the region, including primates, bears, deer, elephants, and squirrels, and bark-feeding may kill individual trees. For 22% of the plant species reported eaten by work elephants in Myanmar, the bark was the only part consumed (Campos-Arceiz et al. 2008a). Although the corky outer bark is among the least digestible of plant foods, the inner layers include the living and much more nutritious phloem and cambium, and it is these that are the target of bark-feeding animals. In some populations of Bornean orangutans (*Pongo pygmaeus*), bark forms a large fraction of the diet when fruit is scarce (Knott 1998; Taylor 2006). However, ketones from the metabolism of fat reserves are present in orangutan urine during these fruit-poor periods, indicating that they cannot maintain body weight on such a diet.

5.2.3 Root-feeders

Twenty to forty percent of plant biomass in a tropical forest is below ground (Mokany et al. 2006) and fine roots probably rival leaves in their inherent vulnerability, but virtually nothing is known about below-ground herbivory in tropical systems (Blossey and Hunt-Joshi 2003). In a secondary forest in New Guinea, more than 90% of the adult chrysomelid beetles feeding on leaves had root-feeding larvae (Pokon et al. 2005). Most of these larvae fed on plants from several different families, which agrees with other studies showing that root-feeding insects have a relatively low degree of specialization. Judging from the tropical crop and non-tropical forest literature, other important groups of root-feeders are found in the insect orders Coleoptera, Lepidoptera, Diptera, Hemiptera (particularly the subterranean larvae of cicadas), and Orthoptera (crickets and mole crickets), as well as a variety of other invertebrate groups, including millipedes and nematodes.

5.2.4 Sap-suckers

The phloem is the nutrient transport system of vascular plants, and phloem sap is a relatively abundant, nutrient-rich plant food that is also, in most plants, relatively free of chemical defences. However, it has several major problems as a food for insects: it is relatively inaccessible, it is very low in essential amino acids, and it has an osmotic pressure 2–5 times that of insect body fluids (Douglas 2006). This may explain why only a single insect order, the Hemiptera (the true bugs), contains species that specialize on this resource. Phloem-feeding hemipterans have specialized mouthparts, symbiotic micro-organisms that probably provide them with essential amino acids, and the ability to transform excess sugars into long-chain oligosaccharides excreted as 'honeydew'. There are also hemipteran species that feed on the xylem fluid and others that feed by emptying individual mesophyll cells (Novotny and Wilson 1997). Xylem fluid is extremely dilute, so large quantities must be processed and, unlike phloem, is under negative tension, so feeding is energetically costly. Hemipterans are extremely abundant in forest canopies, although they are undoubtedly undersampled in studies that collect invertebrates by fogging the canopy with insecticide because the mouthparts get stuck in plant tissues and they do not fall into the traps.

Phloem sap is also consumed by several vertebrates in TEA. The slow lorises (*Nycticebus* spp.) appear to be sap specialists, with the extraction of phloem sap occupying a third of feeding time during a study of *N. coucang* in peninsular Malaysia (Tan and Drake 2001; Wiens et al. 2006). These primates use their lower front teeth to perforate the cambium and then lap up the exposed sap. Other mammals reported to actively consume sap include several species of squirrels and the colugos (*Cynocephalus* spp.). Sap is also consumed by some woodpeckers in the north of the region, although the sap-sucking habit does not appear to be as widespread in TEA as it is in North America (Winkler and Christie 2002). Most reports in TEA refer to the rufous-bellied woodpecker (*Dendrocopus hyperythrus*) (Winkler et al. 1995).

5.2.5 Ants as 'cryptic herbivores'

Ants are the most abundant insects in tropical forests, including the forest canopy. Although many ants appear to be more or less omnivorous, studies of stable nitrogen isotopes have shown that the most abundant species in forest canopies are essentially herbivorous, feeding on extrafloral and floral nectar, wound exudates, and, most importantly, honeydew secreted by sap-sucking hemipterans (Davidson et al. 2003; Blüthgen et al. 2006; Fig. 5.7). So abundant are ants and hemipterans that it is possible that this partnership is responsible for the consumption of more primary production than all other herbivores put together.

Honeydew has more essential amino acids and a lower osmotic pressure than raw phloem sap, so it is a better food, but it is still a high carbohydrate/low nitrogen diet for which ants probably

Figure 5.7 A sap-sucking tree-hopper (*Leptocentrus taurus*, Hemiptera) being tended by an ant (*Camponotus nicobarensis*). It has been suggested that such ant–hemipteran partnerships may be responsible for more herbivory than all other herbivores put together. Photograph courtesy of Yiu Vor.

require special adaptations (Cook and Davidson 2006). The ants protect the hemipterans from natural enemies, including predators, parasites, and pathogens, so the ant–hemipteran relationship is believed to be typically mutualistic. A study in lowland rainforest in Borneo found that the ants were, in general, relatively opportunistic in their choice of hemipterans to defend, while the hemipterans were much more specific in their choice of plants (Blüthgen et al. 2006). In a bizarre exception, the 'migrating herdsman' ant, *Dolichoderus cuspidatus*, has an apparently obligatory partnership with a few species of pseudococcid mealy bug, which in turn feed on young shoots of a wide variety of plants (Maschwitz and Hänel 1985). The ants are constantly moving their mealy bug partners to new feeding sites, providing a mobility that the bugs by themselves would lack.

In at least some cases, the ant–hemipteran relationship is also of net benefit to the plant, despite the carbohydrate cost, because the ants provide protection against other herbivores (e.g. Moog et al. 2005). Overall, the ant–hemipteran partnership can be seen as a 'keystone interaction' that has strong and pervasive effects on the canopy community (Styrsky and Eubanks 2007).

5.2.6 Gall-formers

Some herbivorous invertebrates are able to stimulate undifferentiated plant tissues to produce tumour-like growths that provide food, as well as protection, from natural enemies and environmental extremes. The most abundant gall-inducing insects are members of the dipteran family Cecidomyiidae (gall-midges), but the habit has evolved separately in many unrelated insect groups, including various families of Hymenoptera, Hemiptera, and Thysanoptera, as well as a few Coleoptera and Lepidoptera. Gall-insects tend to be highly host- and organ-specific, but little is known about their ecology in TEA. Casual observations suggest that they are very unevenly distributed among plant species, but that in some cases, gall-formers infect and eventually kill a large proportion of the area of the leaves (or whatever other organ they inhabit). In wet lowland forest in Panama, gall density on leaves was highest in the canopy (Ribeiro and Basset 2007), so observations from the ground may underestimate their importance.

5.2.7 Flower visitors

(See also pollination in 4.3.1.)

Herbivore interactions with flowers range from, at one extreme, 'florivory', where the flower is consumed in the same way as any other plant part (McCall and Irwin 2006), through 'floral larceny', where pollen and/or nectar are removed without pollination (Irwin et al. 2001), to mutualistic interactions where the animal provides pollination services in exchange for a reward (see 4.3.1) (Fig. 5.8). Effective pollinators are probably in a minority among the visitors to most flowers. Note also that many insects that visit flowers as adults (butterflies, moths, flies, wasps, and beetles) are important as herbivores, carnivores, or detritivores in their larval stages, so the impacts of floral food resources may extend beyond their role in pollination mutualisms (Wäckers et al. 2007).

Like fruit pulp (see 5.2.8), floral nectar and, in some plants, excess pollen, are resources produced for consumption by flower visitors. However, although their presence is advertised, these resources are rarely freely available, so some degree of specialized adaptation is usually required to obtain them. For nectar, this protection may be aimed primarily at ants, which rapidly exploit any accessible source of plant secretions (Blüthgen and Fiedler 2004), but protection beyond the needs of ant exclusion is often used to ensure the specificity of pollination (see 4.3.1). Thus a short, narrow corolla tube can mechanically exclude ants, while a long tube will exclude all but a few potential pollinators. Florivores—which include vertebrates such as primates (McConkey et al. 2003; Hanya 2004; Wiens et al. 2006), fruit bats (Nakamoto et al. 2007), and parrots (Walker 2007)—short-cut mechanical defences by eating the whole flower, but many specialized flower visitors, such as *Xylocopa* bees and sunbirds, cheat more elegantly by biting into the base of the corolla tube.

Nectar is basically a dilute sugar solution, dominated by either sucrose or hexoses (glucose and fructose), with small amounts of amino acids and other chemicals. It is primarily a source of energy

Figure 5.8 Bees, such as this *Apis cerana* worker in Hong Kong, depend entirely on floral resources for food. Photograph courtesy of Cristophe Barthelemy.

that can be assimilated quickly and with high efficiency, and nectarivores typically get their protein from other sources, such as pollen (bees) or insects (sunbirds). Secondary compounds in nectar, such as phenolics, may function as attractants to potential pollinators and/or deterrents to nectar-robbers or microbes (Liu et al. 2007a).

Pollen has a highly resistant outer wall and flower visitors vary widely in their ability to extract the contents (Roulston and Cane 2000). The Neotropical hummingbirds, surprisingly, can scarcely digest it at all, while a wide range of other vertebrates do much better. The almost ubiquitous flower-feeding thrips (Thysanoptera) pierce each grain individually with their mouthparts and suck out the contents. Reported protein contents for pollen range from as low as 2.5%, in some wind-pollinated gymnosperms, to as high as 61% (Roulston et al. 2000). Pollen also contains a varying amount of starch and lipid, sterols (which are essential nutrients in insects), vitamins, and phenolic compounds.

The availability of floral resources varies greatly within and between years (see 2.6.2). In the northern half of TEA, there is usually an annual peak between April and June, and a mid-winter trough (Fig. 2.14), while in the equatorial region there can be several years of very low flower-production between 2–3-month episodes of mass flowering. Animals that use floral resources can adapt by synchronizing their lifecycles with the annual patterns of flower availability (e.g. many insects in northern TEA); by rapid population build-up during flowering peaks (e.g. flower-feeding thrips at Pasoh; Appanah and Chan 1981); by diet-switching (e.g. some chrysomelid beetles switch from young leaves to flowers during mass-flowering events at Lambir; Kishimoto-Yamada and Itioka 2008a); by storing food for use during flowering lows (e.g. social bees); by migration (e.g. *Apis dorsata*); or by specializing on the few types of resources that are available year-round (e.g. fig wasps). Fleming and Muchhala (2008) suggest that the low spatio-temporal predictability of floral resources in tropical Asia explains the relatively generalized feeding niches of flower-visiting birds and bats relative to their counterparts in the Neotropics.

5.2.8 Frugivores

(See also seed dispersal in 4.3.2.)

Ripe, fleshy fruits are the only plant part that is consumed largely by vertebrates, rather than invertebrates. Many specialist invertebrates do attack unripe fruits, but ripe fruit pulp is probably too dilute and ephemeral a nutrient source for larval development. Exceptions to this generalization include the many species of 'fruit flies' in the family Tephritidae, whose larvae feed inside fleshy fruits, and scattered members of other invertebrate groups. Ripe and near-ripe fruits are attacked by fruit-piercing moths, in the family Noctuidae, which puncture the skin with a specialized proboscis and feed on the juice (Bänziger 1982). Different moth species vary in their fruit-piercing capabilities, but some can penetrate the intact skin of an orange or longan and can cause severe damage in fruit orchards. Over-ripe or damaged fruits are an important resource for a variety of liquid-feeding insects, including, in tropical forests, a diverse guild of lepidoptera that feed on fruit, rather than nectar, as adults. Fallen figs attract a characteristic assemblage of ground beetles (Carabidae), which feed on fruit pulp, seeds, and/or other insects (Borcherding et al. 2000).

Frugivory was considered from a plant point of view in the previous chapter (4.3.2). For animals, fruit is an easy food: it is conspicuously advertised, it does not fight or run away, it usually has few chemical or physical defences when ripe, and the reward is mostly easily digestible sugars or, in a minority of cases, lipids. It is not surprising, therefore, that most bird and mammal species in TEA seem to eat at least some fruit (Corlett 1998). Living entirely on fruit is not so easy, however, since the nutrient content is diluted by water and the nitrogen content typically very low.

Dependence on fruit varies widely between bird species, for example, with the fruit-pigeons (*Ducula*, *Ptilinopus*, and other members of the same clade) and green pigeons (*Treron*; Fig. 5.9) reported to be 100% frugivorous (e.g. Walker 2007) and high (>80%) dependence reported for a wide range of species, from hornbills and barbets to bulbuls. Dependence may also rise seasonally to near 100% in some species that consume more invertebrates at other times of the year. Green pigeons are known to grind up small seeds within the fruits they eat (mostly figs), but all the other highly frugivorous species appear to pass the seeds undamaged, so they cannot

Figure 5.9 The thick-billed green pigeon, *Treron curvirostrata*, like other *Treron* species, is reported to be 100% frugivorous. Unlike other highly frugivorous pigeons in TEA, green pigeons grind up at least some of the seeds in the fruits they eat. Photograph © Tim Laman.

supplement their diet in this way. Extreme fruit specialists have a variety of adaptations to their diet, including low metabolic rates (Schleucher 2002) and low protein requirements (Pryor et al. 2001). Indeed, low protein requirements seem to be a general characteristic of frugivorous birds, including species that also consume some invertebrates (Tsahar et al. 2005a).

An additional complication for birds is that a clade that includes the Sturnidae (starlings and mynahs) and Muscicapidae (thrushes and Old World flycatchers), is unable to digest sucrose and avoids foods that contain it (Lotz and Schondube 2006). The fruits consumed by frugivorous birds in Hong Kong contain mostly the hexose sugars, glucose and fructose, rather than sucrose, but sucrose is the dominant sugar in some fruits consumed largely by mammals (Ko et al. 1998). Whether or not the same applies in areas with more diverse frugivores and fruits is currently unknown. Starlings, mynahs, and thrushes are probably not important enough as seed-dispersal agents in TEA to be a major selective force against sucrose in fruit pulps, but there are also indications in the literature of a more general preference for hexose sugars among birds (Lotz and Schondube 2006). Throughout the region, the fruits consumed by birds tend to be black (at least to human eyes) or red in colour, although fruits of many other colours are sometimes eaten, and they are either unprotected (i.e. without a protective outer rind or peel) or dehiscent (i.e. they split open when ripe) (e.g. Corlett 1996; Kitamura et al. 2004).

The most fruit-dependent mammals are primates and fruit bats, although fruit also dominates the diet of some carnivores, for at least part of the year (Zhou et al. 2008c). Even the most frugivorous species of gibbons and macaques eat at least some leaves, and all fruit bats that have been studied in detail take both leaves and floral nectar. Young leaves, in particular, can supplement a fruit pulp diet with both extra protein and minerals such as calcium (Nelson et al. 2005). Gibbons prefer large, yellow, thin-skinned fruits with plenty of juicy pulp that are available in large crops, although they also eat some fruits that do not meet all these criteria, either because of seasonal lack of choice or because some other factor is more important (McConkey et al. 2002). Macaques appear to be less fussy about fruit than gibbons—perhaps because of their considerably larger group sizes—while the largely solitary orangutans must be more selective because of their relatively high energy expenditure in moving between trees (Leighton 1993).

Despite the consistent role of colour in fruit choice by primates and, to a somewhat lesser extent, in birds, it seems unlikely that fruit colour *per se* is important to vertebrates, given their learning abilities, but rather it is part of a co-evolved plant–animal signalling system. In a comparison of red and white colour varieties of *Ilex serrata* in Japan (35°N), resident brown-eared bulbuls (*Hypsipetes amaurotis*) ignored colour and based their fruit choice on the sugar concentration in the pulp, but migrant daurian redstarts (*Phoenicurus auroreus*) and orange-flanked bush-robins (*Tarsiger cyanurus*) ate mainly red fruits, suggesting that signal preferences may become more important in the absence of local knowledge of fruit profitability (Tsujita et al. 2007). However, it has recently been shown that birds may select fruits, in part, on the basis of their antioxidant capacity (Schaefer et al. 2008). Anthocyanins are both major fruit pigments and powerful antioxidants, so they are a direct signal of dietary antioxidant reward. Furthermore, the consumption by birds of flavonoids, extracted from fruits, improved the capacity of the immune system to respond to a novel antigen (Catoni et al. 2008).

Although fruit pulp generally has lower concentrations of defensive chemicals than other plant tissues, this does not mean that it is undefended. Anything that is food for vertebrates is also potentially food for microbes, but any defence against microbes risks reducing the palatability of the fruit to vertebrates. There is evidence for a trade-off between attracting dispersal agents and avoiding microbial decay, with plant species spread along an axis from those with highly attractive fruits, which are either removed immediately by frugivores or rapidly decay, to species with unattractive fruits, which can persist on the plant for months until eaten (Tang et al. 2005; Cazetta et al. 2008). The fruits of more than half the species tested in Hong Kong could persist for at least two months if frugivores were excluded by netting. In the genus *Ilex* (hollies), which has relatively long-persistent fruits, the fruit pulp contains phenolics and saponins (Tsang and

Figure 5.10 The ripe fruits of *Ilex* species (hollies), such as this *Ilex pubescens*, contain sugars, saponins, and phenolics. The rate of fruit removal by birds in Hong Kong was positively related to sugar content and negatively related to phenolic and saponin contents (Tsang and Corlett 2005). Photograph courtesy of Billy C.H. Hau.

Corlett 2005) (Fig. 5.10). The rate of removal of these fruits by birds in Hong Kong was positively related to sugar content and negatively to phenolic and saponin content. The ability of birds to detect small differences in the concentrations of nutritional and defensive chemicals shown in this and other studies (e.g. Schaefer et al. 2003; Levey et al. 2007), coupled with the ability of plants to wait for suitable dispersal agents, creates the opportunity for the evolution of plant–disperser specificity. There is, however, little evidence that this has occurred, perhaps because the long life-spans of vertebrate frugivores and the great variation in fruit supply do not permit such specialization.

The variability in the availability of ripe fleshy fruits, within and between years, is probably greater than that of any other major food source. Fruits develop from flowers, but variation in pollination and fruit-set means that fruit resources are less reliable than flower resources. Although individual fruits can persist on the plant much longer than individual flowers (which typically last for only a day or two), and fruit crops generally last longer than flower crops, plants almost always produce more flowers than fruits, and many individual plants flower without fruiting. This variable supply dominates the ecology of frugivores.

Since most frugivores are vertebrates, rapid population build-up is not an option for adapting to the irregular food supply, leaving diet-switching, migration, or specialization on the few fruit species available year-round. Diet-switching to leaves (e.g. many primates; Lucas and Corlett 1991), insects (e.g. many birds), invertebrates or small vertebrates (e.g. some carnivores; Zhou et al. 2008a, 2008b, 2008c), is the commonest response, but short-distance migrations may be under-recorded. Some hornbills and fruit pigeons range daily over large areas in search of fruiting plants (e.g. Leighton and Leighton 1983; Kinnaird et al. 1996; Symes and Marsden 2007). Millions of partially frugivorous birds—mostly thrushes (*Turdus*, *Zoothera*) and chats (*Luscinia*, *Tarsiger*)—make long-distance migrations into the northern half of TEA each winter (e.g. Kwok and Corlett 1999; Wells 2007), with a few continuing further south (e.g. Kimura et al. 2001; Fig. 5.11), but the routes and timing of these migrations are more likely controlled by the 'push' of declining food supplies at higher latitudes than the 'pull' of fruit supplies in the south.

Figure 5.11 The eye-browed thrush (*Turdus obscurus*) breeds in northern Eurasia in the summer and spends the winter in South China and South-East Asia. Photograph courtesy of Owen Chiang.

As discussed in section 2.6.2, the figs (*Ficus* species) are an exception to the dominance of annual or supra-annual fruiting phenologies in the region, with fig crops ripening throughout the year at all latitudes. This dependable supply, coupled with large crops and adequate nutritional value, has lead to the evolution of many more-or-less specialized fig-feeders in TEA, including several primates and fruit bats, as well as some species of hornbills, fruit pigeons, barbets, and other birds (Shanahan et al. 2001; Kinnaird and O'Brien 2005). A large enough territory or a high enough fig tree density can allow sedentary species to feed on figs year-round, although all these species also eat other, non-fig, fruits.

5.2.9 Granivores

(See also seed predation in Chapter 4.)

Seeds are compact packages of nutrients for the young plant and would be an excellent food for animals, if it were not for their physical and chemical defences. The need to overcome these defences restricts seed predation to a few specialized taxa. In TEA, the major vertebrate consumers of seeds are rodents and colobine monkeys, among the mammals, and parrots, some pigeons, and finches, among the birds (Corlett 1998). The other major animal group with generalist seed predators is the ants, while a wide variety of beetles (e.g. Carabidae-Harpalinae; Borcherding et al. 2000) and other insects mostly specialize on one or a few related seed species.

Little is known about how granivores cope with seed chemical defences. Some are probably just very selective, either for seeds with low levels of chemical defences (such as grasses) or for seeds defended by specific toxins that the granivore can overcome. Seeds are usually also defended physically by a varying amount of tissues (endocarp, seed coat, and sometimes other tissues) around the embryo and endosperm, and there may be a trade-off between investment in chemical and physical defences. Granivorous pigeons overcome physical defences by swallowing seeds whole and then grinding them with grit in their muscular gizzards, while parrots, finches, squirrels, and colobine monkeys use their muscular jaws to break them up before swallowing.

Seeds, like fruits, have a strongly seasonal or supra-annual pattern of supply. At Lambir, the populations of the most common species of terrestrial rat (*Maxomys rajah*) and arboreal squirrel (*Callosciurus prevostii*) fluctuated widely in response to the supra-annual fluctuations in fruit and seed availability (Nakagawa et al. 2007). Most granivores can adapt to changes in food supply by changing their diets, but some move when the seed supply is depleted or inaccessible, and migrant seed-eating buntings (*Emberiza*) are prominent in open habitats in the north of the region in winter. Dry seeds are one of the most easily stored of plant foods. Both scatter- and larder-hoarding of seeds by rodents have been reported in TEA (see 4.3.2), but the contribution of this behaviour to food security has not been studied.

5.3 Detritivores

Most dead plant material is processed in a thin layer just above and below the soil surface. Larger fluxes of energy and nutrients pass through this detritus system than through all the herbivores of the canopy layer combined, so there is ample opportunity for specialization in diet among detritivores. Litter starts off as a lower quality food than living plant biomass because herbivores have already eaten some of most nutritious tissues, and plants reabsorb some nutrients during senescence (see 6.3). On the other hand, the chemical defences of living plant tissues are also lost as they senesce and start to decay, so litter-feeders have less need to specialize by plant taxon. After it has been cycled through detritivores one or more times, detritus is no longer recognizable as plant material and it eventually ends up as decay- and digestion-resistant humic compounds (Allison 2006). Detritivores can specialize along this humification gradient from freshly fallen leaves or wood on the soil surface to increasingly refractory plant-derived materials dispersed in the soil (Donovan et al. 2001). Many other organisms in the litter and surface soil layers feed mainly on micro-organisms (most of the Collembola and mites) or on the detritivores (e.g. ants, many beetles, spiders), rather than on the litter itself.

The termites (Isoptera) appear to be the most important group of litter feeders throughout the lowland tropics, although this has rarely been quantified. This dominance in part reflects the phylogenetic and functional diversity of this group of insects, which means that there is a termite capable of feeding on almost any plant or plant-derived material (Davies et al. 2003). The 'lower termites' (Kalotermitidae and Rhinotermitidae) feed on wood, which they digest with the help of symbiotic flagellate protozoa and bacteria. The 'higher termites' (Termitidae) lack the protozoans, and different species feed on a wide range of substrates, from dead wood and leaves to the organic remains dispersed in the mineral soil. The fungus-growing termites (Macrotermitinae) cultivate a fungus (usually a species of *Termitomyces*, Basidiomycotina) on combs built from their faeces and then feed on the older parts of the comb after fungal degradation. This enables them to make extremely efficient use of a wide range of relatively undecomposed plant materials.

Termites as a group compete for detritus with free-living micro-organisms, particularly at continuously wet sites, and with fire at drier and more open sites (Yamada et al. 2007). The proportion of carbon in the annual above-ground litterfall that is mineralized by termites appears to be highest in fire-free tropical seasonal forests, where the fungus-growing termite group is dominant (Yamada et al. 2005b). Among the termites, soil-feeders are in asymmetric competition with the litter-feeders, particularly the efficient fungus-growers, and are relatively more important at sites where these are less active, as in an evergreen lower montane forest in Thailand (Yamada et al. 2007).

On a global scale, the earthworms (Oligochaeta) and the termites have broadly complementary areas of dominance (Donovan et al. 2007), but there has been too little work done on earthworms in TEA to test whether this is true on the regional scale. Earthworms increase in importance, relative to termites, in tropical montane forests (Heaney 2001; Ashton 2003) and in the subtropics. Other important groups of detritivores include millipedes (Diplopoda) and some members of the insect orders Blattodea (cockroaches), Orthoptera, Phasmatodea, and Coleoptera (beetles). Beetles, particularly Cerambycidae, are major consumers of dead wood and are assisted in the digestion process by symbiotic micro-organisms. The ambrosia beetles (Curculionidae: Scolytinae and Platypodinae) take this dependence to the extreme by feeding entirely on the symbiotic fungi that they inoculate into their tunnels. This habit has apparently arisen multiple times and has enabled these beetles to greatly expand their host ranges (Hulcr et al. 2007). At Pasoh, there are distinct communities of ambrosia beetles feeding in recently fallen wood on the forest floor, and in dead and dying branches in the canopy (Maeto and Fukuyama 2003).

Most forests in TEA show a pattern of year-round litterfall with seasonal peaks, although there may also be considerable interannual variation. At Fushan, in north-eastern Taiwan, interannual variation in the frequency of big typhoons caused a more than three-fold interannual range in total litterfall (Lin et al. 2003). A lot of the material brought down by typhoons is 'greenfall'—living green tissue—which is considerably more nutrient-rich than typical detritus.

5.4 Carnivores

Most herbivores and detritivores are invertebrates, so most carnivores are invertebrate-feeders—conventionally and rather misleadingly known as insectivores, although insects do make up a majority of the available biomass. There is no sharp division between the carnivores that prey on invertebrates and those that prey on vertebrates, but diets dominated by one or the other require different adaptations, while animals that consume similar amounts of both, often also eat plant foods and are better classed as omnivores (see 5.6).

5.4.1 Invertebrate-feeders

Most invertebrates are presumably killed by other invertebrates, but this has received relatively little study in TEA, except in agricultural fields, where such predatory arthropods are considered beneficial (e.g. Cai et al. 2007). Judging by their abundance in unselective surveys—such as canopy fogging—ants, beetles, and spiders dominate this feeding guild in tropical forests. As discussed above, most canopy ants appear to have plant-dominated diets, but the resulting high density of patrolling ants must make life extremely difficult for other canopy invertebrates. Down at ground level, TEA has no equivalent of the huge surface-raiding army ant swarms of Africa and the Neotropics, but smaller and less conspicuous species may be important predators. One widespread army ant species in TEA, *Dorylus laevigatus*, lives largely underground where it forms stable trail systems that give quick access to its foraging area (Berghoff et al. 2002). It feeds on earthworms and a wide variety of arthropods. The *Aenictus* army ants, in contrast, are specialized ant-eaters (Hirosawa et al. 2000). Some aspects of the army-ant syndrome, including group-foraging, are also found in unrelated ant genera in TEA, including *Leptogenys* (Ponerinae; Maschwitz et al. 1989) and *Pheidologeton* (Myrmicinae; Moffet 1987, 1988). Other species of soil and litter ants are known to have a variety of more or less specialized invertebrate diets and may form key links in the forest floor food-web (e.g. Masuko 1984, 2008; Wilson 2005).

Predatory beetles include many Staphylinidae and Carabidae (including the subfamily Cicindelinae, the tiger beetles; Fig. 5.12), as well as representatives of other families. Other groups of predatory insects include many Heteroptera (bugs), Mantidae (praying mantids), and Neuroptera (lacewings). Spiders (Araneae) are diverse, abundant and almost all predatory, so they must be important, although there is no quantitative data. There are also other non-insect arthropods, including Chilopoda (centipedes), some Opiliones (harvestmen), and predatory mites.

Many, perhaps most, vertebrates consume some invertebrates, but the specialists are mostly relatively small species. This is clearest among mammals, where there appears to be a size limit of 15–20 kg, above which, relying on small food items becomes energetically impossible, because the maximum rate at which small items can be harvested does not increase with carnivore body size (Carbone et al. 2007). The sloth bear (*Melursus ursinus*, 55–150 kg), which occurs just west of TEA's border with India, is a striking exception to this rule, with its morphological adaptations for the bulk harvest of social insects, including a long tongue, mobile lips, and missing incisor teeth, apparently side-stepping the 'small food items' constraint. The less specialized sun bears (25–65 kg) of TEA, in contrast, do not seem able to maintain body condition on a diet of invertebrates alone (Wong et al. 2002, 2005). TEA also supported a 2-m long giant pangolin (*Manis palaeojavanica*) until the late Pleistocene, which must again have relied on the efficient harvesting of social

Figure 5.12 Tiger beetles (Cicindelinae), such as this *Cicindela aurulenta* in Hong Kong, are predators on other invertebrates. Photograph courtesy of Yiu Vor.

ants and termites. The other specialist mammalian consumers of invertebrates in TEA are all <15 kg and mostly <5 kg. They include two smaller (<10 kg) species of pangolin (*Manis*), four distinctive-looking civets (Viverridae), *Hemigalus derbyanus*, *Diplogale hosei*, *Chrotogale owstoni*, and *Cynogale bennettii*, two species of stink badger (*Mydaus*), now placed in the skunk family, Mephitidae, and several Mustelidae, including the ferret badgers (*Melogale* spp.), and, in the north of the region, the hog badger (*Arctonyx collaris*) and the Eurasian badger (*Meles meles*).

Many smaller species of mammals also specialize on invertebrates, including all members of the order Insectivora (shrews, gymnures, and hedgehogs: 4–1000 g) and many species of rodents (rats, mice, and squirrels). The treeshrews (Scandentia: 40–400 g) feed on both fruit and a wide range of invertebrates, including ants, termites, caterpillars, beetles, cockroaches, crickets, katydids, centipedes, spiders, and earthworms, with coexisting treeshrew species specializing on a few of these taxa (Emmons 2000).

Among TEA's primates, only the tiny (60–140 g) tarsiers (Fig. 5.13) feed exclusively on invertebrates—mostly large insects. The reported importance of invertebrates in the diets of other primates varies between species, site, and season (Campbell et al. 2007), with part of this variation probably reflecting the difficulty in quantifying opportunistic invertebrate consumption in large mammals. In general, colobines, gibbons, and orangutans consume few invertebrates, although some populations are exceptions, while feeding on insects occupies a significant proportion of total feeding time (>10%) in slow lorises and in some populations of macaques. Orangutans in the swamp forest at Suaq Balimbing in Sumatra eat more insects than orangutans elsewhere (12% vs. 1–6%), which seems to reflect both the higher insect abundance at this site and the local invention of tools (modified living branches) for extracting social insects (stingless bees, termites, and ants) from tree holes (Fox et al. 2004).

Little is known about the diets of insectivorous bats in TEA and even less about their ecological impacts. They are certainly the dominant nocturnal predators of flying insects, but leaf-gleaning bats could also be underestimated predators of herbivorous insects in TEA, as has been shown for the Neotropics, where nocturnal exclosures of bats had a bigger impact on both arthropod abundance and herbivory levels than diurnal exclosures of birds (Kalka et al. 2008). Insectivorous bats communities can be incredibly diverse in TEA, with 51 coexisting species recorded in a lowland dipterocarp forest in

Figure 5.13 Tarsiers are the smallest primates in the region and the only ones that feed exclusively on invertebrates, mostly large insects. This is a female of the recently described *Tarsius lariang* in Sulawesi. Photograph © Stefan Merker.

Peninsular Malaysia (Kingston et al. 2003) and 21 persisting even in the highly degraded landscape of Hong Kong (Shek 2006). Coexisting species of insectivorous bats have wing morphologies and echolocation signals that are specialized for the degree of 'clutter' (i.e. obstacles) that they encounter when foraging (Kingston et al. 2003). Three broad guilds can be recognized according to whether they feed: in open spaces above the canopy or in large clearings; in small clearings, along streams, or at the forest edge; or in the highly cluttered space within the forest interior.

The transition from an invertebrate- to a vertebrate-dominated diet occurs at much lower body weights in birds than in mammals, which is hard to reconcile with a general energetic model (Carbone et al. 1999), unless birds are much less efficient at harvesting invertebrates. Among the birds, the 0.5–1.5-kg honey-buzzards (*Pernis*, 2–4 spp.) are the largest invertebrate specialists, feeding largely on the combs, larvae, pupae and adults of social wasps and bees. Invertebrates also dominate the diets of most smaller birds (<200–500 g) and the importance of these visually-hunting predators is apparent in the colours (cryptic or warning), activity patterns (often nocturnal), and other defensive behaviours of herbivorous insects that feed in exposed positions. As mentioned above, however, a study in Panama that excluded birds and bats separately from tree foliage found that, while both had significant impacts on arthropod abundance and herbivory, the impact of bats was significantly greater than that of birds (Kalka et al. 2008). It would be interesting to repeat this in TEA! Exclosures significantly increased insect damage on oil palm seedlings in Sabah, with the density of insectivorous birds explaining 35% of the variation in the effect size (Koh 2008a).

There are numerous ways to be an insectivorous bird: pittas, thrushes, and some babblers feed mainly on the ground; woodpeckers find food on trunks and major branches by gleaning, probing or excavating; many small passerines (e.g. warblers and many babblers) glean insects from leaves; some birds sally from perches to catch insects on the ground, trunk or leaves (e.g. trogons), or in the air (flycatchers); swifts and swallows catch insects while in continuous flight. Each of these categories could be subdivided further, although some bird species use more than one foraging strategy, depending on food availability.

Most species and individuals of invertebrate-feeding birds in forests in TEA join mixed-species flocks for at least part of their foraging time. These flocks are a characteristics feature of tropical and subtropical forests and can make bird-watching in these forests very frustrating, since long periods of apparent birdlessness are interspersed with short periods of frantic activity. In TEA, the flocks are most diverse in the lowland equatorial forests, where they can contain a few individuals each of a dozen or more species (McClure 1967; Croxall 1976), while flocks in the subtropics contain fewer species but more individuals of some of them. In Fushan (24°N), Taiwan, mixed-bird flocks contained an average of 5.8 species and 51.4 birds, and were led and dominated by the grey-cheeked fulvetta (*Alcippe morrisonia*), which averaged 32.5 birds per flock (Chen and Hsieh 2002; Fig. 5.14). There are two basic hypotheses about why birds join these mixed flocks (Primack and Corlett 2005). The predator-avoidance hypothesis suggests that the major benefit is a reduction in the predation risk, largely from raptors, as a result of the collective vigilance of the flock members, while the foraging-efficiency hypothesis suggests that flock members benefit from prey found or flushed out by other members. Both flock composition and bird behaviour in flocks support the predator-avoidance hypothesis, but there is also anecdotal evidence that some species benefit from prey flushing.

Invertebrates also dominate the diets of many small lizards and snakes, and most amphibians. There are no data that can be used to assess their importance relative to other groups of invertebrate-feeders, but there is evidence that densities of frogs and lizards are generally lower on the forest floor in TEA, than in similar climates in the Neotropics (Inger 1980; Huang and Hou 2004). Inger (1980) attributed the low densities he found in Borneo to the supra-annual mast-fruiting cycles in lowland rainforests, reducing the supply of seed-eating insects, but this cannot explain low densities elsewhere in the region. Forest-floor frogs outnumber lizards in sites without a severe dry season, while the pattern is reversed where rainfall is strongly seasonal.

The small body sizes and relatively short lifespans of most invertebrates mean that their abundance is very sensitive to both climate and food supply. Rainfall seasonality is the most important factor in the tropics, although low winter temperatures also have an impact in the north of TEA. In Hong Kong (22°N), the biomass of invertebrates drops drastically in the cool, dry season and the proportion of large individuals declines (Kwok and Corlett 2002). Some insectivorous bats in Hong

Figure 5.14 The grey-cheeked fulvetta (*Alcippe morrisonia*, Timaliidae) dominates mixed-species bird flocks in northern Taiwan. Photograph courtesy of Dr. Yun-Long Tseng.

Kong hibernate, some birds specializing on large and/or flying insects migrate south, and many resident birds eat more fruit. Paradoxically, however, there is also a large influx of tiny (<10 g) insectivorous warblers (*Phylloscopus*) from the north at this time (Dudgeon and Corlett 2004). These winter visitors feed on the small insects that are available year-round. They fly north before the start of the local breeding season, which coincides with an increase in the availability of invertebrates, particularly larger ones (Kwok and Corlett 1999, 2002). This is consistent with the 'breeding currency hypothesis' of Greenberg (1995), which suggests that the density of resident birds is limited by the availability of food suitable for feeding nestlings (particularly large, soft-bodied arthropods) during the breeding season, thus freeing resources for the winter influx of insectivores.

5.4.2 Vertebrate-feeders

The sizes of juvenile frogs, lizards, and snakes overlap with those of insects, so their consumption by primarily insectivorous vertebrates requires no special adaptations. Capturing live, adult vertebrates is a qualitatively different task, in contrast, performed by a guild of animals adapted to subduing large, active prey. In TEA, the killing and eating of vertebrates is dominated by snakes, diurnal and nocturnal birds of prey, and members of the mammalian order Carnivora, although a wide range of other vertebrates, and a few invertebrates (spiders and centipedes) eat at least some vertebrate prey.

Intact mainland forest communities in TEA may support up to 50–65 coexisting species of vertebrate-feeding vertebrates, including 20 or more snakes (Luiselli 2006), 10–12 diurnal birds of prey (Thiollay 1998), up to 8 owls (Francis and Wells 2003), and 15–25 mammals (Rabinowitz and Walker 1991; Johnson et al. 2006; Lynam et al. 2006; Mohd Azlan 2006). Considering just the mammals, different forest sites support up to six sympatric cats, six civets, one linsang (Prionodontidae), three mongooses, eight mustelids (including otters), two canids (dogs), and two bears (although both are largely vegetarian); a diversity in the order Carnivora that is unattached by tropical forests in other regions (Corlett 2007c). The diversity of carnivores declines in more open habitats, with altitude on high mountains, on small islands, or on those that were not connected to the mainland during Pleistocene low sea-levels, and with habitat fragmentation and hunting pressure. Snakes and birds of prey are less sensitive to island effects and some types of human impacts, so large snakes (pythons) or the largest birds of prey (Fig. 5.15) tend to be the top predators on islands (e.g. the Philippines, Sulawesi, and many smaller islands) and in some human-dominated landscapes (e.g. Hong Kong and Singapore).

Prey availability appears to be the main factor controlling carnivore densities. In the order Carnivora, 10,000 kg of prey support only about 90 kg of a given carnivore species (Carbone and Gittleman 2002). Ectotherm carnivores (snakes, lizards, amphibians), which do not use food to maintain body temperature, have much lower food requirements than endotherms, so a given biomass of prey should support a much higher mass of these carnivores. This is consistent with the subjective impression that ectotherm carnivores are far more abundant than endotherms in TEA. In regions with year-round warmth, the energy efficiency and body-form plasticity permitted by ectothermy may be more useful than the capacity for sustained aerobic activity that characterizes modern endotherms. This is particularly likely to be true for ambush foragers, which sit and wait for prey to come within striking distance. No mammal or bird could wait as long for a meal as a python or pitviper can. Endothermy has an obvious advantage only for active foragers and in the subtropics, where most ectotherms are inactive for part of the year.

If prey availability controls predator abundance, then we would expect co-existing carnivores to minimize competition by partitioning the prey resource, by type, size, period of carnivore activity (diurnal, crepuscular, nocturnal), and/or spatial distribution (arboreal vs. terrestrial, forest vs. open, wetland vs. dryland, etc.). Although there is anecdotal support for all these from TEA, no quantitative study has investigated resource partitioning among more than 2–4 species and none has looked at the mechanisms allowing coexistence between members of different classes of carnivores.

Where good evidence for dietary specialization is available, the preferred prey is usually either frogs,

Figure 5.15 The critically endangered Philippine eagle is not only the largest bird of prey in tropical Asia, but also the largest carnivore in the Philippines. Photograph © Eddie Juntilla, Philippine Eagle Foundation.

medium to large snakes (including venomous species), skinks, birds, rodents, or ungulates, presumably because these animals are both abundant and less efficiently captured by generalists. Efficient predation on hard-bodied skinks, for example, seems to require morphological adaptations (Greene 1997), while snakes are protected from generalist predators by their potentially venomous bites, birds by flight, and ungulates by size and speed. The spatial concentration of frogs may also favour specialization. The dominance of rodents in the diet of many species, in contrast, may simply reflect the fact they are the commonest potential prey in their size range. Indeed, rodents are also the commonest vertebrate prey of omnivorous species, such as many civets and foxes, suggesting that few specialized adaptations are needed to harvest them efficiently.

Vertebrates are potentially dangerous prey, particularly for predators that specialize on animals that are a substantial fraction of their own body weight. Among the carnivorous mammals of TEA, the relationship between predator and prey body weights seems to be different for relatively small species (<15 kg), such as small cats, civets, and foxes, and the larger species (>15 kg). The prey of the smaller species consists largely of rodents weighing <300 g, which is typically <20% of predator body weight. In contrast, most of the larger species feed on ungulates or primates (mostly >4 kg) and the largest prey taken is typically around the same weight as the predator itself. Large packs of dholes (*Cuon alpinus*, Canidae), which until recently lived in forests through much of TEA, can kill prey many times the weight of an individual dhole, although less than the combined weight of the pack. Feral dogs (*Canis lupus familiaris*; Fig. 5.16) now do the same in some areas. Both diurnal and nocturnal birds of prey can also take prey approaching their own body mass. The evolution of specialized feet for prey capture and specialized beaks for tearing large prey apart have relieved these birds of the swallowing constraint that limits prey size in birds that must swallow all their food whole (Slagsvold and Sonerud 2007).

Some large reptiles also take prey near their own body mass, including pythons, pitvipers, and crocodiles. Unlike mammals and birds, however, which first become independent foragers when almost fully grown, reptiles are independent at body sizes much smaller than their potential maximum. Changes in diet with development are, therefore, the rule in reptiles because size has such a large

Figure 5.16 Packs of feral domestic dogs (*Canis lupus familiaris*) hunt and scavenge on the outskirts of many tropical Asian cities. Photograph courtesy of Wing-sze Tang, AFCD, Hong Kong.

influence on the ability of an animal to capture, subdue, and swallow prey. This is particularly true in snakes, which swallow all their prey whole, so the maximum prey size is set by the size of the mouth. Interestingly, the largest reptiles—the reticulated python, *Python reticulatus*, the Komodo dragon, *Varanus komodoensis*, and the saltwater crocodile, *Crocodylus porosus*—make the dietary transition from rodents to larger mammals during growth at body masses of around 10–15 kg, which is similar to the size threshold above which mammalian carnivores specialize on larger prey.

In TEA, adult elephants and rhinoceroses have no natural predators, and tapirs seem to be avoided for unknown reasons (Kawanishi and Sunquist 2004). All other animals, including the young of elephants and rhinoceroses, and adult gaur (*Bos frontalis*, >500 kg), are potentially vulnerable. The largest prey (>100 kg) are taken regularly only by the largest mammalian predator, the tiger (*Panthera tigris*; Fig. 5.17) and the dhole packs mentioned above. The adaptable leopards (*Panthera pardus*) feed mostly on smaller prey, although they can kill larger animals. Prey in the 10–50 kg size range is vulnerable not only to these species, and the smaller clouded leopards (*Neofelis nebulosa* and *N. diardi*), but also to large pythons (*Python reticulatus* and *P. molurus*), the saltwater crocodile, and, on the few islands in Wallacea where it occurs, the Komodo dragon. Smaller prey (1–10 kg) is vulnerable to several additional mammalian predators, large individuals of the widespread water monitor (*Varanus salvator*) and false gharial (*Tomistoma schlegelii*), and the largest birds, including the Philippine eagle (*Pithecophaga jefferyi*) and eagle-owls (*Bubo* spp.). The number of potential predators rises steeply for prey <1 kg, and a medium-sized rodent (<200 g) is a potential meal for literally dozens of coexisting mammals, birds, and reptiles.

The relatively flexible behaviour of vertebrate prey means that the impact of predators on a prey species cannot be measured simply by the numbers that are killed. If predation is a major source of mortality, prey species may modify their behaviour to reduce danger, typically at the cost of a less-than-optimum use of their own food resources. Terrestrial herbivores, for example, may forgo feeding in exposed sites. Such indirect impacts of predation pressure may only become apparent when the predators are eliminated and prey behaviour

Figure 5.17 This photograph of the endangered Indochinese tiger (*Panthera tigris corbetti*) in Laos was taken with an infra-red triggered camera. Photograph © WCS.

changes. Unfortunately, there have been no studies of this phenomenon in TEA, where large predators have been eliminated from most of the remaining forest. Have fear-released herbivores (and vulnerable meso-predators, such as civets) changed their behaviour and have these changes had an impact on lower trophic levels? In many areas, human hunters have replaced large predators as a major cause of mortality, but fear of hunters is likely to have very different consequences for behaviour; for example, by encouraging nocturnal activity and the avoidance of roads.

In comparison with flowers, fruits, or insects, vertebrates are a relatively reliable food source throughout the year, although most ectotherm vertebrates—frogs and reptiles—hibernate in winter in the northern TEA, reducing the food supply for any non-ectotherm carnivores that feed on them. Most vertebrate-feeders in TEA apparently lead more-or-less sedentary lives, making, at most, local movements in response to seasonal or other variations in food supply. However, the situation is very different in northern Eurasia, from where the annual decline in food supply drives at least a million raptors (Accipitridae and Falconidae) south into TEA each winter (Bildstein 2006). Most of these follow the East Asian Continental Flyway, which is largely overland, but a smaller number of species make the oceanic crossing to the Philippines.

5.5 Parasites and parasitoids

Parasites are animals that obtain nutrients from one or a few individual hosts, normally causing harm but not immediate death (Begon et al. 2006). Most free-living animals support multiple individual parasites, often of more than one species. This suggests that most individual animals on earth are parasites and, since parasite species tend to be host-specific, probably most species. Apart from micro-organisms, including bacteria, fungi, and protozoa (e.g. Paperna et al. 2005), most parasites are flatworms (Platyhelminthes) (e.g. Wells et al. 2007), nematodes (e.g. Paperna et al. 2005; Wells et al. 2007), mites (e.g. Luo et al. 2007), or insects. Parasitoids are insects with free-living adults, in which the female lays eggs in, on, or near the host, which is then consumed and eventually killed by the developing larvae. Many herbivorous insects are killed by parasitoids rather than predators. Although the mining habit might be expected to provide some protection, leaf-miners suffer very high mortalities (30–60%) from parasitoid wasps (e.g. Lewis et al. 2002). The Hymenoptera includes the most diverse and abundant groups of parasitoids, but there are also many species of Diptera (particularly in the family Tachinidae), and scattered groups in other invertebrate families.

Numerous case studies illustrate the role of parasites in the ecology and evolution of their hosts (Begon et al. 2006), but there is too little information available for the tropics, in general, and TEA, in particular, to make any useful generalizations. Among tropical wild animals, primates have received most attention from parasitologists, but even in this relatively well-studied group, there are far more questions than answers (Nunn and Altizer 2006). Primates also provide the best evidence for the use by wild animals of plants rich in defensive chemicals for self-medication against parasites. Predators and parasites may interact in unexpected ways. In India, the density of sarcocysts of the protozoan parasite *Sarcocystis* in the heart muscles of dhole-killed chital (*Axis axis*) is significantly higher than in animals that die of other causes, suggesting that dhole select infected prey and, since the dhole is an obligatory host for the sexual cycle of the parasite, that the relationship between predator and parasite could be mutualistic (Jog et al. 2005).

The neglect of parasites in regional ecological studies is particularly unfortunate because of their potential interactions with a variety of human impacts, as stressed organisms are more likely to succumb to parasite infections. However, preliminary surveys of gastro-intestinal parasite infections of small mammals, in logged and unlogged forests (Wells et al. 2007), and of blood parasites of birds in forests with various degrees of disturbance (Paperna et al. 2005), did not produce any clear patterns.

5.6 Omnivores

Omnivores are animals that feed opportunistically on a variety of food types, including both plants and animals. No animal species eats all foods in proportion to their availability and nutrient content, presumably because the adaptations to eating, for example, mature leaves, are incompatible with those for overcoming large prey. However, seasonal switches between various proportions of fruit, invertebrates, and small mammals are quite widespread among mammals, including most civets (e.g. *Paguma larvata*; Zhou et al. 2008c; Fig. 5.18) and several mustelids (e.g. Zhou et al. 2008a, 2008b), as well as foxes in the north of the region, and among large birds, including some hornbills, barbets, corvids, and woodpeckers. There is an omnivorous lizard, Gray's monitor (*Varanus olivaceus*) in the Philippines (Auffenberg 1988) and an omnivorous group-foraging ant, *Pheidologeton diversus*, in much of South-East Asia (Moffett 1987). Seasonally changing combinations of fruit and invertebrates

Figure 5.18 The masked palm civet (*Paguma larvata*) is an omnivore, feeding on vertebrates, invertebrates and fruits. Photograph courtesy of Paul Crow, KFBG.

are widespread in smaller birds, including bulbuls, babblers, and thrushes, while combinations of fruit and leaves are common in primates and small ungulates, and have also been suggested for another monitor lizard, *Varanus mabitang*, on the island of Panay in the Philippines (Struck et al. 2002).

5.7 Scavengers

As with live bodies, most dead bodies are invertebrates, but scavenging at this scale has received very little attention in the tropics. The ubiquity, density, and omnivory of the ants may leave little room for specialists in this niche, but facultative scavenging by other predatory invertebrates is likely.

The bodies of vertebrates are usually too large for even cooperating ants to move. Most montane areas have a species of burying beetle (*Nicrophorus*, Silphidae), which buries small (<100 g) carcasses by digging underneath them, but this Holarctic genus does not occur in the lowlands, except in the north of the region and—oddly—on Sulawesi (Hanski and Krikken 1991; Scott 1998). Scavenging is a race against microbial decay, with the added complication that some microbes defend their prey with toxins (Shivik 2006), and it is possible that decay is simply too rapid in the tropical lowlands for the burying strategy to work. The advantages of rapid detection are illustrated by the arrival of carrion-feeding flies at carcasses within minutes of death.

Flies in the families Calliphoridae (blow flies) and Sarcophagidae (flesh flies) are usually dominant on the soft tissues (Hanski and Krikken 1991; Dudgeon and Corlett 2004). Beetles are also important, particularly Scarabaeidae, Hybosoridae, Histeridae, and Staphylinidae, and ants can sometimes dominate to the exclusion of other animals. On Mount Kinabalu, carrion-specialist Scarabaeoidea are absent above 1350 m, perhaps because of competition from the burying beetle, *Nicrophorus podagricus* (Kikuta et al. 1997). Carrion is also attractive to some adult butterflies, particularly males of fast-flying species, which may use it as a source of protein and amino acids (Hamer et al. 2006). Skin, fur, and feathers are consumed by keratin-digesting dermestid beetles (Dermestidae) that arrive as the carcass dries out. After the initial stages, many of the invertebrates on a carcass are there not to consume dead flesh, but to prey on the fly larvae and other scavengers.

For vertebrate carnivores, feeding on vertebrates that are already dead both increases the available food supply, since many vertebrates die for reasons other than predation (perhaps around half; Shivik 2006), and lifts the prey-size constraint, since dead animals do not fight. But large carcasses are rare and must be found before bacteria, or the invertebrates mentioned above, render them inedible. All obligate vertebrate scavengers are large soaring fliers—vultures—presumably because they can search large areas with much greater efficiency than a flapping or walking animal (Ruxton and Houston 2004). Old World vultures rely on their acute eyesight to detect carcasses from the air, so they are confined to open habitats in the TEA, which they presumably moved into after deforestation. By contrast, New World *Cathartes* vultures (Cathartidae) detect carcasses by smell and are the dominant scavengers in Neotropical forests. How would a Neotropical turkey vulture (*Cathartes aura*) fare in a tropical Asian forest? The bone-cracking dentition of hyenas enables them to feed on carcasses that have been stripped bare by vultures, but like the Old World vultures they avoid dense forest. The striped hyena (*Hyaena hyaena*) is widespread in the western half of the Oriental Region, but there have been no hyenas in TEA since the middle to late Pleistocene (Louys et al. 2007). Opportunistic consumption of carrion has been reported for most mammalian carnivores in TEA, many birds (particularly crows and magpies, and the widespread kites, *Milvus migrans* and *Haliastur indus*), and a few snakes. Wild pigs (*Sus* spp.) and rodents, including porcupines, have also been reported to feed on carcasses.

5.8 Coprophages

Vertebrate faeces—dung—vary in composition between animals, depending on their diet, but the bulk of the dung in nature comes from herbivores and thus consists largely of the less digestible components of plant tissues. It is chemically similar to other forms of plant detritus, but physically very different, and, therefore, consumed largely by dung specialists. The invertebrate communities of vertebrate faeces in South-East Asia are dominated by

Figure 5.19 The invertebrate community on vertebrate faeces—dung—is dominated by specialized 'dung beetles', such as these *Catharsius dayacus* (Scarabaeidae). Photograph courtesy of Darren Mann.

the adults and larvae of beetles (mostly Scarabaeidae and related families of 'dung beetles') and flies (particularly Calliphoridae and Muscidae) (Hanski and Krikken 1991; Davis 2000; Dudgeon and Corlett 2004). Flesh flies (Sarcophagidae) can apparently breed on the relatively nutritious faeces of carnivores and omnivores, but not on the protein-poor faeces of herbivores (Bänziger and Pape 2004). Most studies have focused on the composition of the dung beetle community, using unnatural baits (human or cattle dung), and rather little is known about how these organisms deal with vertebrate faeces under natural conditions.

The majority of dung beetles in South-East Asian forests are 'tunnellers', which dig a tunnel below the dung and move some to the bottom for feeding and breeding, but there are also 'rollers' of various sizes, which roll a ball of dung away from the dung pile before burying it (Hanski and Krikken 1991). At Danum Valley, Sabah, human and cattle dung attracted 31 species of dung beetles, among which small diurnal tunnellers (*Onthophagus* spp.) were most diverse and abundant, while large nocturnal tunnellers (mostly *Catharsius dayacus*; Fig. 5.19) accounted for more than half the total biomass (Slade et al. 2007). The large nocturnal tunnellers removed most of the dung and buried most of the artificial plastic seeds incorporated into it, but a full complement of functional groups was needed to maximize both dung and seed removal. Despite their small numerical contribution to dung removal, and a tendency to avoid incorporating seeds into their dung balls, rollers may provide an important service, as they can disperse seeds away from the dung patch. Dung beetles were the first animals to arrive at piles of fresh gibbon dung in central Borneo, but only one seed was removed in this study, while the beetles often removed the dung from around the seeds (McConkey 2005).

CHAPTER 6

Energy and nutrients

6.1 Introduction

Studies of the flows of energy and matter through ecosystems—the traditional tasks of 'ecosystem ecology'—have acquired a new urgency over the last decade in response to the need to understand the potential role of natural ecosystems in the mitigation of anthropogenic carbon emissions (see Section 8.2.3). There is also increasing concern about the impacts of the pervasive enrichment of natural ecosystems by nitrogen and other nutrients (see Section 7.4.12). This renewed interest has revealed that our previous understanding of these processes is in many ways too crude for the level of precision that is now needed. Unfortunately, despite the surge of interest, the demand for novelty in the scientific literature discourages the replication of studies across multiple sites, with the result that different components have been measured at different sites, making a robust synthesis difficult at present.

6.2 Energy and carbon

Energy is captured in photosynthesis and flows through ecosystems in the form of carbon–carbon bonds until it is ultimately released as heat. Energy is used once only, while carbon is cycled; but energy and carbon follow exactly the same path through ecosystems, so it makes sense to consider them together. Carbon, although widespread in the environment, is scarce everywhere. All land plants withdraw carbon dioxide from the same dilute atmospheric pool, causing a decline of 6 ppm in the concentration, every northern-hemisphere growing season, before the concentration increases again each winter, as a result of respiration (Fig. 6.1). All land plants are thus in diffuse competition for carbon.

6.2.1 Primary production

The rate of fixation of carbon (or energy) by photosynthesis is the gross primary productivity (GPP). GPP cannot be measured directly, but it can be estimated from eddy covariance measurements of net ecosystem exchange (NEE) of carbon dioxide between the canopy and atmosphere, and of ecosystem respiration (see 6.2.3). These site-specific estimates can then be extrapolated to broader scales, using predictive models based on variables that can be measured from satellites, plus widely available climatological data (Running et al. 2004; Yuan et al. 2007; Huete et al. 2008). These models are based on the assumption that the potential GPP is directly related to the amount of photosynthetically active radiation absorbed by the vegetation (which can be estimated from satellite-derived spectral indices), but that the efficiency with which this light is utilized may be reduced by suboptimal climatic conditions, such as low temperatures and water stress. However, a recent study found that GPP estimates from eddy covariance measurements at three sites in South-East Asia (in drought-deciduous, dry evergreen, and wet evergreen forest) were more strongly correlated with canopy 'greenness' estimates from satellite data (MODIS EVI), than with the more complex indices (Huete et al. 2008).

Although these results need confirmation from more sites, they do reveal a strong moisture-limitation of monthly GPP throughout the lowlands of continental South-East Asia, with a wet-season peak and dry-season low (Fig. 6.2). In seasonal tropical forests, however, lower dry-season GPP appears to be balanced by higher wet-season GPP, so that estimates of annual GPP are very similar for mixed deciduous forest at Mae Klong (GPP = 32.3 Mg ha^{-1} yr^{-1}) (1 Mg = 1 metric tonne), seasonal evergreen

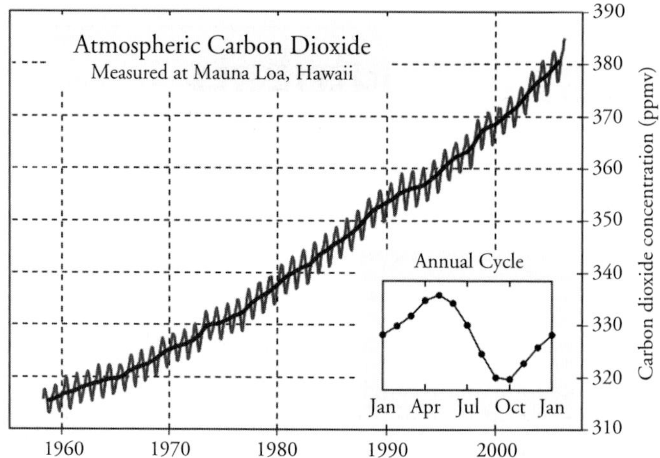

Figure 6.1 The atmospheric carbon dioxide concentration measured at Mauna Loa, Hawaii, showing a decline every northern-hemisphere growing season as a result of photosynthesis, superimposed on the long-term rising trend as result of fossil fuel use and deforestation. Image created by Robert A. Rohde for Global Warming Art.

forest at Sakaerat (37.8 Mg ha^{-1} yr^{-1}), and lowland rainforest at Pasoh (32.2 Mg ha^{-1} yr^{-1}) (Hirata et al. 2008). Across East Asia as a whole, from the subarctic to the tropics, the currently available GPP measurements have a simple linear relationship with annual mean air temperature (Hirata et al. 2008).

Much of the GPP is respired by the plants themselves, leaving the net primary productivity (NPP), representing the net carbon (or energy) gain by the plant community. The ratio of NPP to GPP—the carbon-use efficiency (CUE)—is widely assumed to be around 0.5, but there is evidence to suggest it varies over a wider range (DeLucia et al. 2007). Some, but not all, components of the total NPP can be measured directly. Field studies usually measure the two largest components: the above-ground biomass increment (the increase in the dry mass of above-ground plant parts over a measurement interval of one to several years), and the fine litterfall (the dry mass of above-ground plant parts that is produced and shed during the measurement interval for the biomass increment). Sometimes one of these components is estimated from the other, but a recent review showed that there is no simple relationship between them, so such estimates are very dubious (Shoo and VanDerWal 2008). Another common problem is that the two components are measured over different time periods, ignoring between-year variation.

The proportion of the NPP that goes into the below-ground biomass increment and fine-root turnover is very difficult to measure accurately, and is often estimated as a fixed percentage of NPP. However, the rather limited data available suggest that the proportion of photosynthetic carbon allocated below ground increases with decreasing resource availability (Litton et al. 2007), so assuming a fixed proportion could result in a large underestimate of total NPP at unproductive sites. The reported increase in the root/shoot ratio with altitude on tropical mountains (Leuschner et al. 2007) may be a specific case of this general phenomenon.

Plants also lose significant amounts of carbon in root exudates, by transfers to mycorrhizae, as volatile organic compounds (VOCs, such as isoprene), as organic leachates, and to herbivores, with these additional components either ignored or estimated in most field studies, although they can together amount to as much as 30% of total NPP (Clark et al. 2001; Keeling and Phillips 2007). Additional problems arise from the tendency of litter-trap studies to underestimate litterfall—usually the largest component in NPP—because of losses through decomposition before measurement.

No study in TEA has directly measured all these components, and the many published estimates of NPP from the region have variously ignored, or rather crudely estimated, the components that were

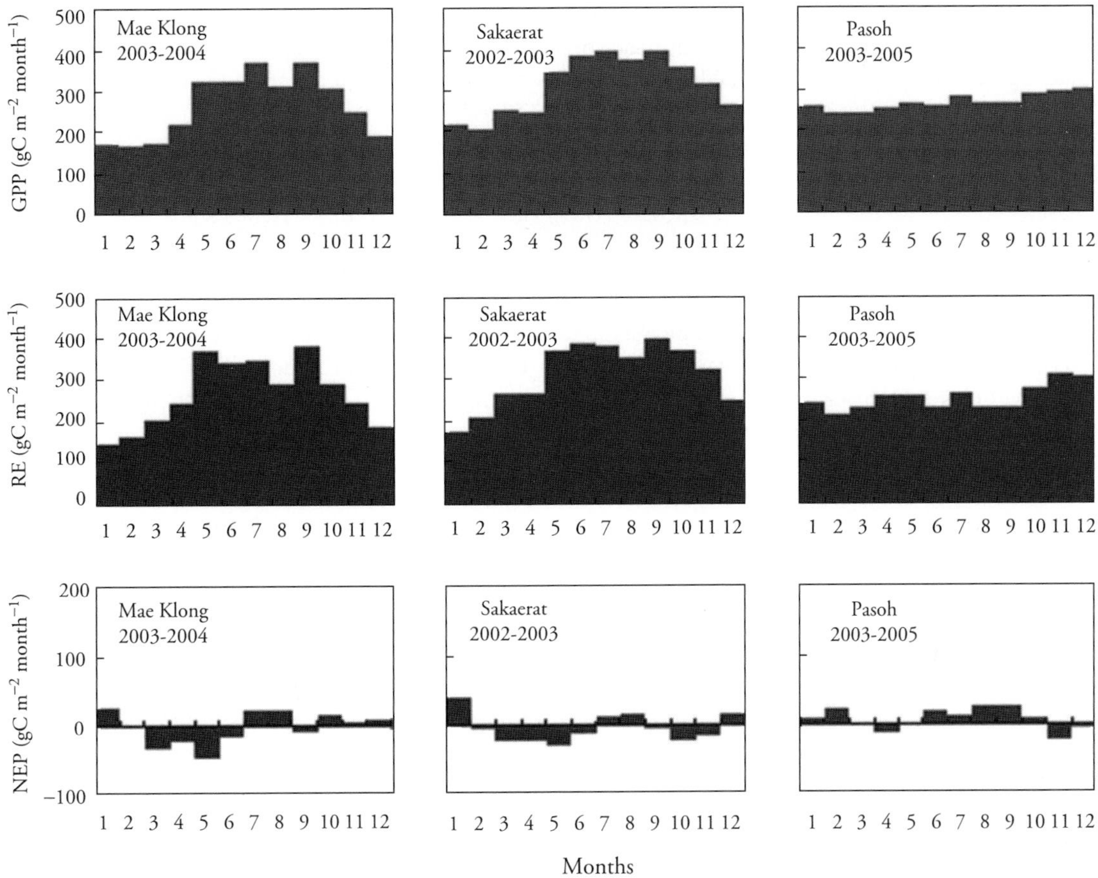

Figure 6.2 Gross primary productivity (GPP), ecosystem respiration (RE), and net ecosystem production (NEP) at three sites in Tropical East Asia: a mixed deciduous forest at Mae Klong, Thailand; a dry evergreen forest at Sakaerat, Thailand; and a tropical rainforest at Pasoh, Malaysia. From Saigusa et al. (2008).

not measured. The largest errors are likely to be in the estimation of below-ground components, so only above-ground NPP (ANPP) can be compared across sites. Allowing for unmeasured components, estimates of total ANPP in lowland tropical rainforests in TEA are mostly in the range 15–30 Mg ha^{-1} yr^{-1} (Clark et al. 2001; Keeling and Phillips 2007; Paoli and Curran 2007). ANPP is lower in the subtropical evergreen broad-leaved forest (<10 Mg ha^{-1} yr^{-1}; Yan et al. 2006; Yang et al. 2007; Zhang et al. 2007), where winter low-temperatures limit carbon fixation. ANPP also declines with altitude (Kitayama and Aiba 2002a; Luo et al. 2002, 2004; Aiba et al. 2005), presumably in response to declining temperature, although many other environmental factors covary along altitudinal gradients (see 2.5.2). Seasonal water-shortage and plant phenology reduce NPP in the drier tropical forest types, since solar radiation cannot be fully utilized when a tree is stressed or leafless; but there are few data from TEA for these ecosystems. Within areas of uniform climate, ANPP is reduced on extreme soil types (Aiba et al. 2005; Miyamoto et al. 2007) and in south-west Borneo NPP was significantly positively related to soil nutrients, especially phosphorus (Paoli and Curran 2007).

After a major disturbance, such as logging, fire, or clearance, NPP peaks at intermediate forest ages and declines in older forests, demonstrating the key role of disturbance history in controlling forest carbon budgets (Pregitzer and Euskirchen 2004). In view of the likely widespread influence of catastrophic disturbances in TEA (Whitmore and Burslem 1998), the extrapolation of NPP measurements from small plots to the landscape or regional level should be done with caution, since different patches in the forest mosaic may have very different productivities, despite uniform climate and soil (Feeley et al. 2007a). In an undisturbed forest in equilibrium at the landscape scale, most of the forest area is expected to be increasing in biomass as it recovers from past gap-forming events, with the offsetting biomass decreases mostly resulting from the deaths of individual large trees, which are rare in space and time.

6.2.2 Biomass

The plant biomass in an ecosystem is a balance between NPP and mortality. Note that, the 'biomass' of a woody plant is a very heterogeneous material. Not only is it conventionally defined to include the whole body of the plant, including dead heartwood and bark, but the living fraction includes both short- and long-lived components. The relationship between NPP and biomass is not simple and the most massive forests, such as the redwoods of northern California, are not the most productive. Within tropical forests, above-ground biomass seems to plateau at high ANPP values (>20 Mg ha^{-1} yr^{-1}), and it may even decline at the most productive sites, where fast-growing, low wood-density species with shorter life-spans may be favoured (Keeling and Phillips 2007).

Most of the highest above-ground biomass values in TEA (>400 Mg ha^{-1}) come from lowland dipterocarp forests, with high densities of huge emergent dipterocarps (Lasco et al. 2006; Paoli et al. 2008), but similar values have also been obtained from small plots in seasonal rainforest in Xishuangbanna, south-west China (Zheng et al. 2006). Values from small plots may be atypically high, however, because of a tendency to site them in the high-biomass parts of a heterogeneous forest mosaic. In these lowland forests, most of the above-ground biomass (>90%) is in the large trees, so half or more of the above-ground carbon can be lost after logging. Post-logging recovery of carbon stocks is slow and, in a lowland dipterocarp forest in Mindanao, only 70% of the original above-ground biomass had been attained by the start of the next cutting cycle, after 35 years (Lasco et al. 2006). Note that plantations of native or exotic tree species can also attain above-ground biomass values >400 Mg ha^{-1} in less than 50 years on good sites (Hiratsuka et al. 2005).

From a greenhouse-gas perspective, the total carbon storage in an ecosystem—in both biomass and soils—is of more importance than biomass alone. The data available from TEA are consistent with the general global pattern of a shift from higher biomass, lower soil-carbon forests in the lowland tropics to lower biomass, higher soil-carbon forests at cooler montane and subtropical sites (Kitayama and Aiba 2002b; Raich et al. 2006). Most of the carbon in lowland tropical forests is in living trees, while most of the carbon in montane and subtropical forests is typically in the soil. If this is generally true, total carbon storage in tropical forests may be more or less independent of temperature, so the predicted warming over the next century could simply transfer carbon stocks from the soil to the biomass, as enhanced decomposition is offset by increased NPP. This scenario, however, ignores concurrent changes in rainfall, carbon dioxide concentrations, and anthropogenic nitrogen deposition, as well as the on-going biomass-reducing impacts of deforestation and logging (see Chapter 7).

6.2.3 Net ecosystem production and exchange

Net ecosystem production (NEP) is the net accumulation of carbon by an ecosystem—the balance between carbon entering the system, mostly by photosynthesis, and leaving it, mostly as carbon dioxide from plant, microbial, and animal respiration, but also as VOCs, methane, and through leaching to groundwater and streams. Accurate estimates of NEP will be essential for any future global carbon-accounting system. At lowland rainforest sites in Costa Rica and central Amazonia, total ecosystem respiration was estimated to be 37–38% from foliage, 14% from woody plant parts, 6–7% from coarse woody debris, and 41% from the soil, with the soil respiration about equally divided between plant roots and soil heterotrophs (microbes and animals) (Chambers et al. 2004; Cavaleri et al. 2008).

Earlier estimates from lowland rainforest at Pasoh, Malaysia, and seasonal rainforest at Khao Chong, Thailand, also suggested that leaves account for at least half of the total respiration by living plants (Yoda 1983).

NEP is a small net difference between two large carbon fluxes, so it tends to fluctuate a lot, both seasonally and between years (Hirata et al. 2008; Saigusa et al. 2008) (Fig. 6.2). The carbon uptake side of the equation—GPP—is a relatively simple process in comparison with the variety and complexity of the processes involved in carbon loss. The NEP of a patch of vegetation is also very sensitive to its disturbance history and successional status, being strongly negative immediately after a major disturbance, then positive during the successional recovery of biomass, and tending towards zero if a steady state is eventually reached.

Net ecosystem exchange (NEE) is the net exchange of carbon dioxide between the ecosystem and the atmosphere—the balance between photosynthesis and ecosystem respiration—and is usually the major component of NEP, at least over short time-scales, when episodic disturbances can be ignored. In practice, NEE and NEP tend to be treated as synonymous, i.e. NEE = -NEP. Unlike GPP, NPP, and NEP, NEE can be measured directly by using the eddy covariance method to estimate the net flux of carbon dioxide above vegetation (Burba and Forman 2008), although there are still some incompletely resolved problems with this technique, which introduce considerable uncertainty into the results (Oren et al. 2006; Kosugi et al. 2008). Total ecosystem respiration can be estimated (with various assumptions) from night-time NEE, so GPP can be calculated. There is now a global network (FLUXNET) of long-term sites measuring carbon dioxide fluxes, including several in TEA, and shorter term measurements have also been made at other sites (Mizoguchi et al. 2009).

The results available so far suggest that the lowland rainforest at Pasoh is a weak carbon dioxide sink throughout most of the year (Kosugi et al. 2008), while tropical seasonal forests in Thailand are sources during the late dry season (February to April) and sinks for most of the rest of the year (Hirata et al. 2008; Saigusa et al. 2008) (Fig. 6.2). A remnant of peat swamp forest in central Kalimantan was a large net carbon dioxide source (3–6 Mg ha^{-1} yr^{-1}) as a result of peat decomposition following lowering of the water table by drainage (Hirano et al. 2007). Conversely, a young, secondary forest, regenerating after fire, was a large net carbon dioxide sink (4 Mg ha^{-1} yr^{-1}), reflecting a relatively low total ecosystem respiration, as a result of low plant-biomass, rather than higher GPP (Hirata et al. 2008).

A characteristic of the currently available NEE estimates from the lowland tropics is their year-to-year variation (Hirata et al. 2008; Saigusa et al. 2008). For example, decreased rainfall in early 2002 significantly decreased GPP, and thus NEE, in the tropical seasonal forest at Sakaerat. Overall, there is increasing evidence that year-too-year variation in GPP and NEE in South-East Asia is strongly influenced by interannual variation in climate in relation to ENSO and other cycles.

6.3 Other nutrients

In addition to carbon, hydrogen, and oxygen, all plants need the macronutrients nitrogen (N), phosphorus (P), sulphur (S), potassium (K), calcium (Ca), and magnesium (Mg), as well as a number of micronutrients. For growth to occur, these must be available in the appropriate forms and supplied at a sufficient rate. N and S are, like C, largely derived from the atmosphere, but most other nutrients originate from minerals in rocks, although dust inputs may also be significant. In contrast to C, however, where most respired carbon dioxide is returned to the global pool and very little is cycled within the ecosystem, the external inputs of these nutrients are generally much smaller than the amounts recycled within the ecosystem, i.e. the ecosystem nutrient cycles are relatively closed. Significant amounts of some nutrients (particularly K) are returned to the soil by leaching from living plant tissues, and a variable proportion of all nutrients (generally 1–10% in forests) enters the herbivore food-chain, but litterfall and below-ground root turnover are the major routes of recycling. The annual nutrient requirements of plants are, therefore, supplied mainly from the decomposition of dead plant materials, rather than external inputs.

Plants can conserve nutrients by withdrawing them from senescent foliage before abscission, with the fraction withdrawn varying considerably between nutrients, plant species, and sites. Resorption is particularly important for N, P, and K, where around half (<80%) is usually resorbed, while Ca and Fe are immobile in the phloem, so cannot be resorbed (Chapin et al. 2002). The resorption of nutrients from senescing leaves means that herbivores, which eat living leaves that still have their full nutrient content, remove around twice as much N, P, and K per unit biomass consumed, as would be lost if the same biomass senesced. This helps explain the large investment plants make in chemical and physical defences, despite the apparently rather minor impact of herbivory (see 5.2). In those parts of TEA subject to typhoons and other wind storms, 'greenfall' of non-senescent leaves may similarly make a disproportionate contribution to nutrient loss from plants and nutrient input to the soil.

The decomposition of dead plant materials and the release of nutrients in plant-available forms are complex processes and the details differ between nutrients, plant materials, and sites (Chapin et al. 2002). The major processes involved are leaching of soluble constituents, fragmentation by soil animals, and chemical alteration by microbes. Decomposition is fastest in continuously warm, damp, aerobic conditions, which helps explain the small litter-pool in lowland rainforests, despite the high ANPP. It is also faster for litter from plants growing on nutrient-rich soils. At Lambir, rates of litterfall are very similar on the fertile shale-derived soils and infertile sandstones, but litter nutrient contents and decomposition rates are lower on the sandstone, resulting in the formation of a distinct humus layer, matted with roots, over the mineral soil (Baillie et al. 2006). At Sepilok, not only the nutrient content and decomposition rate, but also the quantity of litter, decreased along a gradient of soil fertility from alluvial forest through sandstone to heath forest (Dent et al. 2006). The importance of the 'afterlife' characteristics of the dominant plant species in controlling decomposition rates creates the potential for a feedback mechanism, whereby plants on nutrient-limited sites, where leaves tend to be long-lived and heavily defended against herbivores, produce small amounts of low-quality litter that decomposes slowly, thus exacerbating nutrient limitation.

The release of different nutrients is more or less tightly linked with the breakdown of the carbon skeleton, depending on their chemical forms in the plant. N is directly linked to the carbon skeleton by C–N bonds, while P is linked through ester linkages (C–O–P), which can be broken by plant or microbial phosphatases without breaking the carbon skeleton. Ca is a structural component in cell walls and so is also released relatively slowly. At the other extreme, K occurs mostly in the cytoplasm and is easily leached from intact litter.

6.3.1 Nitrogen

Nitrogen is the third-largest component of plant dry matter (after carbon and oxygen) and is incorporated mostly into proteins. The most abundant of these proteins is ribulose-1,5-bisphosphate carboxylase/oxygenase ('Rubisco'), the enzyme that catalyzes the first major step of carbon fixation. Nitrogen gas makes up 78% of the atmosphere, but the N_2 molecule is relatively stable and nitrogen only enters terrestrial nutrient cycles under natural conditions, after biological nitrogen fixation by free-living and symbiotic microbes, or in the rain after it has been converted to ammonia by lightning discharges. Biological fixation is by far the most important of the natural sources, although its role in tropical forests is still poorly understood. In recent years, however, anthropogenic sources of nitrogen have increasingly dominated local, regional, and global N budgets (Galloway et al. 2004; see 7.4.12). In extreme cases, as at Dinghushan in Guangdong, N-deposition from pollutants is so high that there is no net retention by the forest ecosystem and the nitrogen cycle becomes effectively as open as the carbon cycle (Fang et al. 2008).

In the soil, N is available to plants in several different forms (nitrate, ammonium, and dissolved organic matter) that differ in relative availability between ecosystems (Chapin et al. 2002). There is evidence, mostly from studies outside the tropics, that coexisting plant species can reduce competition by specializing on different N sources. Tropical studies have suggested a link with successional status, with pioneers preferentially using nitrate and late-successional species preferring ammonium. However, two abundant, coexisting, late-successional tree species in French Guiana had different N-acquisition strategies, with one apparently foraging mostly in the litter layer for nitrate and the other in the soil for ammonium (Schimann et al. 2008). In contrast, a recent study in tropical forest in Hawaii showed that coexisting species relied on a common pool of inorganic nitrogen, with the preferred form changing with changes in availability, from nitrate at the driest sites to ammonium at the wettest (Houlton et al. 2007).

Nitrogen is thought to be the most common limiting nutrient in extra-tropical ecosystems, including forests in the temperate and boreal zone. In tropical lowland forests on highly-weathered soils, however, there is evidence that N is often in superabundant supply, relative to other nutrients, particularly phosphorus (see below). This evidence is partly direct, from the responses to N-fertilization experiments, but mostly indirect, including the relatively leaky N-cycles and the relatively high N:P ratios in leaves and litterfall in these ecosystems (McGroddy et al. 2004; Townsend et al. 2007).

This relatively high availability of N in tropical forests is presumed to reflect the long-term accumulation of nitrogen fixed by both symbiotic and free-living micro-organisms, although the rates of fixation have not been quantified in TEA. There is evidence to suggest, however, that N may be in short supply in some montane forests (Corre et al. 2006) and on some soil types in the lowlands (e.g. Sotta et al. 2008), and that individual species may be N-limited, even on P-poor soils (Townsend et al. 2007). Moreover, in a Panamanian lowland forest, fertilization with N significantly increased the production of flowers and fruits, but not leaves and twigs, suggesting that N may limit reproduction, even on relatively N-rich soils (Kaspari et al. 2008).

Nitrogen volatilizes in fires at much lower temperatures than most other plant nutrients and nitrate is relatively easily leached from exposed soils. As a result, a large fraction of ecosystem N-stores may be lost during forest clearance. Both agricultural production and secondary succession after abandonment may, therefore, be N-limited on sites where undisturbed vegetation is limited by other nutrients (Gehring et al. 1999; Yan et al. 2006, 2008; Davidson et al. 2004, 2007; Boonyanuphap et al. 2007; Tanaka et al. 2007). In an age-sequence of sites in Amazonia, the leaky-N and conservative-P cycles of mature forests were gradually restored during secondary succession over several decades (Davidson et al. 2007). The frequent fires in dry tropical forests might be expected to result in chronic N-limitation, but the amounts lost in each fire are relatively small because most of the fuel that burns consists of dead grass material from which much of the N has already been translocated below ground (Toda et al. 2007).

6.3.2 Phosphorus

In contrast to N, the largest reservoir for P is in minerals in rocks (e.g. apatite). As soils age, unweathered P-containing rock minerals disappear from the root zone, and most of the soil P becomes bound into forms that are usually considered unavailable to plants. In highly weathered tropical soils (Ultisols and Oxisols; see Chapter 2), reactions with iron and aluminium oxides provide a major sink for P. Unlike N, which is relatively easily replaced from the atmosphere, the replacement of P depends on the very small annual input from dust. Along an age sequence of soils in Hawaii, atmosphere-derived N is limiting on the youngest soils, but gradually accumulates as the soils age, while rock-derived P is abundant in the youngest soils, but becomes increasingly bound in unavailable forms over thousands of years of soil development (Hedin et al. 2003). Note, however, that P is also fixed in unavailable complexes with allophane in young Andisols derived from volcanic ash. Moreover, a study of soils developed since the 1883 Krakatau eruption found low P-availability, limiting to rice growth in a bioassay, despite low allophane contents, perhaps because of rapid accumulation of P in organic forms (Schlesinger et al. 1998).

An increasing amount of evidence has accumulated to suggest that the productivity of tropical and subtropical lowland forests on highly-weathered soils is most often limited by the availability of P. This evidence is partly direct, from the responses to P-fertilization experiments, but mostly indirect, including the efficiency with which plants resorb P from senescing leaves before leaf abscission

(Kitayama et al. 2004; Cai and Bongers 2007; Lovelock et al. 2007), the relatively high C:P and N:P ratios in leaves and litterfall (McGroddy et al. 2004; Townsend et al. 2007), and the covariation between soil P and both plant species composition and productivity (Paoli et al. 2007). Moreover, the amount of stored phosphorus decreases drastically during flowering and fruiting by dipterocarps (T. Ichie, cited in Naito et al. 2008a). As discussed above, similar sorts of evidence suggest that N is often in superabundant supply in these forests. This suggested 'N-rich, P-poor' pattern in tropical forests, contrasts with ecosystems at higher latitudes, where nitrogen is the most common limiting nutrient.

P-limitation is much less likely on younger soils that still contain unweathered primary minerals, including those derived from volcanic ash (but see above) or recent fluvial deposits, as well as soils on the steep, unstable slopes where forest is most likely to persist in human-dominated landscapes. In tectonically active regions, which include much of TEA, a combination of uplift and erosion may continually rejuvenate soils by bringing fresh rock fragments into the root zone, preventing the development of P limitation (Porder et al. 2007).

Despite the growing evidence for pervasive P-limitation on old soils, forests in South-East Asia on soils with very low available P, by global standards, can have very high above-ground biomasses (Kitayama 2005). Paoli et al. (2008) found only a modest increase in above-ground biomass across a 16-fold range of extractable P in a single watershed in Borneo. However, there were big changes in floristic composition and forest structure across this same gradient (Paoli et al. 2007), and a large increase in ANPP, implying, in turn, a large increase in turnover in order to maintain a similar biomass (Paoli and Curran 2007) (Fig. 6.3). It appears that the nutrient gradient is sorting species by nutrient-use strategy, with fast-growing species competitively dominant on the richest soils and species with slower growth, but more efficient uptake and use of nutrients, dominant on poor soils. Interestingly, a comparison of basal areas in mature forests across Borneo shows that biomass is *lower* on the most P-rich soils, suggesting that the productivity gains from increasing nutrient availability may eventually be more than offset by mortality losses, resulting from increasing dominance by fast-growing, shorter-lived tree species (Paoli et al. 2008) (Fig. 6.4).

More generally, the high variability in N:P ratios among co-existing plant species suggests a diversity

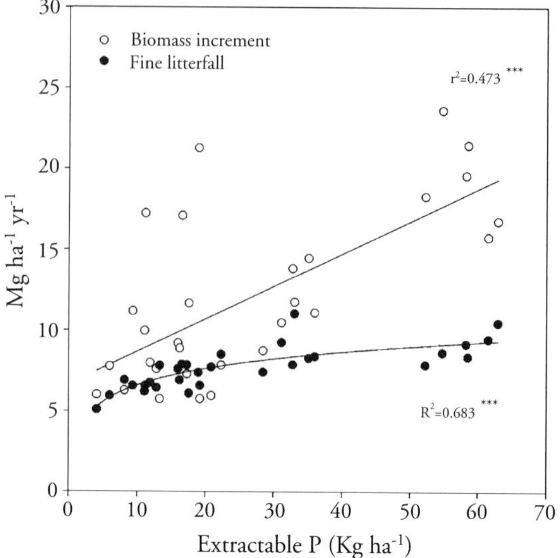

Figure 6.3 The relationship between extractable soil phosphorus and the two major components of above-ground net primary productivity, fine litterfall (•) and stem biomass increment (∘), in lowland tropical rain forest at Gunung Palung National Park, Kalimantan. From Paoli and Curran (2007).

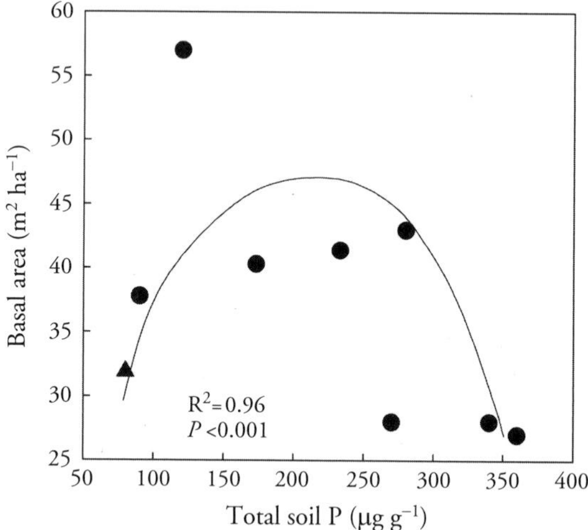

Figure 6.4 The relationship between woody plant basal area and soil phosphorus in mature lowland rainforests in Borneo. The data suggest a significant unimodal relationship (R^2 = 0.96, P <0.001 for a quadratic regression forced through the origin). From Paoli et al. (2008).

of physiological responses to P-poor soils, which may result in different degrees of limitation (Cai and Bongers 2007; Townsend et al. 2007). P deficiency is a major limitation for tropical agriculture, but there are very large differences, both within and between crop species, in their ability to grow on P-deficient soils (Rao et al. 1999). The chemistry of P availability in soils is extremely complex and only partly understood, so it is possible that the 'unavailability' of P in tropical soils has been exaggerated. Dissolved inorganic phosphate, which is the form taken up by plants, is generally found in exceedingly low concentrations in tropical soils, but its turnover rate can be rapid (Turner 2008).

Plants, and/or their mycorrhizae, and/or free-living rhizosphere microorganisms, can potentially influence P availability by secreting extracellular phosphatase enzymes that release inorganic phosphates from organic P forms and by producing organic acids that solubilize iron phosphates (Chapin et al. 2002, Turner 2008). Extracellular phosphatases are N-rich, so the ability to invest N in P acquisition may explain why N-fixing trees are widespread in supposedly P-limited tropical forest ecosystems (Houlton et al. 2008). N-fixation is rare in pioneer trees in TEA, but seedlings of *Pinus massoniana*, the most important pioneer tree in the subtropical evergreen broad-leaved forests of eastern China, greatly increased their secretion of low molecular weight organic acids in solution culture under P-deficient conditions (Yu et al. 2008b). Moreover, soils in the humid tropics can provide very heterogeneous environments, varying in critical physico-chemical properties, over short spatial and time-scales, with consequences for the mobilization of P from forms that have usually been considered unavailable (Chacon et al. 2006). Turner (2008) argues that the variety of organic and inorganic compounds of P in the soil, coupled with the diversity of ways in which plants can access these compounds, provides an opportunity for resource partitioning and thus species coexistence on P-limited soils.

The role of the almost ubiquitous mycorrhizal associations may be critical here (Alexander and Lee 2005). However, while the mechanisms of inorganic P-capture by mycorrhizal fungi are well understood, the ability of tropical mycorrhizae—both arbuscular (AM) and ectomycorrhizal (ECM)—to access the much larger pools of organic P in soils, is less clear at present. Moreover, there is evidence for considerable functional diversity among mycorrhizal fungi and this could interact with the diversity of host plants in complex ways. Plants pay for their mycorrhizae with significant amounts of fixed carbon, but it is important to remember that enhanced access to P is not the only potential benefit of this relationship, with both increased N-capture and non-nutritional benefits, such as defence against root pathogens, among other possible benefits.

P availability tends to covary with the availability of other rock-derived nutrients (e.g. John et al. 2007), as well as other soil parameters, so field correlations alone cannot easily distinguish the effects of P from a general influence of soil fertility. Unfortunately, P-fertilization experiments are very difficult to perform in intact forest, while experiments on seedlings, particularly when done in pots, can be difficult to interpret. In general, however, the results of P-fertilization experiments in TEA do not provide support for widespread P-limitation of seedling growth, at least among mycorrhizal dipterocarps (Burslem et al. 1994, 1995; Bungard et al. 2002; Palmiotto et al. 2004; Brearley 2005; Brearley et al. 2007b).

6.3.3 Essential cations

The attention given to N and P in studies of tropical forests largely reflects theoretical expectations of an N–P dichotomy among ecosystems in nutrient limitation. However, the neglect of other nutrients in many studies risks making the N–P dichotomy a self-fulfilling prophecy. The essential cations, K, Ca, and Mg are, like P, derived from primary rock minerals, but they are more mobile in soils than P, and thus more easily depleted, and are required in larger amounts by plants. Near the coast, the input in rainfall from marine aerosols may be sufficient to prevent nutrient limitation by these cations (Chadwick et al. 1999), but this needs more study in TEA. Associations between tree-species distributions on various scales and the availability of one or more of Ca, K, and Mg, have been widely reported (Baillie and Ashton 1983; Amir and Miller 1990; Potts et al. 2002; Paoli et al. 2006; John et al. 2007), as have responses in seedling-fertilization experiments (Brearley 2005; Kaspari et al. 2008). Clearly it is very important that both observational and experimental studies include more than just N and P.

6.3.4 Micronutrients

The essential micronutrients, boron (B), copper (Cu), iron (Fe), manganese (Mn), molybdenum (Mo), and zinc (Zn) are required in tiny amounts, but deficiencies of each of them have been reported in agricultural systems and plantations in TEA. The large agricultural literature on micronutrient-limitation of crop yields in the tropics, contrasts with the almost total silence on this topic from ecologists. A recent study in the Neotropics, however, found that associations between species-composition and variation in soil nutrients were just as strong for B, Cu, Fe, and Zn, as they were for N and P (John et al. 2007). Moreover, at a site in Panama, micronutrient addition (B, Ca, Cu, Fe, Mg, Mn, Mo, S, and Zn) increased litter-decomposition rates by 81%, compared with 33% for P addition (Kaspari et al. 2008). At this site, Mo limits N-fixation by free-living bacteria (Barron et al. 2009).

6.3.5 Aluminium, manganese, and hydrogen

Aluminium makes up 7% of the Earth's crust, mostly as harmless oxides and silicates. In soils with a pH <5.2, however, Al^{3+} and, less commonly, Mn^{2+} can reach toxic concentrations, with plants varying widely in their degree of tolerance, both within and between species (Kochian et al. 2004). Again, this is an issue that has received more attention in the agricultural, than the ecological, literature. In the Neotropical study mentioned above, Al concentrations had as strong an influence on plant distributions as soil nutrients, but in a different direction (John et al. 2007). Interestingly, while most plants that grow on soils with high concentrations of available Al exclude it from their above-ground parts, a minority of species are accumulators, with Al concentrations in their leaves that are as high as, or higher than, those of the major nutrient cations (Jansen et al. 2002). Examples from TEA include many Melastomataceae (e.g. *Melastoma* [Fig. 6.5], *Memecylon*, *Pternandra*) and Rubiaceae (e.g. *Urophyllum*), and some or all species in a scattering of genera in other families (e.g. *Adinandra*, *Anisophyllea*, *Aporusa*, *Baccaurea*, *Camellia*, *Eurya*, *Helicia*, *Litsea*, *Pentaphylax*, and *Symplocos*). Accumulation of a range of other metals is also known (Cd, Cr, Cu, Mn, Ni, Se, and Zn), with a role in defence against herbivores or pathogens the most plausible of several theories to explain this behaviour (Boyd 2004).

Figure 6.5 The common pioneer shrub, *Melastoma malabathricum*, not only tolerates the high levels of aluminium in acid soils, but appears to require aluminium for optimal growth and accumulates more than 10 mg per g dry weight in its leaves and roots (Watanabe et al. 2008). Photograph courtesy of Hugh T.W. Tan.

It is important to note, however, that the strong influence of soil pH on the availability of both nutrient and toxic cations can make it difficult to identify the specific factor or factors that are most important in any particular case. Indeed, in soils that are not buffered by Al or Fe oxides (such as many Spodosols and Histosols; see Chapter 2), pH can fall so low (<4.2) that hydrogen H^+-ion toxicity may become the major growth-limiting factor, as has been suggested for heath forests (Luizão et al. 2007).

6.4 Research needs

There are big gaps in our current understanding of the movements of energy and nutrients through regional ecosystems. Below-ground stocks and processes have been largely ignored and other minor, but potentially significant, components, such as herbivory and VOCs, have rarely been measured. Many measurements come from small plots, which may not be representative of the forest as a whole. Nutrient studies have focused on N and P, or on these plus K, Ca, and Mg, while the potential roles of micronutrients, aluminium, and hydrogen ions have been neglected. Those data we do have were obtained by multiple different methods from many different sites, making it impossible to identify which results are of general significance and which are site-specific.

Given the potentially important role of ecosystem processes in mitigating—or exacerbating—the impacts of anthropogenic climate change, air pollution, and a host of other human disturbances (see Chapter 7), there is an urgent need for a more coordinated approach to data collection. Plot-scale studies, replicated across climatic, soil, and disturbance gradients, need to be combined with both shoot- and root-scale studies of representative plant species, and landscape- to regional-scale studies using satellite data. This may sound impossibly ambitious, but the AsiaFlux network of eddy flux towers (Huete et al. 2008), provides a possible model for regional collaboration, and carbon offsets may provide part of the finance (see Section 8.2.3). Making use of the CTFS network of large plots (see 4.2) would allow ecosystem processes to be linked to forest dynamics—particularly important for understanding carbon budgets and the impacts of climate change.

Another neglected area is the influence of plant nutrition on herbivores and higher levels in the food web. Terrestrial herbivores have much (>5–10 x) lower carbon to nutrient ratios than most plant materials (Elser et al. 2000). This suggests that herbivores, in general, will be nutrient-limited, although the massive energy, and thus carbon, requirements of endotherms may reduce the impact of this apparent disparity, as long as the excess carbon is metabolizable (Klaassen and Nolet 2008). Animals also have very different micronutrient requirements. In particular, sodium (Na) is not an essential nutrient for most plants, while it is for animals, particularly vertebrates, some of which show an innate Na hunger (Karasov and Martínez del Rio 2007). Many vertebrate herbivores visit so-called 'salt licks' and supplementation of a mineral-poor plant diet with sodium, calcium, and other minerals is the most plausible explanation for this behaviour (See 5.2.1). Social bees can also be attracted by salt solutions (Roubik 1996).

Differences in herbivory in the tropics have usually been attributed to differences in the contents of phenolics and other carbon-based defences, but the possible role of element ratios in food choice has rarely been investigated. One exception is the evidence that fruit bats eat leaves as a source of calcium (Nelson et al. 2005). It has also been suggested that Borneo's nutrient-poor soils may be the ultimate explanation for both the island's relatively few species of large mammals (e.g. no tigers or leopards) and the generally smaller size of the large-mammal species present, than their closest relatives elsewhere, but the actual mechanism is unclear (Meiri et al. 2008). Herbivory studies across soil-fertility gradients could be very interesting, as would the impact of long-term supplementation with mineral nutrients.

CHAPTER 7

Threats to biodiversity

7.1 Introduction

Our views of the impact of people on biodiversity depend very much on what baseline is used for comparison. Human impacts are often underestimated in TEA because comparisons are made with the earliest written descriptions, which are most often from the twentieth century. But even in those parts of China where written records extend back more than a millennium, few managers of nature reserves recognize elephants and rhinoceroses as among the former inhabitants. As a result of the region's complex history (see Chapter 1), any baseline is fairly arbitrary, but the middle Holocene, 5000 years ago, is a reasonable compromise: changes in the region's biota before this time cannot be confidently attributed to people, however suggestive the circumstantial evidence, while the spread of agriculture and agricultural populations afterwards resulted in the ecological transformation of a steadily increasing proportion of the land area. It is largely an inferred baseline, since we have no detailed information from this period, but historical records from sparsely populated areas are consistent with this picture.

Five thousand years ago, TEA was almost entirely forested and had a total human population of only a few million people. These forests supported at least 30 large mammal species (>45 kg), with up to a dozen coexisting in one area, and a host of other plant and animal species—in total, around 15–25% of the entire global terrestrial biota (see 3.6). Today, one billion people inhabit the region, more than half of the forest has gone, more than half of what remains has been logged or otherwise degraded, and the great majority of forest, logged or unlogged, has lost most or all of its large mammal fauna to hunting. The non-forest areas are increasingly dominated by monoculture crops and plantations, or by urban areas and their associated infrastructure, all of which are hostile environments for forest-dependent plants and animals. There have been very few recorded extinctions, but much of the native biota is confined to rapidly shrinking areas, with more or less intact forest communities; while a few native species, and an increasing number of exotics, have massively expanded their populations.

7.2 Don't trust the numbers!

This chapter quotes a lot of different numbers from a variety of historical sources and global databases. Beware! Some of these statistics are probably as good as they could be, but they can still mislead, as much as they inform, particularly when 'sanitized' by being quoted outside their original context. For a start, statistics are only available for things that are fairly easily measured or estimated, and these are often not the variables of greatest ecological interest. They may serve as reasonable proxies for the variables of interest, but this is usually impossible to prove. For example, the total forest area can be measured with a fair degree of accuracy from satellite data, but the area of forest with intact communities of large vertebrates cannot.

A second major problem is that most of the recent statistics are gathered by national agencies in the countries in question. Despite the best efforts of international agencies, the definitions of the variables and the methods of data collection often differ widely among countries, as does the degree of accuracy. Data for rates of change within one country are likely to be more reliable than comparisons between countries, but this is by no means always true.

Finally, the best statistics are usually only available at the whole-country level, which is rarely a unit of ecological interest. Countries are an ecological factor in TEA, as is obvious from the changes in forest cover when certain borders are crossed, such as that between Myanmar and Thailand, or Brunei and the Malaysian state of Sarawak. But countries are also by no means uniform. China is vast and mostly outside the tropics; Thailand and Vietnam spread over very broad latitudinal ranges; and Malaysia, Indonesia, and the Philippines are divided between islands. Rural–urban contrasts are also huge, with the former dominating direct ecological impacts but the latter dominating most national statistics.

7.3 The ultimate causes

7.3.1 The growth of human populations

Human population growth (Fig. 7.1) is the ultimate driver of most adverse impacts on biodiversity. In theory, however, human impacts could increase, even with stable or declining populations, if the amount each individual consumes increases and/or the technologies used to acquire each unit of consumption have a greater impact. In simple terms, the environmental impact of a human population can be expressed as the product of these three factors: the population size, the per capita consumption, and the impact per unit of consumption (Ehrlich and Goulder 2007).

Hunter-gatherer population densities in tropical forests were probably always below one person per square kilometre, except in exceptionally productive areas, such as coastal and estuarine habitats, and most were probably much lower than this (Robinson and Bennett 2000). This suggests that TEA could have supported, at the most, a few million people before the introduction of agriculture. Hunter-gatherer populations can increase only by invading new lands, as happened 40,000–50,000 years ago, or through improvements in technology, such as the ability to target arboreal prey with blowpipes or bows, or the ability to process otherwise toxic plants. Ultimately, however, the inaccessibility or inedibility to humans of most primary production in tropical forests, and the meagre secondary production, sets a limit that could not be exceeded even with the best modern technology.

Agriculture can support much higher population densities and agricultural populations can, in theory, continue to expand as long as there is additional land to clear. Except in a few favoured areas, however, population growth in TEA remained very slow and much potential agricultural land remained unexploited until recently. Disease and warfare seem to have been the main factors limiting growth in agricultural populations in TEA, although hard evidence is in short supply (Perkins 1969; Reid 1987). By 1600, there may have been around 80 million people in the region. Most were

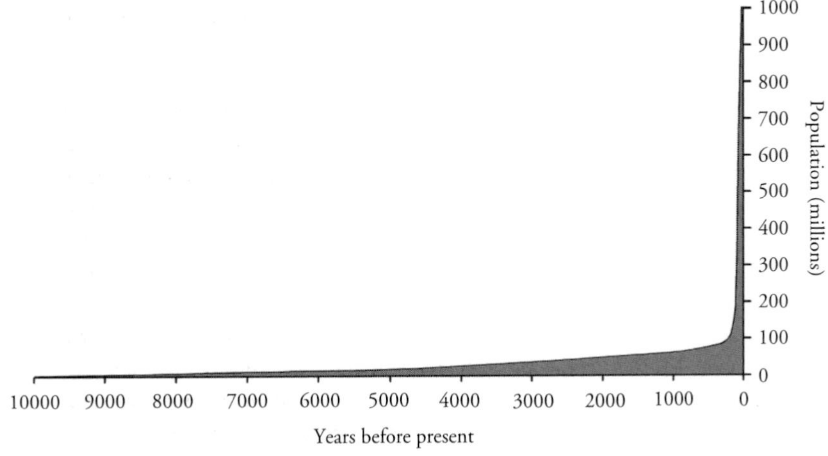

Figure 7.1 The human population of Tropical East Asia from 10,000 years ago until the present day. Extrapolated from multiple sources.

in the densely populated parts of south-eastern China, while South-East Asia's population was concentrated in a few large trading cities and in scattered pockets where soil, climate, and history favoured intensive rice-cultivation. These included the Red River (Sông Hóng) delta, in Vietnam, and parts of central Myanmar, central and east Java, Bali, south Sulawesi, and central Luzon (Reid 1987). In between these areas of dense settlement, the forest was largely intact, occupied by low-density populations, living by shifting cultivation and/or by collecting forest products for trade.

Population growth remained low in South-East Asia in the seventeenth and eighteenth centuries, but then increased dramatically from the nineteenth century. The reasons for this are not entirely clear, but probably include the end of the incessant warfare that characterized much of the region in pre-colonial times. China's population followed its own path during this period, however, controlled largely by internal political upheavals, with population growth concentrated in the previously under-populated south-west. Growth rates finally exploded everywhere in the twentieth century, particularly after 1950, as death rates rapidly declined, following the widespread adoption of public health measures, vaccinations, and modern medicine, while birth rates remained high. In some areas, such as the Philippines and Malaysia, the population quadrupled in only 50 years. Today there are approximately one billion people in TEA—15% of the world population.

This period of explosive population growth is now over and birth rates are declining everywhere, although life-expectancies are still increasing (Husa and Wohlschlägl 2008). The total fertility rate—the average number of children that would be born to each woman if age-specific fertility rates remained as they are at present—is still relatively high in Laos, Cambodia, the Philippines, and Malaysia, but has fallen below long-term replacement (2.1–2.15 children per woman) in China, Vietnam, Myanmar, Thailand, Singapore, and Brunei, with Indonesia not far behind (Table 7.1). Population growth rates will remain high for several decades, however, as the hundreds of millions of young people, who are the legacy of the explosive growth phase, have children themselves. The population of TEA will, therefore, continue to rise until at least 2050, when projections suggest it may have reached 1.2 billion. Moreover, there is no likelihood of a deceleration in the growth of the second component in the impact formula, per capita consumption: an aspiration to the middle-class life-style seen on television is one of the unifying features of modern TEA.

There are few data on the third component—the impact per unit of consumption. Crop yields per unit area are increasing, so less deforestation is required per unit of consumption, but other inputs—pesticides, fertilizers, fossil fuels—are also increasing, so the overall impact is hard to assess. The carbon intensity of the global economy—the carbon dioxide emissions to produce a unit of economic activity—has been declining for decades, but has recently started to increase (Canadell et al. 2007).

Overall, it is clear that TEA has a long way to go before human impacts stabilize and, at least initially, this will be at a considerably higher level than today. Things will get considerably worse for biodiversity before they start getting better.

7.3.2 Poverty

TEA as a whole is getting rapidly wealthier, but the distribution of this wealth is highly uneven, both among and within countries (Table 7.1). All countries except Brunei, Malaysia, Singapore, and Thailand still have substantial numbers of people living on less than US$1 per day and all but Brunei, Indonesia, Malaysia, Myanmar, and Singapore report that more than 10% of their populations are undernourished. The relationship between poverty and biodiversity is complex. On the one hand, poor people in rural areas are more likely to harvest wild species for food, medicine, firewood, timber, and other resources, both to meet subsistence needs and to raise cash for other necessities, such as school fees (e.g. Yonariza and Webb 2007; Zackey 2007). On the other hand, poor people consume fewer natural resources per capita than richer people and lack both the financial capital and the political influence needed for commercial logging and large-scale forest clearance. Poor people are more likely to hunt, but rich—at least, relatively rich—people provide the major market for those wildlife products whose high prices reflect rarity and/or illegality (Corlett 2007a).

Table 7.1 Statistics relating to human populations in Tropical East Asia (TEA), by country, for the region as a whole, for the region without China, and for other representative tropical countries.

Country	Population (million)	Population density (million)	Urban Population (%)	Population growth rate (%)	Total fertility rate	Projected Population (million)	GDP per capita (PPP)	Corruption perceptions index (high = good)	Proportion of undernourished people (%)
Year(s)	2005	2005	2005	2000–5	2007	2050	2006	2006	c.2004
China (all)	1313	137	40	0.7	1.8	1409	7800	3.5	9
China (south)	438	273	–	0.7	1.8	470	–	–	–
Brunei	0	65	74	2.3	2.0	1	25,600	NA	5
Cambodia	14	77	20	1.8	3.1	25	2800	2.0	26
Indonesia (all)	226	119	48	1.3	2.4	296	3900	2.3	17
Indonesia (west)	210	156	–	–	–	275	–	–	–
Laos	6	24	20	1.6	4.6	9	2200	1.9	19
Malaysia	26	78	67	2.0	3.0	40	12,800	5.1	5
Myanmar	48	71	31	1.1	2.0	59	1800	1.4	19
Philippines	85	282	63	2.1	3.1	140	5000	2.5	16
Singapore	4	6336	100	1.5	1.1	5	31,400	9.3	–
Thailand	63	123	32	0.8	1.6	67	9200	3.3	17
Vietnam	85	256	26	1.5	1.9	120	3100	2.6	14
TEA (ex China)	541	140	–	–	–	740	–	–	–
TEA	989	180	–	–	–	1200	–	–	–
Brazil	187	22	84	1.4	1.9	254	8800	3.5	6
DR Congo	58	25	32	3.0	6.4	187	700	NA	76
PNG	6	13	13	2.1	3.8	11	2700	2.0	–
Madagascar	19	32	27	2.8	5.2	44	900	3.2	37

Data from the on-line databases of the Food and Agriculture Organization of the United Nations (FAO), the United Nations Population Division, the International Monetary Fund, and Transparency International.

7.3.3 Corruption

In much of TEA, corruption—the illegal use of public office for private gain—has helped create societies in which wealth and power are synonymous, to the detriment of both development and biodiversity (Table 7.1). In some parts of the region, a 'culture of impunity' puts wealthy people, with the right political connections, effectively above the law. In Cambodia, for example, large-scale, illegal logging operations have been dominated by relatives and friends of the prime minister and other senior officials (Global Witness 2007). In Laos, corrupt military and local officials facilitate illegal log exports to Vietnam (EIA/Telepak 2008), while in parts of Indonesia, military involvement in logging has undermined the rule of law (Human Rights Watch 2006). The flagrant disregard for laws shown by corrupt elites is then used by people at the other end of the power spectrum to justify their own illegal behaviour (e.g. Zackey 2007).

7.3.4 Globalization

Globalization is a two-edged sword, in relation to biodiversity. On the one hand, the demands of global markets are increasingly the major drivers of deforestation and logging in the region; but, on the other hand, increasing environmental awareness in the developed-world markets may create a demand for better environmental behaviour among producers (Nepstad et al. 2006; Butler and Laurance 2008). Big corporations also provide a far smaller number of targets for pressure from conservation groups, than do the tropical rural poor. Unfortunately for this optimistic view, the major markets for many tropical products, including timber and palm oil, are now in the rapidly developing economies of China and India, where environmental standards are still low and from where pressure on producers is currently less likely (see also Section 8.2.7). In general, globalization is most advanced in commodities, such as timber, biofuels, and animal feeds, where a number of alternative, more or less substitutable, sources are available. In such cases, changes in the prices paid for the end product can have large and often unpredictable impacts on deforestation choices in the tropics, within months.

Figure 7.2 Schomburgk's Deer, *Rucervus schomburgki*, inhabited open, swampy plains in central Thailand until its extinction in 1932 as a result of hunting and the loss of its habitat to commercial rice production. This photograph was taken in 1911 in Berlin Zoo and was obtained from Wikipedia.

7.4 The major threats to biodiversity

7.4.1 Habitat loss

Only the most adaptable of wild species can survive the complete destruction of the habitat to which they have become adapted over evolutionary time. Habitat destruction on this scale is, therefore, an almost unselective threat to biodiversity. The best example of the near total loss of a habitat in TEA is probably the replacement of the grass-dominated vegetational mosaic of seasonally inundated, river floodplains, by agriculture—principally rice (Dudgeon 2000). Although these sites must have attracted attention from the beginning of agriculture, their complete conversion to cropland has occurred only in the last 400 years, leading to the extinction of endemic deer taxa, including Schomburgk's Deer, *Rucervus schomburgki* (Fig. 7.2), in Thailand and Père David's deer, *Elaphurus davidianus* (Fig. 8.13), in eastern China, and drastic reductions in the populations of numerous other species. Riverine grasslands are exceptional in TEA, however, where almost everywhere has a climate that can support some form of forest (see Chapter 2).

7.4.2 Deforestation

For the great majority of TEA, therefore, the major threat to biodiversity comes from the conversion of forest to non-forest habitats, including crops, anthropogenic grasslands, and urban areas. Overall, TEA today is only around 40% forested (32% in southern China, 44% in South-East Asia), meaning that more than half of the original forest cover has gone (Table 7.2; Fig. 7.3). On a country basis, forest loss ranges from 30–40% in Laos, Cambodia, and Malaysia, to 70–80% in China, Thailand, and the Philippines. These figures underestimate the extent which the original forest cover has been lost, however, since a wide range of anthropogenic ecosystems, including regrowth, logged forests, and plantations, are included as 'forest' in the available statistics and it is not possible to consistently disaggregate them. Even if it were possible to get reliable estimates of primary forest loss, these would still greatly understate the pressures on lowland forests, which in the equatorial region at least, support the great majority of the biodiversity. Vast areas of lowland TEA have no remnants of the original forest at all. Most surviving forest is at high altitudes—with the lower limit set by local agricultural practices—or, to a lesser extent, on extreme substrates, such as limestone, sand, or deep peat, which have limited agricultural potential.

Table 7.2 Statistics relating to forests in Tropical East Asia (TEA), by country, for the region as a whole, for the region without China, and for other representative tropical countries.

Country	Land area (1000 km²)	Forest area (1000 km²)	Forest (%)	Annual forest loss (%)	Annual log production (million m³)	Population (million)	Population density (million)	Road density (km/1000 km²)
Year(s)	2005	2005	2005	2000–5	2006	2005	2005	c.2005
China (all)	9326	1973	21	−2.2		1313	137	201
China (south)	1620	518	32	(−2.2)	3.3	438	273	400
Brunei	5	0	53	0.7	NA	0	65	693
Cambodia	177	104	59	2.0	0.1	14	77	216
Indonesia (all)	1826	885	49	2.0	26.0	226	119	202
Indonesia (west	1342	564	42	(2.0)	0.1	210	156	NA
Laos	231	161	70	0.5	NA	6	24	135
Malaysia	329	209	64	0.7	27.0	26	78	301
Myanmar	658	322	49	1.4	4.1	48	71	41
Philippines	298	72	24	2.1	0.9	85	282	671
Singapore	1	0	3	0.0	0.0	4	6336	4734
Thailand	512	145	28	0.4	5.2	63	123	112
Vietnam	325	129	40	−2.0	NA	85	256	683
TEA (ex China)	3878	1706	44	1.2	c. 64.0	541	140	–
TEA	5500	2224	40	0.4	c. 67.0	989	180	–
Brazil	8457	4777	57	0.6	22.9	187	22	207
DR Congo	2268	1336	59	0.2	0.1	58	25	67
PNG	453	294	65	0.5	2.2	6	13	44
Madagascar	582	128	22	0.3	0.1	19	32	86

Data from the on-line databases of the Food and Agriculture Organization of the United Nations (FAO) and ITTO (2006).

Estimates of the current rates of forest loss (Table 7.2) suffer from the same problems as the figures for total forest cover, with the recent statistics confirming *increases* in the forest cover of China and Vietnam, despite continued loss of natural forests in both countries (Meyfroidt and Lambin 2008). Cambodia, Indonesia, and the Philippines show the highest rates of loss (*c*.2% per year), despite the great differences between these countries in both the current extent of forest and the human population density. The high percentage loss from Indonesia, which has a quarter of the region's forest, is particularly disturbing, as is the continued loss in the Philippines, where past deforestation and high endemism (see 3.9.9) means that numerous forest-dependent species are under threat of global extinction. Within these countries, there are 'hotspots' of particularly intense clearing activity, including Riau province in Sumatra and the area in Cambodia along its border with Thailand (Hansen et al. 2008).

The overall rate of forest loss in South-East Asia is currently the highest in the global tropics (Laurance 2007a), although per capita rates of loss are relatively low (Wright and Muller-Landau 2006a). Note also that the very high biomasses of lowland dipterocarp forests in South-East Asia (see 6.2.2), means that carbon emissions per hectare, as a result of deforestation, will be higher than elsewhere (Paoli et al. 2008). The great majority of forest clearance is initially for agriculture, with urbanization spreading later into previously agricultural areas. Forest is cleared by individuals, families, villages, and local, national, regional, and global businesses, but the main driver in most areas is now the planting of cash crops on an industrial scale. The most important of these are oil palm (e.g. Fitzherbert et al. 2008; Koh and Wilcove 2008) (Fig. 7.4) and rubber (e.g. Li et al. 2007; Stone 2008b), although a wide range of other crops, including bananas, cashew nuts, cassava, cocoa (Siebert 2002), coconuts, coffee (WWF

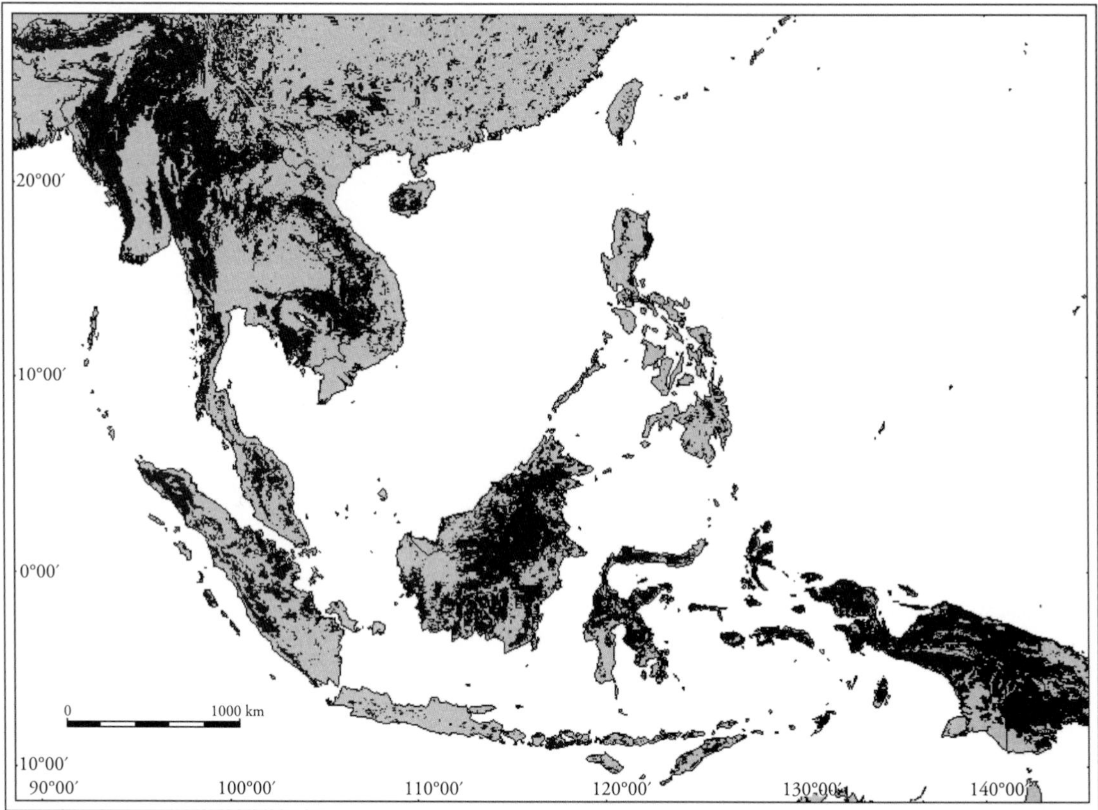

Figure 7.3 Forest cover in Tropical East Asia in the 1990s. Note that Indonesia, in particular, has lost a considerable amount of forest since then. Map from the World Conservation Monitoring Centre, reproduced from Corlett (2005).

2007), sugarcane, tea, and pulp-wood plantations are locally important.

Conversion of forest to agriculture has a massive impact on forest biodiversity, but the cleared areas retain an ability to regenerate by secondary succession, with the rate of recovery dependent on the spatial extent of the cleared area and the degree of damage to the soil (see Chapter 4). The slowest recovery is on land that has been degraded by prolonged cultivation. In Singapore, such sites still support a floristically impoverished forest a century after abandonment (Turner et al. 1997). The present and future value of these secondary forests, as a substitute or supplement to primary forest in the maintenance of native species, is one of the key questions for predicting the future of tropical-forest biodiversity and one that cannot be answered satisfactorily from existing data (e.g. Dunn 2004; Wright and Muller-Landau 2006a, b; Barlow et al. 2007; Bowen et al. 2007; Gardner et al. 2007).

7.4.3 Fragmentation

If the surviving forest in each country was in a single, large, undisturbed block, the ultimate impact on biodiversity would be relatively easy to assess. In reality, however, deforestation has been much less tidy than this, with much of the remaining forest cover in scattered patches of widely varying sizes, shapes, and degrees of isolation and disturbance. Small forest remnants in deforested landscapes have allowed the survival of local-endemic, forest-dependent species that would otherwise have been lost, but this fragmentation of the remaining forest area also brings a host of additional problems. Research, principally in the

Figure 7.4 The replacement of hyper-diverse lowland rainforest by a monoculture of oil palm in Sumatra. Photograph © Ardiles Rantes, Greenpeace.

Neotropics, has identified a wide range of adverse impacts, associated with distance from the forest edge (Fig. 7.5), the total area of the fragment, and/or its degree of isolation from larger areas of forest (Turner 1996; Laurance et al. 1997, 2002; Sodhi et al. 2007; Laurance 2008c). There has been much less research in TEA, but this has tended to confirm the major general conclusion from the Neotropics, that fragmentation greatly adds to the adverse impacts of deforestation (e.g. Zhu et al. 2004; Benedick et al. 2006). Small fragments (<100 ha) can support a surprising proportion of their original flora and fauna for decades after isolation, and are thus well worth conserving, if there is no other forest (Turner and Corlett 1996). But even relatively large fragments (>10 km^2) lose sensitive species (e.g. Brühl et al. 2003). Moreover, forest fragments are a relatively new feature of the landscapes of TEA, so the long-term fate of their biota is unknown.

The impacts of deforestation and fragmentation on biodiversity are less severe for those species that can make use of the non-forest matrix, than for those that cannot (Sekercioglu et al. 2002; Ewers and Didham 2006). For species that can move through the matrix, either under their own power or dispersed by some vector, fragments are less isolated and effective population sizes are larger. The ability to forage in the matrix further reduces the adverse impacts of fragmentation, while for the few forest species that can successfully live and reproduce entirely within the matrix, reduced competition may permit an explosive increase in population size and expansion beyond the original geographical range. All this depends on the nature of the matrix habitat, however, with its ability to support forest species likely to increase with its similarity in structure to natural forest. Similarity in structure between the matrix and the fragments also reduces the severity of edge effects and thus increases the effective area of the fragments.

Surveys of the mosaic of villages, gardens, orchards, regrowth, and forest remnants, which characterize areas recently cleared by peasant farmers, may give an optimistic impression of the ability of deforested landscapes to support many elements of the forest biota (e.g. Thiollay 1995; Sodhi et al. 2005a; Round et al. 2006), but this is probably misleading. Clearance continues, forest fragments lose species with time since isolation (with the rate of loss highest in smaller fragments), and remnant native trees are rarely replaced when they die (Corlett 2000). There have been few studies in long-settled landscapes, but these all suggest that only the most adaptable of forest species persist. There is also an

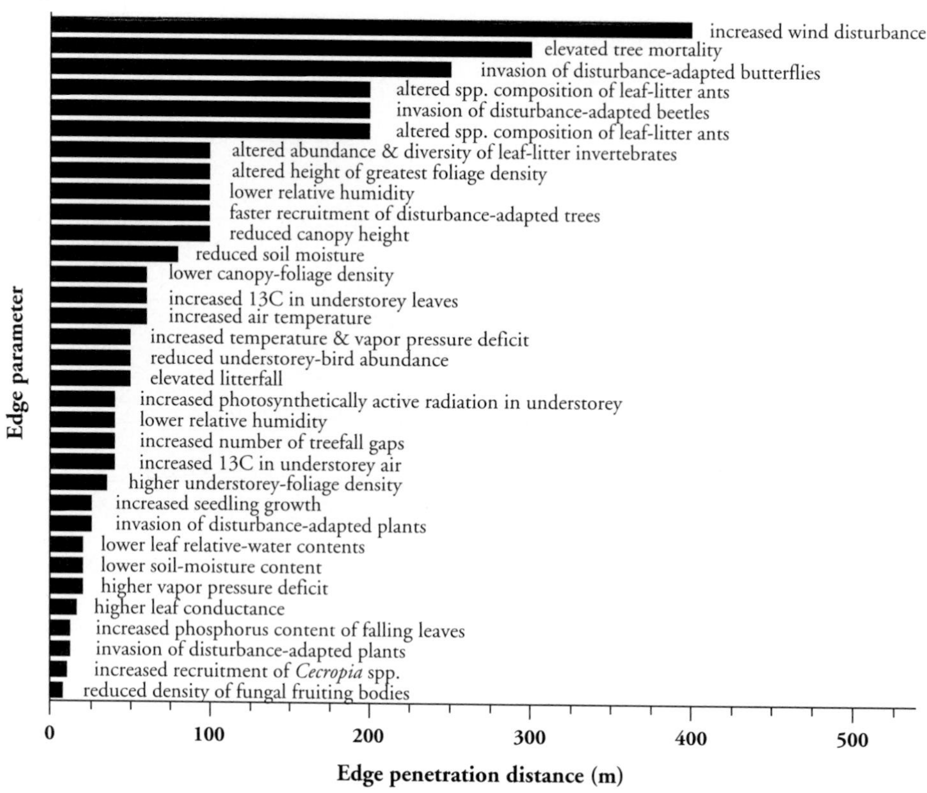

Figure 7.5 Penetration distances of different edge effects into the forest remnants of the Biological Dynamics of Forest Fragments Project in Central Amazonia (Laurance et al. 2002).

increasing tendency, in much of TEA, for traditional complex farming systems to be replaced by simplified monocultures (e.g. Siebert 2002; Hu et al. 2008). Moreover, peasant farmers are no longer the major agents of forest clearance in much of the region, and the vast plantation monocultures that result from modern deforestation support few forest species from the start (Chung et al. 2000; Brühl et al. 2003; Donald 2004; Sodhi et al. 2005a; Aratrakorn et al. 2006). Anecdotal information suggests that the increasingly heavy use of agrochemicals by farmers of all types is also a significant threat to biodiversity in agricultural landscapes, but this problem has received very little attention in TEA.

7.4.4 Mining

Although the total area impacted by mining is much smaller than that impacted by agriculture, the extreme nature of mining impacts, particularly where little or no attempt is made at post-mining rehabilitation, means that the damage is cumulative (Laurance 2008d). Mining also finances roads into areas that were previously protected by inaccessibility. Moreover, the processing facilities associated with many mining operations are often situated next to rivers or on the coast, leading to potentially massive water-pollution problems.

Mining operations in TEA range from large-scale mines using the latest mining technology, through medium-scale mines, which are mechanized but often relatively unsophisticated, down to small-scale 'pick-and-shovel' operations. In general, environmental standards decline along with size and the smallest operations are often illegal and, therefore, completely unregulated (e.g. for Indonesia: McMahon et al. 2000). A major current concern is the potential for the expansion of open-cast mining of Tertiary coals in the rainforest areas of Borneo and Sumatra. Indonesia is already the world's biggest exporter of coal.

7.4.5 Urbanization

The impacts of urbanization are at least as severe as those of mining and cover much larger areas. An estimated 70,000 km^2 is covered by artificial impervious surfaces in TEA—1.3% of the total land area (Elvidge et al. 2007). Urbanization has only recently become extensive enough in TEA to directly threaten biodiversity and no examples of global extinctions resulting from the spread of urban areas have been reported, although damage to coastal, estuarine, and river floodplain habitats has been massive. Urban areas also impact biodiversity indirectly through their roles as a source of air pollution and invasive exotic species, and as a market for forest products (see below). An estimated 45% of the regional population lives in urban areas, and this percentage is growing fast. Three cities—Shanghai, Manila, and Jakarta—have populations exceeding 10 million and several others are approaching this size.

7.4.6 Logging

Official statistics on the timber trade are readily available (ITTO 2006; Table 7.2), but these are an unreliable measure of logging impacts. Most timber harvest in TEA is at least partly illegal (illegal sites, species, tree sizes, logging practices, and/or exports), large volumes of illegally exported timber are 'laundered' through neighbouring countries, and growing internal markets for timber are monitored poorly or not at all. Illegal logging is a massive problem in TEA, particularly in Cambodia, Vietnam, Indonesia, Myanmar, and Laos. Estimates of the percentage of logs harvested illegally in Indonesia, the country with the largest remaining forest area, range from 50 to 80% (Turner et al. 2007). The uncertainties in these and other estimates come partly from the legal ambiguities surrounding the forest industry in many countries in the region. The fact that a logging operation has documents that apparently authorize its operations, says little about its legality. Increasing efforts are being made to remove the ambiguities in the law and to control this illegal harvest and trade, at the national and international level, but the problem is likely to continue in the immediate future.

Indonesia and Malaysia are by far the largest producers in the region, with similar 'official' log-production figures in 2006 (Table 7.2), although these numbers are certainly an underestimate for Indonesia. Myanmar is next in terms of timber production from natural forests, while Thailand produces more from rubber and other plantations. China is also a significant producer of tropical timber, mostly from plantations. However, China's domestic production has been dwarfed by the exponential increase in imports over the last decade: half of all timber traded in the world is now destined for China (Laurance 2008b). China, India, and Japan are the major Asian (and, indeed, global) importers of tropical timber, followed by Thailand and Vietnam, whose domestic wood industries can no longer be supported by declining domestic production.

The harvesting of trees for timber has had both direct and indirect impacts on forests in TEA (Fig. 7.6). The indirect impacts—through fire, hunting, and deforestation—are often much greater, but understanding the direct impacts is important because it is these that determine the potential for sustainable timber production from managed natural forests. In species-rich tropical forests, only a few tree species are usually recognized by the international timber trade, so logging intensities are low. In South-East Asian lowland dipterocarp forests, however, many species are grouped into a few market categories, resulting in more intense logging. The most productive forests in Borneo can have >20 saleable large trees per hectare, most of which are dipterocarps (Sist et al. 2003a). Moreover, domestic markets in the region are typically far less fussy about the species, size, and quality of the logs, which can greatly increase initial logging intensities in accessible forest areas and also encourages re-logging of previously logged areas for smaller, less desirable trees.

The amount of direct damage done to a forest by logging depends on many factors, including topography, soil type, and the adoption (or not) of a variety of techniques intended to reduce damage (see Section 8.5.1), but the most important variable is the

Figure 7.6 A logging concession in Peninsular Malaysia.

logging intensity (i.e. the number of trees harvested per hectare). At high logging-intensities (>*c*.8 trees ha^{-1}), massive damage to the remaining stand appears to be inevitable (Sist et al. 2003a, b). This damage includes dead and injured residual trees, large canopy openings that encourage the growth of climbers and pioneers, as well as increasing fire risk, and soil compaction along the skid trails that can cover a third or more of the total area, leading to decreased infiltration, increased erosion, and slow regeneration. Note that high logging-intensities and high rates of collateral damage mean that carbon dioxide emissions from logging operations in lowland dipterocarp forests will also be much higher than those from logging other tropical forests (Lasco et al. 2006; Paoli et al. 2008).

As mentioned in Chapter 2, most studies of the botanical impacts of logging in lowland dipterocarp forests have been done only 1–6 years after logging, or in older forests that received post-logging management that is rarely carried out today, so the long-term impacts of present-day logging practices are largely unknown (Bischoff et al. 2005). Logging impacts on animals have been better studied. Most animal species survive through a conventional logging cycle, although there can be large changes in community composition (Meijaard et al. 2005, 2006; Cleary et al. 2007; Wells et al. 2007; Meijaard and Sheil 2008). Some taxa and feeding guilds, however, appear to be particularly sensitive to logging, such as terrestrial insectivores, some babblers (Fig. 7.7), hornbills, and most woodpeckers, among the birds, and ground squirrels, sun bears, and some civets among the mammals. Recovery of the affected species can be very slow. Logging impacts in other forests types, including the extensive peat-swamp forests in the Sundaland region and the more seasonal forest types in continental Asia, have received far less attention and there is currently insufficient information available to make any generalizations, except that logging intensities tend to be higher in species-poor forests.

The major types of indirect damage as a result of logging—provision of access to hunters, increased risk and intensity of fires, and deforestation—are covered in separate sections. These indirect impacts are, in theory, preventable, but their frequent association with logging is not a coincidence. Hunting

Figure 7.7 Terrestrial insectivores such as this black-capped babbler, *Pellorneum capistratum*, in Sabah, appear to be particularly sensitive to logging impacts. Photograph courtesy of Jon Hornbuckle.

is often tolerated or even actively encouraged, as a source of meat at remote sites (Bennett et al. 2000). Logging may promote deforestation by providing access to small farmers, but, in Indonesia, the same industrial conglomerates control much of the logging, wood processing, and plantation industries, so the link between logging and deforestation is often more direct, with logging just the first stage in the conversion of rainforest into a plantation monoculture (Barber et al. 2002; Curran et al. 2004). In an additional twist, however, proposals for new oil palm plantations in Indonesia are often used as an excuse for logging in areas that are unsuitable for oil palm (Sandker et al. 2007).

7.4.7 Collecting of non-timber forest products

A huge variety of non-timber forest products (NTFPs) have been harvested from the region's forests for thousands of years and several have been important trade items for much of this time. Bamboos, rattans, resins, and mushrooms are probably the most widely harvested plant NTFPs in TEA, but numerous other items are important locally (Table 7.3). The collection of NTFPs is still a significant source of employment and income for many people in the region, but rising populations and shrinking forest areas have put many of these livelihoods under threat. Conservationists have viewed the harvest of NTFPs as a relatively benign operation, which can produce a cash income from the forest without the adverse impacts associated with timber harvest, but recent reviews have concluded that this optimism is not, in general, justified (Kusters et al. 2006; Belcher and Schreckenberg 2007). Increasing demand leads either to depletion of the resource or to intensification of its production, with an associated increase in adverse impacts on biodiversity. The potential for sustainability will depend on the plant part harvested and the damage done during harvest, but lack of management can undermine almost any production system.

7.4.8 Hunting

As with the harvesting of non-timber plant products, the hunting and trapping of vertebrates could, at least in theory, be a sustainable source of food and income from the forest. However, this is probably only possible in practice where the hunted area is surrounded by a large, unhunted refuge for prey populations: a situation that exists almost nowhere in TEA today. Several groups of nomadic hunter-gatherers persisted into the twentieth century in

Table 7.3 Examples of non-timber forest products in Tropical East Asia—excludes plant species used for firewood and as fodder for livestock.

Type and major uses	Major taxa
Bamboos construction, scaffolding, baskets, handicrafts, edible bamboo shoots	Many genera
Rattans furniture, matting, baskets, etc.	*Calamus, Daemonorops, Korthalsia*
Fibres textiles, cordage, etc.	*Pandanus* spp. and many minor taxa
Exudates latex, resins, gums with many uses	*Agathis, Canarium, Dipterocarpus, Dyera, Palaquium, Payena, Pinus, Sindora*
Non-seed carbohydrates food	*Amorphophallus, Dioscorea* + the stems of various palms
Fruits and nuts food	*Aglaia, Antidesma, Artocarpus, Baccaurea, Canarium, Castanopsis, Durio, Garcinia, Mangifera, Nephelium, Salacia, Syzygium*
Vegetables food	*Champereia manillana, Claoxylon longifolium, Ficus* spp.
Honey food, medicine	*Apis dorsata, Apis* spp., *Trigona* spp.
Edible fungi	Many species
Medicinal plants	Numerous species

Sources: Verheij & Coronel 1991; Dransfield & Manokaran 1993; Dransfield & Widjaja 1995; Flach & Rumawas 1996; Boer & Ella 2000; Brink & Escobin 2003.

TEA and forest farmers have always hunted to supplement their crops, but over the last 50 years, hunting for subsistence has been increasingly overshadowed by hunting for the market (Bennett 2007; Corlett 2007a). Market access encourages hunting beyond subsistence needs and of species for which there is no local use. The hunted biomass today is still dominated by the same species (pigs, deer, and primates) as in the past, sold mostly for local consumption, but a huge range of additional species are targeted to feed a colossal regional trade in wild animals and their parts, for luxury foods, medicines, trophies, raw materials, and pets. Urban wealth is increasingly replacing rural poverty as the main driver for hunting in TEA.

Although trade has encouraged the hunting of a few species, particularly rhinoceroses, for hundreds of years, the breadth and volume of the modern Asian wildlife trade are unprecedented (Corlett 2007a; Singh et al. 2007a, b; Venkataraman 2007) (Fig. 7.8). The use of wildlife in medicinal products is particularly important in TEA because of the wide range of species involved. China has been the main market over the last 20 years and continues to be very important (Zhang et al. 2008a), but there is also a huge demand in other countries in the region, particularly in the increasingly prosperous urban areas (e.g. Venkataraman 2007).

Most mammal species weighing more than 1–2 kg are widely hunted (Corlett 2007a), including large fruit bats (Struebig et al. 2007), and even rats and squirrels are targeted in some areas (Khiem et al. 2003; Wattanaratchakit and Srikosamatara 2006). Indeed, with the decline of larger mammals, squirrels are now the most hunted and traded mammals in Laos (Timmins and Duckworth 2008). There is less information available on the hunting of birds, but large species, such as pheasants (Brickle et al. 2008), pigeons (Gibbs et al. 2001), and hornbills (Kinnaird and O'Brien 2007), are widely hunted and many smaller species are targeted locally for food, medicine, or pets (BirdLife International 2001). Crocodiles and turtles are hunted almost everywhere, while large snakes and monitor lizards are

Figure 7.8 Pangolins (*Manis* species) are being traded illegally in huge numbers from all over South-East Asia to China, where they are sold as a health food and medicine. This photograph shows part of a large shipment of frozen pangolins seized in Hong Kong in 2005–6. Photograph © AFCD, Hong Kong.

also heavily exploited in many areas (Keogh et al. 2001; Shine et al. 1998, 1999; Zhou and Jiang 2005).

All studies of recent hunting in TEA report declines of large-bodied species and all have concluded that current harvest rates are unsustainable for most species (Bennett 2007; Corlett 2007a). The elimination of many large mammals from South China over the last 1–2 millennia was mentioned in Chapter 1, but the same process has been compressed into 50 years or less in much of TEA, and is still continuing with the elimination of progressively smaller species. It should not be surprising that a recent assessment of the conservation status of the world's mammals found that by far the highest density of globally threatened species was in TEA (Schipper at al. 2008). Morrison et al. (2007) estimate that only 1% of the Oriental Region still supports all the large (>20 kg) mammals that it once held. Many areas now support none. The existence of large areas of suitable, but unoccupied, habitat for most of the species that have been eliminated ('empty forests'), is strong evidence that hunting, rather than habitat loss, was the major factor in the decline of these species.

The focus on relatively large-bodied species is likely to maximize the secondary impacts of hunting. The elimination of all large carnivores might have resulted in an explosion of prey populations, were it not that these too are also heavily hunted. The elimination of all the dispersal agents for large fleshy fruits—large birds, primates, large fruit bats, civets, and terrestrial herbivores—from many areas, is probably already having an unrecorded impact on plant regeneration and forest succession (Kitamura et al. 2005; Corlett 2007a, b; Terborgh et al. 2008). Impacts of hunting on seed and seedling predation might seem less worrying but, as discussed in Chapter 4, patterns of seed and seedling mortality are central to current theories of plant species coexistence.

The overall impact of hunting is likely to be an increase in the relative abundance of small-fruited, small-seeded, fast-growing and relatively palatable plant species at the expense of species with large fruits and seeds, and a large investment in anti-herbivore defences (Corlett 2007a, b). This change will be slow in intact but 'empty' forests, since individual plants and local plant populations may persist for decades or centuries without vertebrates, but it will be faster in degraded areas where seed dispersal is essential for establishment. At the landscape level, the result will be a 'pioneer desert' (Martinez-Garza and Howe 2003), with pockets of plant diversity where the original forest was not completely cleared. This is the situation in Hong Kong, where most large vertebrates have been eliminated and the vegetation is now recovering from centuries of human impacts (Hau et al. 2005). Although the flora is still very diverse, most species are rare, and both the seed rain and woody secondary vegetation are dominated by a few plant species that are well dispersed by small passerines (Corlett 2002; Au et al. 2006; Weir and Corlett 2007).

7.4.9 Fires

Fires are a major primary human impact in seasonally dry forests; while in areas without a regular dry season, most fires are in forest that has already been logged or otherwise degraded. Fires can be detected from satellites, so the fire-detection density is a useful indicator of the pervasiveness of human impacts in the region. A comparison between fire-detection densities inside tropical forest reserves and paired contiguous buffer areas found a great deal of variation within TEA, with protected areas in Cambodia and Indonesian Borneo (Kalimantan) particularly ineffective at reducing fire occurrence, while those in Malaysia were largely fire-free (Wright et al. 2007).

As discussed in Chapter 2, the role of natural fires started by lightning in seasonal TEA may have been underestimated, but anthropogenic fires are as old as human occupation of the region. Relatively open, more or less deciduous forests, with a grassy understorey, form a fire-climax over large areas of continental South-East Asia. The near annual fires lit by forest users maintain these forests as flammable, relatively species-poor, low-carbon ecosystems. However, an assessment of the adverse impacts of these fires, which burn millions of hectares each year, is hindered by the lack of control areas with long-term protection from anthropogenic fires. Short-term protection simply increases fuel and fire risk, but long-term protection is expected to shift the structure and species composition in favour of a less flammable, more evergreen, closed-canopy forest.

Lack of control areas for comparison is not a problem in wetter parts of the region, where fires were very rare until recently. ENSO-associated droughts provide conditions suitable for forest fires over much of Indonesia, but the disastrous rainforest fires of 1982–83 and 1997–98 resulted from an interaction between climate, population growth, and public policy (Goldammer 2007; Murdiyarso and Adiningsih 2007; Tacconi et al. 2007). Fire is the cheapest way to clear land for agriculture and is used for this purpose by both smallholders and big corporations (Fig. 7.9). In 1982–83, fires affected around 5 million hectares of land on Borneo, much of it previously logged forest, but also some drought-affected unlogged forest. The 1997–98 ENSO was less severe, but the area burned was larger, as a result of expanded clearance for plantations, totalling around 11 million hectares on Borneo, Sumatra, Java, and Sulawesi. Smoke haze resulting from the 1997–98 fires affected millions of people in the region, with peat fires contributing most of the particulates. Estimates of the carbon dioxide produced vary widely, but plausible assumptions suggest that more than one gigatonne of carbon was released to the atmosphere (Page et al. 2002; Murdiyarso and Adiningsih 2007; Tacconi et al. 2007)—about one-seventh of the global anthropogenic carbon emissions in that year. The smoke haze also reduced photosynthetically active radiation and thus primary production (Hirano et al. 2007). Fires and smoke haze are now an annual phenomenon in the region, but the moderate 2006 ENSO produced another peak in activity.

Whether fire is seen as a cause or a consequence of rainforest degradation is largely a matter of perspective, since the two are so closely intertwined in South-East Asia (Langner et al. 2007). Most fires occur in degraded forests but, in drought years, they may spread further. Severe impacts from a single fire have been recorded for plants, birds, and butterflies in TEA (Adeney et al. 2006; Cleary et al. 2006; Cleary and Mooers 2006; Slik et al. 2008), but the worst damage is done by repeated fires and combinations of fire with other impacts.

Figure 7.9 Clearance fires in Riau, Sumatra, during dry periods contribute to carbon emissions and regional air pollution. Photograph © Vinai Dithajohn, Greenpeace.

7.4.10 Invasive species

Invasive exotic ('alien' or 'introduced') species are not yet a major threat to native biodiversity in TEA, except in human-modified habitats and on some oceanic islands (Corlett 2009a). However, once a species is established in the wild, outside its native range, it is rarely possible to eliminate it, so both the numbers of introduced species and their impacts on native species can only increase. A successful alien survives in the same way as a successful native. Introduced species may, however, be recognizable long after their initial establishment: by being confined to disturbed, anthropogenic habitats, since species adapted to these habitats are most likely to be introduced; by their dispersal-limited spatial distributions, since they have had less time than natives to occupy suitable habitats; by their lower genetic diversities, as a result of their populations being established recently from a few founders; by suffering less damage from species-specific natural enemies (herbivores, pathogens), since these are usually left behind during the introduction process (e.g. DeWalt et al. 2004); and/or by their possession of character traits that do not exist in the native biota, as a result of their selection from a larger, global pool of species. The last two of these—escape from natural enemies and sampling from a larger species pool—could potentially give aliens a competitive advantage over otherwise similar natives.

The invasive-species problem often looks far worse than it really is, because exotic plants are such a conspicuous component of the vegetation of chronically disturbed urban and agricultural habitats, which are the ones we most often see. In most cases, however, exotic plant species are invading natural or semi-natural vegetation only along roads and tracks, and, less commonly, after natural disturbances, such as landslides or typhoons. These species are typically eliminated—though often very slowly—after the disturbance ceases. Even in the highly degraded upland landscape of Hong Kong, exotic plant invasions are largely confined to road- and stream-sides, and unshaded areas disturbed by feral cattle (Leung et al. 2009).

A small minority of plant species, however, have invaded minimally disturbed natural vegetation. In TEA, for example, only a single exotic plant species, *Clidemia hirta* (Melastomataceae) (Fig. 7.10), has invaded lowland rainforest in Singapore (Teo et al. 2003) and at Pasoh, Malaysia (Peters 2001). Both sites are surrounded by exotic-rich anthropogenic vegetation, so lack of propagule pressure is

Figure 7.10 *Clidemia hirta*, a small, Neotropical shrub in the family Melastomataceae, is one of the few alien species that has invaded undisturbed rainforests in South-East Asia. Photograph courtesy of Hugh T.W. Tan.

unlikely to explain the absence of other invaders. A few other plant species have invaded more or less intact natural vegetation elsewhere in the region, notably *Chromolaena odorata* (Asteraceae) in the seasonal tropics (Kriticos et al. 2005), *Eupatorium adenophorum* (Asteraceae) in the subtropics (Zhu et al. 2007b), and *Mimosa pigra* (Mimosaceae) in open freshwater wetlands, although all of these are most abundant at disturbed sites.

There is less information for animals, but the pattern appears to be essentially similar, with exotic ants, earthworms, amphibians, reptiles, birds, and mammals conspicuous in urban and agricultural areas (e.g. Dudgeon and Corlett 2004; Pfeiffer et al. 2008), but rare—or, at least, rarely reported—from most natural vegetation (e.g. Bos et al. 2008). Again there are exceptions, but these are almost all from disturbed, fragmented, and/or peri-urban forests (e.g. Hong Kong: Dudgeon and Corlett 2004) and there have so far been no reports of exotic animals at remote rainforest sites in South-East Asia. An American ambrosia beetle, *Euplatypus parallelus*, has been found in primary forest at Pasoh (Maeto and Fukuyama 2003), but there are disturbed areas nearby. Ambrosia beetles, with their typically generalist feeding habits (see 5.3) and tolerance of extreme inbreeding, are prime candidates for invaders of undisturbed primary forests (Kirkendall and Ødegaard 2007), so further observations on this group would be valuable.

The clearest exception to this general pattern is the widespread invasion of subtropical pine forests in northern TEA by the exotic pinewood nematode, *Bursaphelenchus xylophilus* (Shi et al. 2008b). Of North American origin, this pest is spread by native bark beetles and can kill susceptible pine species, such as *Pinus massoniana* in southern China, within months. Most economic damage has been to plantations, but this species has also spread into natural forests in the region. A range of other exotic forest pests are established in southern China, including the loblolly pine mealy bug (*Oracella acuta*) and the pine needle scale (*Hemiberlesia pitysophila*), but their ability to invade natural forests is unrecorded.

The apparent resistance of intact continental communities to exotic invasion, contrasts with the apparent susceptibility of equivalent communities on oceanic islands. The high diversity of species and functional groups on continents and the resulting efficiency of resource capture provide the most plausible general explanation for this difference, while freeing of resources by disturbance can explain the invasibility of chronically disturbed urban and agricultural habitats (Denslow 2003;

Teo *et al.* 2003; Daehler 2006). TEA has no really remote oceanic islands, but the Ogasawara (Bonin) Islands on the north-east fringe of the region lack native frogs, snakes, and non-flying mammals, as well as many other poorly dispersed groups of animals and plants (see 3.9.4). This has apparently resulted in both vacant niches for invasive species and a native biota that is highly vulnerable to competition, predation, and disease. Introduced tree species threaten the native forest by competitively replacing native trees (Yamashita et al. 2003); introduced honeybees threaten native-pollination systems (Abe 2006); introduced rats predate fruits and seeds (Abe 2007); introduced predators threaten native land-snails (Sugiura et al. 2006; Chiba 2007); an introduced lizard, *Anolis carolinensis* (Fig. 7.11) threatens many native insects (Karube and Suda 2004; Abe et al. 2008); feral cats predate native birds (Kawakami and Higuchi 2002); and feral goats threatened native vegetation until they were controlled (Shimizu 2003).

The other oceanic islands in TEA are less isolated and invasive impacts are generally less obvious. Most examples of actually, or potentially, harmful vertebrate introductions involve the movement of species within TEA, rather than introductions from outside the region. Thirty individuals of the mongoose, *Herpestes javanicus*, were released on Amami Island, in the Ryukyus, in 1979, to control a native poisonous snake (*Protobothrops flavoviridis*), but the expanding population now threatens the endangered endemic rabbit, *Pentalagus furnessi*, as well as native birds, reptiles, and amphibians (Watari et al. 2008). Two species of civet, *Viverra tangalunga* and *Paradoxurus hermaphroditus* are established on Sulawesi and several islands in the Moluccas (Flannery 1995), and the latter species was a Neolithic introduction to Flores (van den Bergh 2008b). Sulawesi already had a native civet, but the introduction of mammalian carnivores to islands that previous lacked them altogether, must have had significant impacts, which have not been studied. The many translocations of snakes within the region (Brown and Alcala 1970; Lever 2003; de Lang and Vogel 2005) may also have had an impact, particularly on smaller islands with an impoverished native snake fauna. The region's largest snake, *Python reticulatus*, for example, is probably introduced in most of the Philippines (Brown and Alcala 1970).

Herbivorous vertebrates have also been widely introduced in TEA. Pigs, macaques, and porcupines were pre-colonial introductions to Flores (van den Bergh et al. 2008b). Within the last century, three

Figure 7.11 The Carolina anole, *Anolis carolinensis*, is native to the south-eastern United States, but has been widely introduced on oceanic islands, including the Ogasawara Islands, where it threatens many native insects. Photograph by Robert Michniewicz, Wikimedia Commons.

species of deer and the Asian elephant have been introduced to the Andaman Islands, while pigs were probably introduced in prehistoric times. Unlike Flores, which had stegodonts until the late Pleistocene, the Andamans had no large native herbivores and the introduced species are having a significant impact on forest structure and species composition (Ali 2004, 2006). Translocations within the region are difficult to detect, unless they occurred in historical times, and it is likely that many more island populations of carnivorous and other vertebrates are introductions than is currently recognized. The evidence for deliberate vertebrate introductions over a period of at least 7000 years on Flores, which is not the most accessible of islands, should serve as a warning (van den Bergh 2008b).

7.4.11 Diseases of wildlife, people, and plants

Pathogens and parasites are an essential component of any natural ecosystem, but emerging infectious diseases (EIDs) are diseases that have recently increased their incidence, geographic range, or host range (Cunningham et al. 2006). Although usually applied to diseases of people, the same criteria can be used to identify EIDs, in both wild and domesticated animals and in plants.

Many of the most spectacular recent EIDs of people originated from wild-animal populations in the tropics and subtropics (Jones et al. 2008), including HIV/AIDS, Ebola, Nipah, SARS, and probably the H5N1 highly pathogenic avian influenza. The emergence of most of these diseases can be attributed, at least in part, to increased human impacts on natural systems: hunting (HIV/AIDS in Africa), the wildlife trade (SARS in China), and the expansion of intensive agriculture into previously forested areas (Nipah in Malaysia). The Nipah virus spread from *Pteropus* fruit bats through pigs to people in Malaysia in 1998–99, killing 105 people, prompting the slaughter of more than a million pigs, and costing the Malaysian economy more than US$500 million (Halpin et al. 2007). The SARS virus outbreak in 2002–3 originated in China, probably also in bats, and spread through crowded wildlife markets to people, eventually killing 774 people worldwide and costing more than US$30 billion. The recent outbreak of highly pathogenic avian influenza appears to have originated from migratory waterfowl populations in East Asia and has subsequently killed >200 people and tens of millions of wild and domestic birds, as well as prompting the culling of hundreds of millions more. Another virus from within the region, the Ebola Reston strain, was detected in long-tailed macaques (*Macaca fascicularis*) exported from a facility in the Philippines, but, although it can infect humans, it does not cause symptoms. The wild reservoir of this virus is unknown, although bats have again been suggested.

There is a risk that pathogens and parasites that spread outside their native range may encounter hosts with fewer parasite-specific defences and, thus, become increasingly invasive (Tella and Carrete 2008). EIDs have not so far been recognized as an extinction threat for wild species in TEA, although the examples of Ebola in West African apes, avian malaria in Hawaiian birds, the fungus *Batrachochytrium dendrobatidis* in amphibians in many parts of the world, and the fungus *Phytophthora cinnamomi* in Australian plants, illustrate the risks. Many less dramatic cases have probably been overlooked. Throughout TEA, deforestation and fragmentation have brought wild populations of native species into close proximity with human-dominated landscapes, where exotic plant and animal species provide both new hosts for native diseases and new diseases for native hosts. The exchange of malarial parasites (*Plasmodium* spp.) between people and non-human primates, when they are brought together, is a good example of this process and its potential dangers (Jongwutiwes et al. 2004; Reid et al. 2006). Bidirectional transmission of other pathogens is also possible where wild macaques interact with people in a recreational setting (Fuentes et al. 2008).

7.4.12 Air pollution and nutrient enrichment

Every organism in TEA today is exposed to an atmosphere that differs significantly in composition from any to which its ancestors would have been exposed. Changes in the concentration of the major greenhouse gases (CO_2, CH_4, N_2O) are considered separately below, while other air-borne pollutants, with potentially significant impacts on biodiversity, are considered in this section. Unfortunately, air-pollution studies in TEA have focused on urban areas and very little is known about pollution in rural areas. As a result, the adverse impacts of air pollution on agriculture, forestry, and natural ecosystems have probably been underestimated. Indeed, there is a global gap in knowledge of pollution impacts on tropical forests (Zvereva et al. 2008).

The major source of air pollution in TEA is the burning of fossil fuels, particularly coal (Fig. 7.12), and the most important primary pollutants are sulphur dioxide (from sulphur impurities in the fuel) and nitrogen oxides (from oxidation of atmospheric nitrogen). Ozone, which is produced from hydrocarbons and nitrogen oxides in the presence of sunlight, is the most important secondary pollutant (i.e. one formed in the air by reactions between primary pollutants). Particulates (aerosols) are derived from a variety of primary and secondary sources. A huge variety of volatile organic compounds (VOCs) are also present, particularly in urban areas, but their impacts on natural systems are unknown.

Sulphur dioxide emissions have risen rapidly in China and the more developed parts of South-East Asia over the last 20 years, largely because of the rise in coal combustion. There is very little data from outside the major urban areas, but total sulphur-deposition levels are in the range where negative impacts on agriculture and natural ecosystems are to be expected in southern China, north-eastern Vietnam, much of Java, and around major cities elsewhere (Siniarovina and Engardt 2005; Aas et al. 2007). Symptoms ranging from decreased tree growth to leaf damage and mass die-back have been attributed to sulphur dioxide emissions in the Pearl River Delta region of Guangdong (Kuang et al. 2008).

Emissions of NO_x (NO and NO_2) from burning fossil fuels and NH_3 from intensive agriculture have also increased greatly over recent decades (Schlesinger 2009). The main impact on natural systems is through wet and dry deposition, which can acidify the soil and change nutrient cycles (see 6.3). Current levels of N deposition in southern China are already very high (10–50 kg N ha^{-1} yr^{-1}) by global standards (Chen and Mulder 2007; Fang et al. 2008) and similar levels probably occur in other densely populated regions of TEA. Studies of the impacts in

Figure 7.12 The major source of air pollutants in most of Tropical East Asia is the burning of fossil fuels, particularly coal.

TEA have only recently started, but these levels are above the threshold for damaging impacts in temperate forests.

Background levels of ozone are rising throughout the region. Again there is very little relevant rural data, but extrapolations from studies elsewhere in tropical Asia suggest that ozone has already reached damaging levels in many areas and will get worse (Ishii et al. 2007; Wang et al. 2007e). There is an urgent need for more research on ozone impacts on natural systems, including free-air fumigation studies of entire communities (Morrissey et al. 2007).

Particulates (aerosols) in the atmosphere have multiple different origins (industry, agriculture, forest fires, and a variety of secondary chemical processes in the atmosphere) and properties, and their direct impacts on plants have received little attention. Whatever the source, however, aerosols tend to reduce the total amount of photosynthetically active radiation (PAR, 400–700 nm) reaching the top of the vegetation, while increasing the fraction of diffuse radiation, and thus enhancing the transmission of radiation within the canopy. The short-term impact on plant growth can be positive or negative, with long-term changes in species-composition likely, as a result of changes in competitiveness. The impacts of aerosol emissions on climate are considered below. The apparent paradox that aerosols can cause large reductions in solar radiation at the surface while, in some cases, contributing to increasing air temperatures, is explained by the fact that aerosols do not just reflect radiation, but also absorb it.

7.4.13 Climate change

TEA is both a major contributor to global climate change and a major victim. On the one hand, rapid industrialization in China means that it will soon surpass the United States as the world's largest emitter of carbon dioxide, if it has not done so already, and the pace of deforestation in Indonesia, particularly the destruction of carbon-rich peat swamps, probably raises that country to third in the global emission ranks. Riau Province, Sumatra, alone has been estimated to produce 0.22 gigatonne of carbon dioxide per year, which is more than the Netherlands (Uryu et al. 2008). On the other hand, the cumulative contribution of TEA to current atmospheric carbon dioxide levels is much smaller than that from the developed world, carbon emissions per capita are still much lower, and a significant fraction of the emissions attributed to countries in the region are a consequence of consumption outside the region.

Global atmospheric carbon dioxide levels have risen from around 280 ppm, before the industrial revolution, to 385 ppm in 2008 (Fig. 7.13), with the rate of increase accelerating sharply since 2000. It has been estimated that tropical deforestation contributed around 20% of the rise in the 1990s. Currently, around half of the anthropogenic carbon emissions remain in the atmosphere, while the rest is absorbed by terrestrial and oceanic sinks. There are still very large uncertainties in the overall carbon budget for the tropics (Baker 2007). The remaining undisturbed tropical forests have been considered to be a major sink because of increased growth resulting from rising carbon dioxide levels. Although this may be the case for the Amazon region (Lloyd and Farquhar 2008; Phillips et al. 2008), the global pattern is less clear (Chave et al. 2008). A study at Pasoh showed that stem growth rates have significantly *declined* over the past two decades, possibly in response to rising temperature (Feeley et al. 2007b). Moreover, even if there is a global tropical forest sink, deforestation is reducing its size, while logging, fire, and other disturbances are releasing additional carbon dioxide.

Carbon dioxide is not the only anthropogenic greenhouse gas. Methane (CH_4) is second in

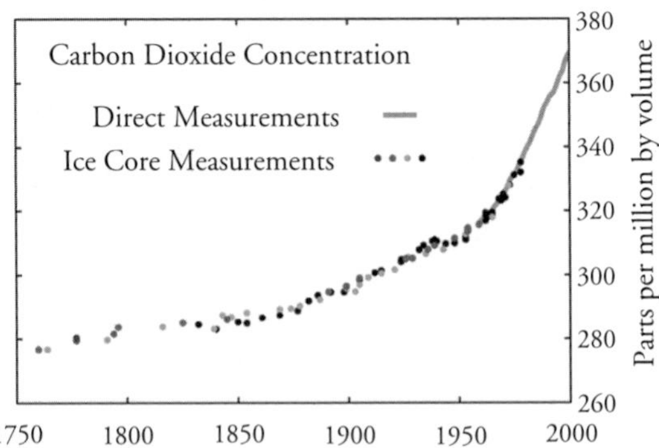

Figure 7.13 Carbon dioxide concentration in the atmosphere over the last 250 years based on both direct atmospheric measurements and sampling of gases trapped in ice cores. Image created by Robert A. Rohde for Global Warming Art.

importance, with an overall warming impact of between a fifth and a half of that of carbon dioxide, with the large uncertainty reflecting its contribution to the formation of ozone, another potent greenhouse gas, in the troposphere. The major anthropogenic sources of methane within TEA include wet-rice cultivation (estimated to produce 26% of China's emissions; Streets et al. 2003), cattle (from gut fermentation and manure), landfills, wastewater treatment, extraction and use of fossil fuels, and biomass burning. Nitrous oxide (N_2O) is third and of particular concern because of its long life (around a century) in the atmosphere, high warming potential (310 times that of CO_2), and rapid increase in atmospheric concentration. Major regional anthropogenic sources include agricultural activities, particularly excessive use of nitrogen fertilizers, and probably also emissions from natural ecosystems subject to excessive nitrogen deposition (Zhang et al. 2008a).

The impact of aerosol emissions on global, regional, and local climates is complex and there is considerable uncertainty in published estimates (IPCC 2007; Streets 2007; Ramanathan and Carmichael 2008). Black carbon (soot), produced from burning fossil fuels and biomass, is the dominant absorber of visible solar radiation in the atmosphere and may be the second strongest contributor to current global-warming after carbon dioxide (Ramanathan and Carmichael 2008). On the other hand, other aerosols, including organic carbon and sulphates, scatter more than they absorb and have a significant cooling effect. Furthermore, aerosol particles can influence the formation and properties of clouds in complex, non-linear ways. Different mixes of aerosol emissions can, therefore, have very different impacts on climate. Aerosols have a much shorter lifetime than greenhouse gases (days rather than decades), so future changes in emissions—which are highly uncertain—may have large and relatively rapid impacts on climate (Levy et al. 2008).

Deforestation also impacts climate through other mechanisms, including changes in evapotranspiration, surface albedo (reflectivity), and aerodynamic roughness (Bala et al. 2007; Pielke et al. 2007). While these changes are relatively easy to estimate if forest is replaced by uniform pasture, the impacts of conversion to tree crops or a mosaic of different land-uses are less clear. Overall it appears that the impact of any increase in albedo, as a result of deforestation in the tropics, is more or less cancelled out by the decrease in cloudiness that results from reduced evapotranspiration, so carbon dioxide emissions are the major influence of deforestation on climate.

Since 1970, TEA has experienced a general increase in temperature of 0.2–1.0°C, with the largest increases in the north. The median warming predicted for the twenty-first century by global climate models, ranges from 2.5°C in South-East Asia to 4.5°C in the north of TEA (IPCC 2007; Fig. 7.14). Rainfall projections have a greater degree of uncertainty, but most current climate models predict a moderate (0–15%) increase in annual rainfall in most of the region, with the strongest increases broadly following the seasonal movement of the ITCZ (see Chapter 2), while dry seasons will generally be more severe. Note, however, that current global-climate models do not adequately reproduce the effects of topography on spatial patterns of rainfall, or the interannual variation, so these rainfall projections should not be taken as definitive. Some models predict a decrease in ENSO frequency and increase in amplitude with rising global temperatures (Bush 2007), but this is far from certain. The IPCC projected a sea-level rise of less than 60 cm over the twenty-first century, but this is probably conservative, and a rise of several metres cannot be ruled out (Hansen 2007; Hansen et al. 2007). Indeed, the potential for unpredicted, non-linear, more or less rapid, and catastrophic changes in processes that are currently not fully understood, applies to all aspects of climate change; it is important not to

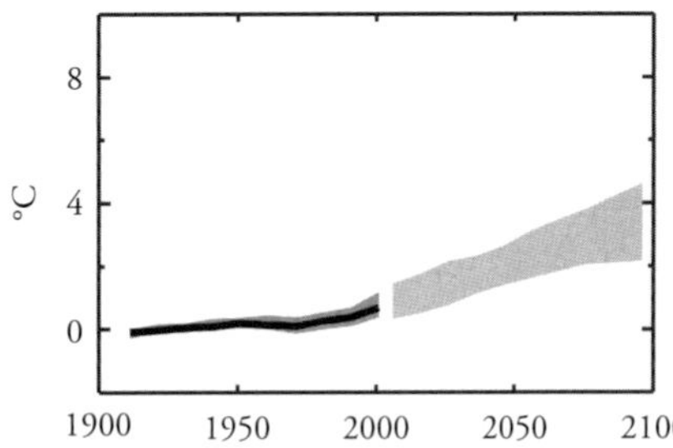

Figure 7.14 Observed and projected temperature changes for South-East Asian land regions relative to 1901–50. Figure reproduced courtesy of the IPPC Secretariat. More details on the construction of this figure are given in box 11.1 and section 11.1.2 in IPCC (2007).

be lulled into a false sense of security by the smooth projections in the IPCC reports (Lenton et al. 2008).

Predicting impacts on biodiversity from climate change is especially difficult in TEA because of the lack of information on how present-day climates influence the distributions and ecologies of organisms, the uncertainties in the rainfall projections, and the ubiquity of other human impacts. The region's biota has survived previous periods of climate change (see 1.5), but the predicted changes over the next 100 years will be much more rapid than past natural changes and, for temperature at least, will probably exceed the extremes of the last two million years. At the same time, the fragmentation of natural ecosystems and the increasing 'unfriendliness' of the matrix between fragments will make response by migration difficult for all but the most mobile species, while logging, hunting, fires, and pollution will cause additional stress.

Adverse impacts are most easily predicted for mountainous areas. Considering just the effects of temperature, a species could avoid thermal stress by dispersing uphill, as moth species on Mount Kinabalu have since 1965 (Chen et al. 2009). A warming of 3°C could be compensated for by a vertical movement of around 500 m, which in typical topography might require actual dispersal distances of 1–3 km in a century. In practice, outliers in the current distributions of species will reduce the distances that need to be moved, suggesting that dispersing uphill is a viable response to warming for most plant and animal species. Exceptions could include plant species that rely entirely on short-distance dispersal by ants, rodents, or wind (see 4.3.2), and plant or animals species confined to particular types of soil or topography. With long-lived trees, there is also the additional problem that previously established plants may persist for a century or more, despite climate change, thus blocking the establishment of species from lower altitudes.

A more general problem, however, is that the area available declines with altitude, eventually reaching zero at the summit. Even where additional, cooler habitat exists in the region, dispersal distances will often be too great for successful dispersal (e.g. Weir and Corlett 2007). Cool-adapted montane species will thus be pushed off the summit by the warming climate. Montane areas in TEA support many endemic species of animals and plants, apparently adapted to a narrow range of temperatures and often inhabiting a single mountain (e.g. Balete et al. 2007), so many extinctions can be expected. Additional complications arise in montane areas from the complex spatial patterns of rainfall and the importance of water from clouds at many sites (see 2.2.4). Predicting changes in rainfall over complex topography is beyond the capability of current climate models, but a rise in the basal altitude of cloud is expected and could have a severe impact on drought-sensitive montane plants and animals (Williams and Hilbert 2006).

Predicting impacts in the lowlands is a lot more difficult. The spatial and temporal patterns of rainfall are at least as important as temperature in controlling current distributions in the lowlands, and their changes are less predictable. Moreover, the dispersal distances required to compensate for these changes will be impossibly large for many—probably most—species, particularly in the typically highly fragmented, lowland landscapes of TEA, and multiple additional human impacts (fragmentation, logging, hunting, fire, N-deposition, etc.) will intensify over the same period.

Perhaps the greatest single worry is that there may be a general upper limit to thermal tolerance, as a result of two million years of relatively cool climates, which, once exceeded, could lead to reduced plant growth, widespread die-offs of plants and animals, and the release of huge amounts of additional carbon dioxide into the atmosphere (e.g. Colwell et al. 2008). The recent declines in stem growth at Pasoh, mentioned above, could be an early sign of this (Feeley et al. 2007b). Data for tropical ectotherms (insects, amphibians, reptiles) suggests that most species are currently living near their physiological optimum temperature, so that any warming will reduce their fitness (Deutsch et al. 2008; Tewksbury et al. 2008). Tropical-forest species may be particularly vulnerable because of the narrow temperature range they experience in the constant shade and their lack of behavioural options for thermoregulation. This contrasts with the situation at higher latitudes where, although temperature increases over the next century are expected to be greater, most species have broader thermal tolerances and are currently living in temperatures below their physiological optimum. It is important to note, however, that there is currently very little information available on the heat sensitivity of tropical lowland species, so predictions of 'lowland biotic attrition' remain largely speculative.

Global-vegetation models, such as BIOME (Prentice et al. 1992), can be used to predict changes in the potential natural vegetation as a result of climate change (e.g. Weng and Zhou 2006). These show a northward expansion of forest belts in TEA, as low-temperature constraints are lifted, with other changes depending on the details of the rainfall scenario used. For example, using two regional climate scenarios, the modified BIOME model of Weng and Zhou (2006) predicts an expansion of dry forests and savanna, at the expense of broad-leaved evergreen forest, in south-west China (and, presumably, the adjacent part of Myanmar). This is a region of exceptionally high biodiversity today, so such drastic predictions are worrying. However, it is also a region with extremely complex topography, where current climate models are expected to do a poor job of predicting rainfall.

These are equilibrium-vegetation models, based on the physiological tolerances of the component plants, and they ignore the processes by which the predicted changes in vegetation will take place. These processes include seed dispersal, mentioned above, and the complex interactions involved in replacing one vegetation type by another, particularly if both are dominated by long-lived trees. In the next few decades at least, the most obvious changes in plant communities are likely to be in the reproductive and vegetative phenologies of existing individuals, in growth rates, and probably in the climate-sensitive lifecycle stages between seed dispersal and seedling establishment (see Chapter 4). Many animals, however, are more mobile and the type of range shifts already observed in numerous species of birds, butterflies, and other organisms in the temperate zone, will become increasingly frequent in TEA.

Interactions with other anthropogenic stresses may be more of a threat to biodiversity than the direct impacts of climate change. The impacts on compensatory dispersal of the fragmentation of natural ecosystems and the simplification of the matrix are obvious. But the elimination by hunters of the dispersal agents for large, large-seeded, fleshy fruits will limit dispersal for many species, even where continuous habitat is available. Fragmentation and logging will interact with any increase in dry-season severity to increase fire frequency and intensity. Even in the least disturbed sites, climate change, rising carbon dioxide levels, ozone and particulate pollution, and enhanced nitrogen-deposition, will interact to produce unpredictable changes in species composition.

7.5 Predicting extinctions

Most of the major threats described in the previous section are already well advanced, yet the number of known extinctions in TEA is still extremely low. The 2007 IUCN Red List includes only 10 plant species from TEA that are believed to be extinct in the wild, of which 4 survive in cultivation, and only 9 animal species (plus 11 species from the oceanic Ogasawara Islands), most of which are problematic from a taxonomic and/or historical point of view. A number of additional species, including the Javan lapwing, pink-headed duck (Fig. 7.15), and white-eyed river-martin, are probably extinct, but conservationists have been wary of committing the 'Romeo error', of giving up on a species that may still survive (Butchart et al. 2006). The Cebu flowerpecker was rediscovered on the Philippine island of Cebu in 1992 after being assumed extinct for 40 years. The rapid pace of lowland deforestation in many parts of the region suggests that these extinction numbers may be underestimates, since numerous species in

Figure 7.15 The pink-headed duck, *Rhodonessa caryophyllacea*, used to inhabit lowland marshes in Myanmar, India, and Bangladesh, but is probably now extinct, as a result of hunting and habitat loss. This picture, obtained from Wikimedia Commons, is of a painting from A.O. Hume and C.H.T. Marshall, 1879–81, *The game birds of India, Burmah and Ceylon*, published by the authors in Calcutta.

groups that are less well-studied than the birds, lack recent positive records; but it is clear that the predicted mass extinctions have not yet happened.

A major reason for the low number of extinctions so far is undoubtedly the time-lag between loss of habitat and the death of the last individual of a species. An individual tree may persist in a tiny forest fragment for a century or more without reproducing: not extinct but 'committed to extinction'. Janzen (2001) memorably referred to these individuals as the 'living dead'. The same will apply to any small population of plants or animals confined to a habitat fragment. They may reproduce, but if they cannot expand their population, then eventual extinction is inevitable, as a result of random variation in reproduction and mortality, and the loss of genetic variation. Studies of species losses from isolated forest fragments confirm this inevitable decline and suggest that the time-lag may range from months to centuries for different groups of organisms in forest fragments with areas of 10–1000 hectares (Corlett and Turner 1997; Brooks et al. 1999; Laurance et al. 2002; Ferraz et al. 2003; Sodhi et al. 2005a).

Extinctions as a result of habitat loss can be predicted by using the well-studied relationship between species number and habitat area ($S = cA^z$, where S is the number of species, A is area, and c and z are constants). Note that this approach can only predict extinctions for species that are threatened largely by habitat loss, rather than direct exploitation. For birds and mammals, predictions of extinctions from past forest-loss more or less match the numbers considered threatened or already extinct in the IUCN lists (Brooks et al. 1997, 2002). Since threatened species are defined as those with 'a high risk of extinction in the wild', this suggests that the use of the species–area approach to predict threats is reasonable. However, it is not expected that all species listed as threatened will go extinct, so these species–area predictions may be overestimating the eventual number of extinctions. Indeed, a recent study of the rates of movement of bird species between IUCN Red List categories found much lower extinction rates among critically endangered species than predicted, with much of this shortfall attributable to active conservation measures (Brooke et al. 2008).

Brook et al. (2003) used species–area relationships to extrapolate from the actual numbers of extinctions that have occurred as a result of deforestation in Singapore since the nineteenth century to estimates of possible future extinctions in the whole of South-East Asia as a result of projected forest losses. Their estimates range from 2–25% in reptiles, which seemed relatively insensitive to forest loss, to 21–58% for freshwater fish (Table 7.4). Other studies have used varying estimates of future forest cover

Table 7.4 Observed and inferred extinction rates during >95% deforestation in Singapore, between 1819 and 2002, and projected percentage losses for South-East Asia by 2100 (Brook et al. 2003).

Taxon	Singapore		SE Asia
	Recorded losses by 2002	Inferred losses by 2002	Projected losses by 2100
Vascular plants	26%	74%	12–44%
Freshwater decapod crustaceans	30%	82%	14–50%
Phasmids	20%	67%	9–38%
Butterflies	38%	73%	19–43%
Freshwater fish	43%	87%	21–58%
Amphibians	7%	71%	3–41%
Reptiles	5%	49%	2–25%
Birds	34%	59%	16–32%
Mammals (excluding bats)	42%	78%	21–48%
Weighted mean	28%	73%	13–42%

Inferred losses are estimated on the assumption that pristine habitats in Singapore supported the same species as similar habitats in nearby Peninsular Malaysia. Projected extinctions for South-East Asia are based on both observed and inferred losses in Singapore, the species–area relationship for each taxon, the proportion of the forest lost already, and the present rate of deforestation.

to come up with overall projections for species loss in the Asian tropics of 20–30% (Wright and Muller-Landau 2006a, b; Brook et al. 2006).

Species–area curves are crude tools, so these estimates could be very wrong, even if the predictions of future forest cover are correct. Many—probably most—medium and large vertebrates are threatened by hunting, so their population sizes are either unrelated to the area of remaining habitat (e.g. the Javan Rhinoceros, which has been extirpated from almost all of its huge historical range) or are much lower—and thus much more vulnerable to extinction—than area alone would predict. More generally, the carrying capacity of much of the remaining forest for a wide range of species is being reduced by logging, fires, air pollution, and nitrogen deposition. Brook et al. (2008) suggest that synergies among extinction drivers—habitat loss, fragmentation, overharvesting, fire, climate change, etc.—will make predictions based on habitat loss alone overly optimistic. On the other hand, many forest plant and animal species can persist in landscapes that would be considered completely deforested in a satellite-base assessment, making use of tiny forest fragments, patches of secondary forest, and trees in the agricultural mosaic. Moreover, the time-lag between habitat loss and extinction raises the possibility that species reduced to small populations may be 'rescued' by the expansion of secondary forests, if human impacts decline after reaching a peak in the next few decades.

The biggest unknown is the impact of anthropogenic climate change. Currently, the uncertainties in the climate projections for the region are compounded by lack of information on the responses of species and ecosystems. The climate projections will continue to improve in both accuracy and detail, but the gaps in our ecological knowledge are unlikely to be filled in the foreseeable future, so predictions of extinctions as a result of climate change will continue to be relatively speculative.

7.6 Satellites and cryptic threats

The use of satellite-based sensors to detect changes has dramatically improved our understanding of the threats to regional biodiversity from deforestation, fragmentation, and fire. It is important to realize, however, that areas which appear pristine in even the highest resolution satellite data, have often had their animal communities drastically altered by hunting. Other potential 'cryptic' threats include air pollution, the harvesting of non-timber forest products, and the spread of emerging pathogens and exotic invaders (Laurance et al. 2006). Logging impacts are on the margins of detectability, depending on their intensity and recentness. Climate change will cause both satellite-detectable changes in phenology and generally undetectable changes in species-composition. Moreover, the focus on detectable threats has resulted in a tendency to downplay the importance of synergies between different types of impacts. Almost everywhere in TEA suffers from multiple threats and, in many cases, the combined impact of two or more of these is likely to be greater than the sum of their individual effects: logging and hunting, logging and fire, climate change and fragmentation, roads and invasive species, and so on.

CHAPTER 8

Conservation: saving all the pieces

8.1 Introduction

How can the biodiversity of TEA be saved from the threats outlined in Chapter 7? In a region that supports one billion people and 15–25% of global terrestrial biodiversity, conflicts between the two are inevitable. There will be local win-win scenarios, but on a regional scale, crops and urban areas occupy land that is needed by wildlife, and even the best-managed resource extraction reduces the carrying capacity of the natural habitats that remain. There have been recent suggestions that human welfare—in particular, the welfare of poor rural people—should be placed at the heart of conservation (e.g. Kaimowitz and Sheil 2007). Others have argued that movement to cities will continue to be the main pathway out of rural poverty, despite attempts to improve rural living conditions (Fuentes 2008). Whoever is right, it seems likely that the necessary compromises between conservation, and the livelihoods and aspirations of people will best be made when the needs of both are better understood. This chapter, therefore, focuses on the needs of biodiversity conservation. The needs of the rural poor cannot be ignored, however, even when the focus is on biodiversity.

8.2 Who should pay and how?

8.2.1 Who should pay?

Ultimately, all conservation is local and it is rarely practical, even if it were morally justifiable, to enforce conservation regulations in the face of widespread local hostility. There has been considerable debate in recent years about the need for the developed 'north' to pay more of the costs of tropical conservation, which are currently paid largely by the undeveloped 'south' (e.g. Balmford and Whitten 2003; Whitten and Balmford 2006). North–south rhetoric makes rather little sense in TEA, which encompasses the global extremes of wealth and poverty, but the basic issue is the same.

At first sight, an increasingly wealthy region could easily afford to trade-off some economic development for better conservation, but it is not the wealthy that pay most of the price in practice. Much of the surviving biodiversity in TEA is concentrated in remote areas, where the local populations are on the geographic, social, and economic margins of society (Fig. 8.1). The direct costs of planning, setting-up, and managing protected areas in such places are usually met by the national government and/or international donors. Local communities, however, pay the typically much larger 'passive' costs that result from controls on exploitation, lost opportunities for expansion of agriculture, increased crop damage, and attacks by wildlife on livestock and people. If alternative sources of cash income are not available, some degree of 'illegal' activity is almost inevitable (e.g. Yonariza and Webb 2007).

Of course, local people may also benefit from conservation activities, through the maintenance of local ecological services, such as a reliable, clean, water supply, but the major benefits of tropical conservation are spread much more widely. These benefits include nature-based tourism for national and international participants, carbon fixation, and the preservation of global biodiversity. Both equitability and practicality suggest that it is the broader national and international community that should pay the bulk of the costs for tropical conservation, including compensation to local communities for the passive costs mentioned above. Greatly increased funding from national governments and/or international sources will not solve all the

Figure 8.1 Penan man and child in Sarawak. The Penan are nomadic hunter-gatherers in Borneo whose traditional way of life has been devastated by logging and deforestation. Photograph © Dang Ngo, Greenpeace.

problems of either tropical conservation or rural poverty, but it can help reduce the conflicts between the two. Possible mechanisms for achieving this are considered below. One general issue is the need for external funding to be sustained, since erratic funding is unlikely to have any lasting impact.

The local costs of tropical conservation are by no means the only example of environmental injustice in the twenty-first century. United Nations estimates suggest that the poorest billion people on the planet contribute only 3% of total global carbon emissions, with the emissions of the richest individuals exceeding those of the poorest by a factor of thousands (UNDP 2007). Again, this is often presented as a north–south issue, but the extremes exist within every country in TEA. It is, therefore, of critical importance to ensure that the costs of mitigating climate-change impacts on biodiversity are met, as far as possible, by those responsible for these impacts, without placing an additional burden on the poor.

8.2.2 Payments for environmental services

The need to transfer more of the costs of conservation from the rural poor to national and international donors has resulted in calls for new approaches to conservation. The best developed of these new ideas is the concept of direct payments for environmental services (PES) (Wunder 2007; Goldman et al. 2008; Wunder et al. 2008). The basic idea is straightforward: a voluntary transaction in which the beneficiaries of environmental services (water, carbon storage, biodiversity, etc.) pay local land-users for the continued provision of these services, or for their restoration. The voluntary, negotiated nature of PES agreements distinguishes them from command-and-control approaches to the same environmental problems. In developing countries without a strong legal framework, payments need to be contingent upon the continued provision of the service, so they are made periodically in response to monitored compliance.

The environmental service that is most relevant here is the protection of biodiversity. Ideally, the land owner—or, at least, a land user or users with the capacity to enforce the deal—would be paid to preserve or restore some component of biodiversity, such as an endangered species or ecosystem (see also 8.2.4). In practice, however, the linkage is usually less direct than this, with the payment being made for the preservation of forest cover, the control of hunting, or some other restriction on land-use that is believed to favour biodiversity conservation. Indeed, there may be more money available for payments for other environmental services, particularly watershed protection (Fig. 8.2) and carbon offsets (see below), but the conservation benefit can be similar as long as restrictions on hunting and other forms of exploitation are included in any agreement. Water payments are already being used to support protected areas in Laos and Vietnam (McNeely 2007).

China has two major PES schemes—the Natural Forest Conservation Program (NFCP), which pays forest enterprises to stop logging natural forests and to increase the plantation area; and the Grain to Green Program (GTGP) (also known as the Sloping Land Conversion Program), which pays farmers in key river-catchments to convert cropland on steep

Figure 8.2 Payment by downstream-users to owners of upstream forests for the continued provision of a reliable supply of clean freshwater is an example of a payment for environmental services (PES) with potential biodiversity benefits. Photograph © AFCD, Hong Kong.

slopes to grassland or forest (Liu et al. 2008). Both schemes have already had large impacts on vegetation cover and the biodiversity benefits are likely to increase as more encouragement is given to the use of diverse native species in forest rehabilitation.

The explosion of both practical trials and theoretical studies of PES over the last few years precludes a synthesis at this stage, but a number of general problems are apparent (e.g. Wunder 2006, 2007; Ferraro 2008; Wunder et al. 2008). PES can work, but it is only one conservation tool of many and it is not necessarily either simpler or cheaper than the alternatives. In particular, unless PES schemes are very carefully designed and implemented, there is a risk that most payments will be made to land-users to continue not doing things that they were not going to do anyway, so the buyer is getting little or no additional conservation for the money spent. This will happen if, for example, the user accepts payments for preserving forest cover on land that was not going to be cleared (e.g. Sánchez-Azofeifa et al. 2007; Muñoz-Piña et al. 2008). The problem of assessing 'additionality' (i.e. the extent to which the payments result in environmental benefits in addition to 'business as usual') is much less severe when the service provided is active restoration, rather than maintaining the status quo; but many PES schemes are likely to be of the latter type.

Lack of additionality may be considered less of a problem when the major aim of the scheme is to reward stewardship, rather than change behaviour, but this sort of scheme is not PES in the strict sense, since the buyer does not receive any service as a result of the payment. Indeed, the confounding of two objectives—the supply of environmental services at least cost and the provision of extra income to the rural poor—makes the conservation effectiveness of many so-called PES schemes very difficult to evaluate. At the other extreme, PES is unlikely to be effective when the rewards of the unwanted behaviour are high, unless combined with effective enforcement and/or the provision of attractive alternatives. PES makes most sense at the margins of profitability, when small payments can tip the balance in favour of conservation (Wunder 2007). Even then, the prospect of continuing monitoring and payments into the indefinite future may make alternatives, such as land purchase or legal enforcement, appear more attractive.

It is easy to find problems with PES, but most of these apply also to all the alternatives, along with additional ones. Most experience so far comes from the Neotropics, and there is a huge scope to try out PES schemes in TEA. It would be particularly interesting to run trials with payments that are linked directly to components of biodiversity that are relatively easy to monitor, such as the densities of diurnal primates, the camera-trap success rates for nocturnal terrestrial mammals, or the extent of forest clearance as assessed from satellite imagery. The additional costs of monitoring compliance could potentially be balanced by a large increase in efficiency over less direct schemes. It may also be easier to raise funds for PES when the linkage with target species is clear. Moreover, PES payments need not be in cash, which may not bring long-term benefits to recipients, but could be in the form of schools, healthcare, training, or development assistance.

8.2.3 Carbon offsets

No potential mechanism for funding tropical conservation has received as much recent attention as carbon offsets. The principle is simple. Tropical deforestation accounts for at least a quarter of all anthropogenic carbon emissions, with both the highest rates of deforestation (see 7.4.2) and, as a result of high forest biomass, the highest carbon emissions per hectare cleared (see 6.2.2), in TEA. It, therefore, makes sense for the wealthy industrialized countries, which are responsible for most cumulative emissions, as well as major emitters within the region, to fund conservation actions that reduce carbon emissions from tropical forests. Potentially, everyone benefits: the industrial emitters can meet their emissions targets more cheaply than alternatives, while the tropical nations receive much-needed funds for conservation and other uses. Currently, most carbon-offset schemes in the tropics are voluntary, but future international agreements are very likely to impose binding cuts in emissions that will make such schemes a necessity. Carbon offsets are thus a major opportunity for forest conservation in the TEA.

Credits for reducing emissions will be traded by nations and companies, with the price determined by a global carbon market. Reducing emissions from deforestation and degradation ('REDD') does not count as a source of carbon credits under the current Kyoto Protocol, but it is almost certain to be included in any future agreement (Laurance 2008a). A variety of practical problems must be overcome to make this possible, including establishing an appropriate baseline against which any reductions in deforestation will be measured, assessing emissions per hectare of clearance, and ensuring the permanence of cuts in deforestation (Fearnside 2006; Laurance 2007b). Poor governance in most tropical forest countries—particularly on the forest frontier—is likely to be a widespread problem (Ebeling and Yasué 2008). There are also concerns that the impacts of climate change on forest growth, litter decomposition, and fire frequencies may have unpredictable effects on carbon storage. Most of these uncertainties, however, can be fairly easily dealt with by initially giving less credit per tonne of carbon kept out of the atmosphere by avoided deforestation, than is given for alternative, better-quantified, offsets. This discount will then, in turn, create an incentive to improve mechanisms for monitoring and enforcing cuts in deforestation. In any case, most tropical forest countries will need a large infusion of resources from the developed world in order to set up credible mechanisms for quantification (Laurance 2008a). Note that the drainage of carbon-rich peat swamps in equatorial South-East Asia produces far greater carbon emissions than deforestation alone (Uryu et al. 2008; Wösten et al. 2008), making the protection of peat swamps a particularly attractive target for carbon offsets.

Avoiding deforestation is not the only mechanism for reducing emissions in the tropics, although it is the one with the greatest immediate benefit for both biodiversity conservation and carbon mitigation (Nabuurs et al. 2007). Other sources of carbon credits with conservation benefits include reforestation of degraded lands (8.9.1), reducing logging impacts (8.5.1; Putz et al. 2008b), changes in agricultural practices (8.6), and restoring the water table in drained peat swamps. Reforestation has the advantage that it removes carbon dioxide from the atmosphere, rather than just not putting it there. It is often assumed that the greatest carbon benefits will come from plantations of fast-growing acacias, pines, and eucalypts, with few or no biodiversity benefits, but both theory and some experimental evidence suggest that more diverse plantations may fix more carbon. This is currently being tested in the massive Sabah Biodiversity Experiment. Recent estimates suggest that a combination of avoided deforestation and reforestation, both largely in the tropics, could offset 2–4% of projected global emissions by 2030 (Canadell and Raupach 2008).

The costs of carbon offset by reforestation depend largely on growth rates and land costs, so it is likely to be an economically competitive land use only in low-income tropical areas with low population densities, such as parts of Indonesia. Reduced impact logging (RIL, see 8.5.1) may pay for itself, but the added incentive of carbon credits should not only encourage its uptake, but will also give an advantage to legal over illegal operations. Changing agricultural practices can not only increase carbon sequestration, particularly in the soil, but also reduce methane and nitrous oxide emissions.

8.2.4 Biodiversity offsets

Carbon-offset schemes have the advantages for biodiversity conservation that they are relatively straightforward (i.e. a tonne of carbon is a tonne of carbon, whatever and wherever the source) and that there are potentially huge amounts of money involved. However, pure carbon-offset schemes do not necessarily require much native biodiversity (see above). Biodiversity offsets involve the same basic principle—a business or government funds conservation actions that compensate for unavoidable harm to biodiversity, so as to ensure no net loss—but they focus directly on biodiversity, or a proxy such as habitat. However, they are far less straightforward, since there is no single 'currency' to be traded. Many existing biodiversity offsets are voluntary, but there are also a variety of more or less compulsory schemes in various parts of the world, such as the requirements for 'no net loss' of wetlands from developments in the USA and for 'net conservation benefits' in Western Australia (ten Kate et al. 2004; Gibbons and Lindenmayer 2007; Latimer and Hill, 2007; Blundell and Burkey 2008; Stokstad 2008).

Biodiversity offsets normally try to achieve equivalence—like for like—so funding would generally come from companies within the region that are involved in plantations, mining, or other activities that cause unavoidable harm to biodiversity, largely through habitat loss. For example, a company planning to convert 500 hectares of lowland rainforest to oil palm, might offset both the carbon and biodiversity impacts of this by protecting an equivalent area from clearance. This would only contribute to 'no net loss', however, if the areas were truly equivalent, if the protected area would otherwise be cleared, and if there was no 'leakage' (i.e. no additional area cleared elsewhere as a consequence; see 8.10). An alternative would be rehabilitating or re-creating 500 hectares of rainforest on degraded land, but this could take centuries to reach equivalence, assuming that this goal is even possible. Yet another alternative would be to pay for improvements to existing protected areas. Although at first sight this third alternative does not represent a true biodiversity offset, it may actually achieve more conservation than the equivalent expenditure on the protection of additional areas. Whatever the conservation action, more conservation will usually be achieved if the biodiversity offsets from several projects in a region can be aggregated into a single large area, either by collaboration between multiple companies or through a 'conservation bank' set up by an NGO, government department, or private company, which then sells on biodiversity credits to companies that need them.

There are numerous potential problems with biodiversity offsets, including the difficulties of enforcement in countries without a strong legal system, the risk that offsets will create a 'license to trash' elsewhere (ten Kate et al. 2004), the difficulties of ensuring equivalence and additionality, the need for local support, and the possibility of unforeseen social impacts. As with so many new ideas in conservation, we urgently need rigorously monitored pilot studies within TEA to assess the practicality of introducing biodiversity offsets on a larger scale and, indeed, making them compulsory. There is likely to be far more money available for carbon than biodiversity, so the major role for biodiversity offsets in TEA will probably be in strengthening the conservation benefits of carbon-offset schemes—both avoided deforestation and reforestation. Hopefully, for many buyers of carbon credits, the added attraction of 'carbon + biodiversity' offsets will compensate for considerably greater costs. The Climate, Community, and Biodiversity Alliance (CCBA), which has a membership of both NGOs and companies, has created standards for carbon credits that also bring both biodiversity *and* social benefits.

8.2.5 Tourism

> Come face to face with turtles and orangutans!
>
> See gibbons singing duets as the mist burns off the rainforest!
>
> A chance of glimpsing Asian elephants at the river's edge.
>
> Fierce Komodo dragons, stunning coral reefs, isolated beaches, and volcanoes....
>
> Sulawesi has the highest number of endemic bird species of any of the Indonesian islands with 70 species found nowhere else!
>
> Mount Kinabalu is the highest peak in South East-Asia. Guided walks take you through this botanical paradise.
>
> See the rare sight of fiery Rafflesia blooms, world's largest flower.
>
> See the world's smallest fly-eating plant, rare orchids, and exotic medicinal plants.
>
> In October, the springs at Poring attract many of Borneo's most beautiful [butterfly] species and the flowers at Danum a variety of stunning birdwings and both dragontails.
>
> (Quotations from websites offering tours in TEA.)

Tourism is the world's fastest-growing industry, so it is not surprising that it is seen as a potential source of funds for conservation. Malaysia alone recorded 21 million tourist arrivals in 2007, a 20% increase over the previous year (Tourism Malaysia website). Domestic tourism is often bigger still, with Chinese-speakers making up more than 98.5% of tourists in China (Zhong et al. 2007). The involvement of local people in tourism could offset some of the passive costs of conservation (see 8.2.1) and increase local support for protected areas. Note that I have avoided the term 'ecotourism' because of its many definitions, although within TEA this term is often applied to any recreational visit to a natural area. Most 'ecotourism' in TEA involves visits to minority ethnic groups and most of this has little or no conservation benefit (Zeppel 2006).

Local visitors to protected areas are most often on day-trips from home, so ease of access is crucial and sites within 3–4 h travel time of urban areas can be overcrowded at weekends and on public holidays (Fig. 8.3). Accessibility is also crucial for the highly organized tour groups that typify multi-day tourism within the region, because a natural area visit is just one attraction in a tight itinerary. Tourists from outside TEA are drawn largely to a small number of well-known sites, such as Komodo Island and Mount Kinabalu, but smaller numbers visit even the most remote and inaccessible areas.

There has been no review of nature tourism in TEA, but some general patterns can be derived from websites that advertise such tours, and from my personal experiences in many parts of the region. Most visitors visit protected areas for the opportunity for open-air recreation in an attractive setting, rather than biodiversity as such. With a few exceptions, including the Komodo dragon and, at some sites, primates, TEA lacks the large, easily watchable, wildlife of the African savannas. Organizers of tours in TEA are usually careful not to promise sightings of specific large mammals. Most tourists see only domesticated elephants, for example, although the last wild elephants in China are a significant tourist attraction in Yunnan and 'Wild Elephant Valley' was named one of China's 50 most recommended destinations for foreign tourists in 2006. Most tours include cultural and scenic attractions, as well as wildlife.

The major exceptions to this general pattern are tours for birdwatchers (Mollman 2008), for whom the opportunity to see new species over-rides most other considerations. Several companies advertise bird tours to various parts of TEA and many other tours are organized by birdwatching societies, both within and outside the region. Diversity and endemism are the chief draws for this group and, although the spectacular hornbills are a major attraction, birdwatchers are probably unique in their desire to see every bird species. Birders are thus more evenly spread through the region than generalist nature-lovers or specialists on other groups. Plant diversity also attracts specialist visitors,

Figure 8.3 The heavy use of a walking trail at the weekend in Pat Sin Leng Country Park, Hong Kong, illustrates the pressure put on protected areas that are easily accessible from major urban centres. Photograph © AFCD, Hong Kong.

particularly in south-west China, which was an important source for temperate garden plants, and on Mount Kinabalu, with its huge range of different plant communities. The giant flowers of *Rafflesia* feature in many tours of Borneo. Smaller numbers visit TEA to see butterflies, and individuals or private groups come for other taxa, such as dragonflies or beetles. Increasingly, specialist groups within the region—particularly lepidopterists—organize their own expeditions.

The diversity of origins and motivations among visitors to protected areas in TEA makes generalizations difficult, but research elsewhere in the world has shown that the impacts of tourism on conservation are mixed (e.g. Stronza 2007; Turton and Stork 2008). In TEA, adverse direct visitor impacts, such as noise, litter, and path erosion, are highly localized at present, although this situation could change as visitor numbers and adventurousness increase. Indirect impacts from the expansion of infrastructure, including roads, hotels, shops, and other facilities, are obvious at many popular sites. There is very little information available on economic or other benefits to local people, but the overall impression is that the provision of services to tourists at the more popular sites is dominated by outsiders from the urban areas where most visitors originate. Local communities are most likely to benefit if the provision of some services (e.g. accommodation, guides) is restricted to them or if there is some mechanism by which income from entry and other charges is distributed.

Increased employment and income from tourism may reduce adverse impacts from subsistence hunting and farming, but could also increase such impacts by financing purchases of new technology (guns, chain-saws, outboard motors) or by attracting outsiders. At present, the major conservation benefit from nature tourism is probably the impression it gives to the urban elite that conservation activities can be justified on economic grounds. The national tourist authorities of Malaysia and Indonesia, in particular, promote their countries as destinations for nature lovers and tourism is recognized in both countries as a major source of money and employment.

There is evidence from the developed world that biodiversity interpretation in protected areas can increase both visitor satisfaction and knowledge (e.g. Hill et al. 2007; Pearce 2008). In my experience, no protected area in TEA has sufficient biodiversity information available for an interested, but non-specialist, independent visitor and most have little or no such information. An even more serious deficiency is the absence of interpretative material (signboards, leaflets, maps) targeted at domestic visitors, thus greatly reducing the educational potential of protected areas. The provision of biodiversity information and conservation education in popular protected areas is an obvious role for local, regional, and international NGOs.

8.2.6 Voluntary conservation measures

In many parts of the world, private companies and wealthy individuals make a significant contribution to conservation by acquiring land of high conservation-value and protecting it from incompatible developments. Large-scale land acquisitions by foreigners are likely to be viewed as 'neocolonial' in TEA, but the protection of small areas of critical habitat would be a useful outlet for local philanthropy. An alternative would be to 'adopt' an existing protected area, providing field uniforms, equipment (binoculars, cameras, GPS; Fig. 8.9), and transport, or funding a field-studies centre for schools. In countries with an appropriate legal framework, the establishment of a conservation trust fund can ensure sustained funding into the future. Major banks within the region provide 'philanthropic services' to wealthy clients, which include the identification of projects that match each client's particular philanthropic vision, so it may not be necessary for a project proponent to approach individual donors directly.

8.2.7 Certification

The idea of certification (often called 'eco-certification') is that it gives consumers the opportunity to choose products from environmentally and socially responsible producers. Normally this will mean that the consumer pays more and the producer receives more than for otherwise similar products from irresponsible producers. In theory, if not always in practice, this will then mean that wealthy

Table 8.1 The Forest Stewardship Council's 'Principles and Criteria for Forest Stewardship' (FSC 2004).

1. Compliance with laws.
2. Tenure and use rights to the land.
3. Respect for indigenous peoples' rights.
4. Maintenance or enhancement of the social and economic well-being of forest workers and local communities.
5. Sustainable economic, environmental, and social benefits.
6. Conservation of biodiversity and ecosystem services.
7. A management plan.
8. Monitoring and assessment of economic, social, and environmental impacts.
9. Maintenance of high conservation value forests.
10. Plantations should reduce pressures on natural forests.

consumers of tropical products contribute more to biodiversity conservation in the tropics.

By far the best-developed certification systems are those that apply to forest products. The Forest Stewardship Council (FSC) was established in 1993 by environmental and social groups to certify forests and forest products (timber, furniture, paper, etc.) that meet a range of stringent, but practical, criteria, relevant to environmental and social responsibility (Table 8.1). The intention of the standards is that they should be tough enough to make a difference, while not being so burdensome that the costs of compliance exceed any possible gains. The FSC does not issue certificates itself, but accredits certification organizations. These issue two types of certificates: Forest Management (FM) Certificates, for responsibly managed forests, and Chain of Custody (COC) Certificates, which allow products to be labelled with the FSC logo (Fig. 8.4).

The FSC now competes with a range of more business-friendly certification programmes initiated by the forest industry. The prior existence of the FSC benchmark has made it difficult for these competitors to establish themselves and none currently have the same credibility, although some have undoubtedly led to improvements in forest management. FSC itself has been criticized for certifying timber extracted from old-growth forests,

Figure 8.4 Brazilian timber marked with the logo of the Forest Stewardship Council (FSC), showing that it has been produced in a way that meets a range of environmental and social criteria (Table 8.1). Photograph © Daniel Beltra, Greenpeace.

but the sustainable management of natural tropical forests is greatly preferable to their replacement by cash-crop monocultures, which is usually the most probable alternative. The possibility of corruption and fraud are other concerns for this and other forms of certification (Butler and Laurance 2008).

Roughly 10% of the world's production forests have FSC certification, but these are largely outside the tropics and the direct impacts of FSC certification in TEA are still small. Although all countries except Cambodia and Myanmar have some FSC certificates, these are mostly for plantations and very little natural forest is certified. A major reason for this is that most timber and forest products produced in the region are sold in the relatively environmentally insensitive markets of Asia (particularly China [Laurance 2008b], but also Japan and South Korea), where the costs of FSC certification are not currently justified by increased market access or price premiums (Cashore et al. 2006). However, China is also the major single exporter of wood products to Europe and North America, so there are opportunities for consumers in these countries to insist on higher standards. The major impact of the FSC in TEA so far has been through the development of competing national schemes in the major exporting countries, Indonesia and Malaysia, which, despite some major problems, are under pressure to meet many of the FSC principles.

It is not easy to transfer the lessons of FSC to other products whose production threatens biodiversity. A certified chain of custody up to the point of sale to consumers is more difficult to achieve for palm oil or natural rubber, for example, since both are typically relatively minor (and often anonymous) components in consumer products. The Roundtable on Sustainable Palm Oil (RSPO) is currently introducing a certification system for palm oil, in the face of considerable scepticism from conservationists, but it is too early to judge its success. Most attention has focused on relatively high-value products for which consumers already recognize many categories, differentiated by price. The standards for Fair Trade certification (Fairtrade Labelling Organizations International) include reducing impacts on biodiversity, although as a fairly minor component; a small number of producers in TEA of coffee (Indonesia), rice (Laos, Thailand), and tea (China, Laos, Vietnam) are currently certified (Fig. 8.5). There have also been attempts by the International Federation of Organic Agriculture Movements (IFOAM) to ensure that products labelled as 'organic' come from farms that are biodiversity-friendly (McNeely 2007).

Figure 8.5 Tropical East Asia produces a variety of products that are certified as 'organic' or Fair Trade, with criteria that include a biodiversity component. Currently, the great majority of these products are sold outside the region, like this Sumatran coffee, but the potential regional market is huge. Photograph courtesy of John Corlett.

The greatest scope for expansion of biodiversity-friendly certification in TEA is probably at the national and regional level, making use of the increasing environmental awareness of wealthy urban consumers. For many export products, however, direct pressure by major buyers from outside the region, who fear embarrassing publicity, is likely to be more effective than certification targeted at consumers. Butler and Laurance (2008) argue that public pressure on these corporations and on industry trade-groups could be an important new weapon in the fight to save tropical forests.

8.2.8 The role of NGOs

The big international conservation NGOs (BINGOs; particularly Worldwide Fund for Nature, Wildlife Conservation Society, Conservation International, and the Nature Conservancy) are the most visible face of tropical conservation for people living outside the tropics and have a combined income exceeding US$1 billion per year. In theory, these organizations could help funnel funds from wealthier nations and individuals to meet the needs of tropical conservation. In practice, perceptions of these organizations within the tropics are often less than positive. Many perceive a large gap between the huge resources available to these NGOs and their relatively modest lasting conservation achievements on the ground.

The BINGOs have also been variously criticized for their dependence on huge cash incomes, for their restricted project time-frames, for their generalized, top-down approach to conservation, for their slick marketing that oversimplifies complex problems, for their fuzzy objectives, for their focus on charismatic megafauna to the exclusion of other taxa, for competing rather than cooperating with each other and with local NGOs, and for their partnerships with governments at the expense of local people (Chapin 2004; Dowie 2006; Halpern et al. 2006; Rodríguez et al. 2007; Stone 2007; and personal communications from many other people). Moreover, although accurate estimates are not available, it is widely believed that a large proportion of the money they raise is spent on expensive international experts and equipment, rather than on cheap local services.

Much of this criticism may be undeserved. As outsiders, operating only with national government approval, the BINGOs must usually concentrate on 'big conservation', such as the design of protected area networks (8.3.2) and they cannot get involved in local conservation battles. WWF's recent 'Heart of Borneo' initiative illustrates the BINGO advantage—the ability to get three countries (Brunei, Indonesia, and Malaysia) to agree to a massive project for the conservation of forests in 220,000 km^2 of Borneo. On the other hand, its vagueness has raised fears that the main impact will be as a distraction from the real needs of conservation in Borneo (ATBC 2007; Stone 2007). In particular, the project so far lacks clear, measurable, targets, so success will be hard to assess.

The most pressing conservation problems in TEA are often the pervasive but unspectacular erosion of biodiversity through encroachment on protected areas, hunting, and illegal logging. These problems may be better tackled by local NGOs, with less funds and expertise, but more local knowledge and support. Unfortunately, there has been no detailed study of local NGO effectiveness in TEA, so their importance is hard to evaluate. There are also large differences between countries, with some NGOs in the Philippines, for example, apparently particularly effective (Posa et al. 2008). An alternative model, combining the advantages of global reach with local knowledge, is provided by BirdLife International, which is a global partnership of independent local NGOs. BirdLife offers training and other assistance to member organizations, as well as an overall 'seal of approval', which can be very useful in negotiations with national governments.

8.3 What should be protected?

8.3.1 Surrogates in conservation planning

The resources available for conservation are limited, so the allocation of these resources requires careful planning. A variety of methods have been developed for doing this (see 8.3.2), but they all depend on adequate data about the spatial distribution of biodiversity. In practice, however, such data are always incomplete, so conservation planning is necessarily based on the use of surrogates for which we do have data (Rodrigues and Brooks 2007). The most widely used surrogates are better-known taxa, such as vertebrates and vascular plants, or vegetation. The underlying assumption is that plans made on the basis of these surrogates will also be effective for the conservation of taxa for which we have no, or inadequate, data—usually the overwhelming majority. Is this assumption justified?

While many unrelated taxa share broadly similar patterns of diversity (see Chapter 3), the evidence suggests that using one taxon as a surrogate for others does not work well on the fine spatial scale at which most conservation decisions are made (Das et al. 2006; Grenyer et al. 2006; Rodrigues and Brooks 2007; Kremen et al. 2008). Surrogacy appears to be particularly ineffective in human-dominated landscapes, where the differential response of different taxa to disturbance can result in very different patterns of diversity (Yip et al. 2004, 2006). It is also ineffective for taxa, such as land snails, where diversity is concentrated in a particular subset of habitats (Fontaine et al. 2007). It is true that there are also examples in the literature of strong correlations between particular taxa at particular spatial scales (e.g. plant and vertebrate diversity across China's 28 provinces; Qian 2007), but unless we can predict which taxa will be correlated and at which scales *before* the data are collected, these findings are of little practical use. Moreover, successful surrogates for diversity patterns will not necessarily be useful for setting priorities, where vulnerability and irreplaceability are key issues (see below), or for predicting the persistence probabilities for poorly known taxa (Rodrigues and Brooks 2007).

Although more research is needed, a reasonable conclusion at this stage is that conservation decisions should be based on as wide a range of taxonomic groups as possible (e.g. Kremen et al. 2008). In TEA, data for birds and large diurnal mammals are widely available and relatively cheap to collect, since most can be identified in the field from standard manuals. Data for trees are probably the next most widely available, despite the need for relatively rare field-skills, backed up by a good herbarium. Data for all other taxa are patchy, reflecting both a lack of expertise and a lack of demand. An inevitable consequence of this has been that conservation planning in the region has been based almost entirely on large mammals, birds, and/or trees, with input for other taxa of only local importance. Conservation funding from international donors is increasingly concentrated on large areas supporting charismatic large vertebrates, so there is a real risk that small sites with high plant and/or invertebrate diversity will be overlooked.

In some cases, decisions will have to be made without adequate data on any group of organisms; for example, when establishing protected areas at remote and inaccessible sites. In such cases, vegetation will often be the best surrogate available, although remote sensing from satellites offers a variety of other possible indices (Ranganathan et al. 2007). The use of remote-sensing data as a surrogate for biodiversity is likely to be misleading on a regional scale, particularly where hunting has created 'empty forests', but vegetation maps can potentially be useful for extrapolating data from small sample plots to the surrounding landscape. It is also possible to use environmental data alone (e.g. temperature, rainfall, and geology), on the assumption that selecting sites covering a wide range of environmental conditions ensures comprehensive coverage of species. While this approach is likely to improve on random site-selection, simulations and some real-world data suggest that its effectiveness is much improved when the relationship between environmental and community gradients can be calibrated with species-distribution data (Arponen et al. 2008).

It is important, however, not to let concerns about the effectiveness of conservation surrogates delay effective conservation action. A recent study that looked at the optimum allocation of global conservation spending concluded that these allocations were considerably less sensitive to the taxonomic group assessed than to the socio-economic factors that determined threats and costs (Bode et al. 2008; Polasky 2008). Indeed, when these factors were taken into account, the use of information on endemicity in vascular plants and vertebrates resulted in very similar funding allocations, i.e. the choice of surrogate did not matter. It is not clear if the same would be true on the finer scale at which real conservation happens, but it is always useful to remember that, while the goals of conservation are biological, the means are largely socioeconomic (Polasky 2008).

8.3.2 Setting priorities

It will not be possible to protect all the surviving biodiversity in TEA immediately, so some species or areas will be protected first and some later, or not at all. Ideally, priorities for conservation action would be determined by high vulnerability (i.e. a high likelihood of being lost in the short term) and, for areas, high irreplaceability (i.e. a lack of alternative options). In practice, even after these filters, more species and areas are in need of immediate conservation action than there are resources available. The limiting resources may not only be money and land, but also trained conservation professionals, and both public and political support. We, therefore, must decide not only what *needs* saving, but also, within this list, what we will try to save with the resources we have.

The IUCN's Red List of Threatened Species is the most authoritative source of information on the global conservation status of species and recognizes >15,000 species as threatened (IUCN 2004; Rodrigues et al. 2006). Unfortunately, the coverage of the IUCN lists is limited by lack of data and resources. Less than 3% of known species have been evaluated so far, with a strong bias towards vertebrates and plants. Only amphibians, birds, mammals, conifers, and cycads have been comprehensively assessed, but turtles and tortoises are almost done, the assessments of the legumes and palms are in process, and the assessment of all plants is being given a high priority. The IUCN Red List categories and criteria (IUCN 2001, 2005) are intended for the assessment of global status, but guidelines have also been developed for assessments at regional and national level (IUCN 2003) (Fig. 8.6). The regional guidelines include an extra category—regionally extinct—and an assessment of the potential for rescue of regional populations by immigrants from outside the region. The application of the criteria to small countries can result in a disproportionately high number of species under threat (Miller at al. 2007), but this will cause problems for prioritization only if the global status of species with tiny national populations is ignored when setting national priorities. The development of national red lists using the standard IUCN criteria would be a useful contribution to prioritization, but it would be even more useful to develop red lists for the region as a whole or for sub-regions, such as Sundaland and Indochina (see 3.8).

Which species should have the highest conservation priority in TEA? If coverage in the media was taken as a guide to priorities, then the answer would be tigers, rhinoceroses, elephants, and orangutans. To some extent, this reflects the use of these large and charismatic taxa as fund-raising 'flagships' for biodiversity as a whole, but, as mentioned above, there has also been a tendency to use these taxa as surrogates for biodiversity when planning

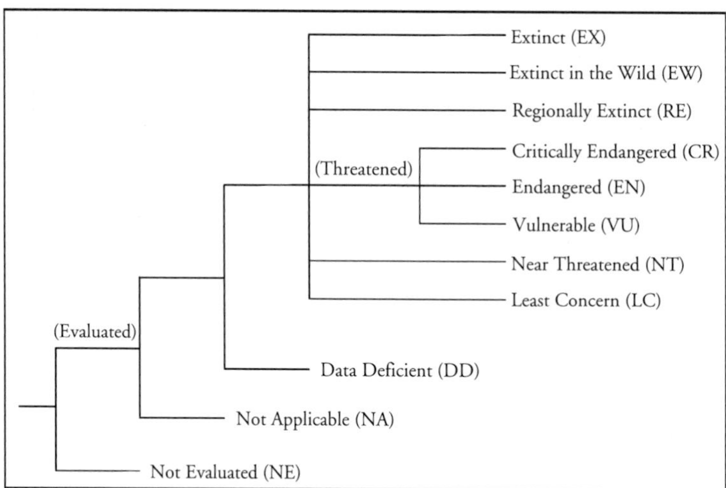

Figure 8.6 IUCN categories for use at the regional and national level. Re-drawn from IUCN (2003).

for conservation. Charisma, however, is not the ideal way of prioritizing endangered species for conservation action. Among the alternatives that have been suggested are: ecological uniqueness (i.e. prioritize species with a unique role in ecosystem function), potential economic value (e.g. the wild relatives of crop plants and domestic animals), and phylogenetic distinctiveness (i.e. prioritize species with no close relatives).

The EDGE (Evolutionarily Distinct and Globally Endangered) programme of the Zoological Society of London prioritizes mammals and amphibians based on the amount of unique evolutionary history each species represents (Isaac et al. 2007). This is then combined with a global endangerment score derived from the IUCN Red Lists to give an EDGE score for each species. The top 100 EDGE mammals include, in TEA, the Sumatran and Javan rhinoceroses (ranked 6 and 11, respectively), the Asian elephant (12), and the orangutan (97), but also the less well-known Sumatran (10) and Amami (44) rabbits, Vietnam leaf-nosed bat (15), Hainan gymnure (45) and Dinagat and Mindanao gymnures (both 47), bumblebee bat (49), Muennink's spiny rat (51), Malayan and Sumatran water shrews (both 58), Senkaku mole (68), and Philippine flying lemur (71). The next update is likely to include the recently discovered Laotian rock-rat, *Laonastes aenigmamus*, because of its ancient divergence (*c*.44 million years ago) from the nearest living relatives (Huchon et al. 2007) (Fig. 8.7). Nobody is suggesting that phylogenetic distinctiveness should be the only basis for prioritization—ignoring, for example, the key ecological roles of big cats, pigs, rats and monkeys. However, the argument that the loss of millions of years of evolutionary history is a greater disaster than losing a recently diverged species with non-threatened relatives seems incontrovertible. The currently identified EDGE species deserve priority and we need to extend the concept to other taxa as phylogenies improve.

In theory, when priorities have been assigned to species, the occurrence of these could then be used to prioritize areas, although the data incompleteness,

Figure 8.7 The recently discovered Laotian rock-rat, *Laonastes aenigmamus*, diverged tens of millions of years ago from its nearest living relatives (Huchon et al. 2007). It is an example of a species that deserves conservation priority because of its phylogenetic distinctiveness. Photograph © Uthai Treesucon, the David Redfield Expedition.

mentioned in the previous section, means that there is usually a degree of subjectivity in this. Threatened ecosystems, such as beach forests, may also merit special protection, even if they lack priority species. Three international NGOs have independently developed global priority schemes. Conservation International's 34 biodiversity hotspots each have more than 1500 endemic plant species and have lost at least 70% of their original habitat (Mittermeier et al. 2004). The 34 hotspots originally covered 15.7% of the Earth's land surface, but with 86% of their habitat already destroyed, the intact remnants now cover only 2.3%. WWF's Global 200 Eco-regions includes 142 'irreplaceable' terrestrial eco-regions covering 40% of all land on Earth, although intact remnants make up far less (Olson and Dinerstein 2002). Both these schemes include most of TEA (Fig. 8.8), so, although they may be useful in attracting conservation funding to the region, they are little use for setting priorities within it.

WWF's eco-regions, however, can usefully be ranked in other ways in order to identify regional priorities. Fa and Funk (2007) used endemic vertebrates to rank 796 global eco-regions by their importance for vertebrate conservation. TEA includes 16 of the top 100, with the Andaman, Ogasawara, Ryukyu (Nansei), Palawan, Nicobar, and Mentawai Islands all in the top 50, and much of the Philippines and Sulawesi in the next 50. Krupnick and Kress (2003) ranked eco-regions in the Indo-Pacific

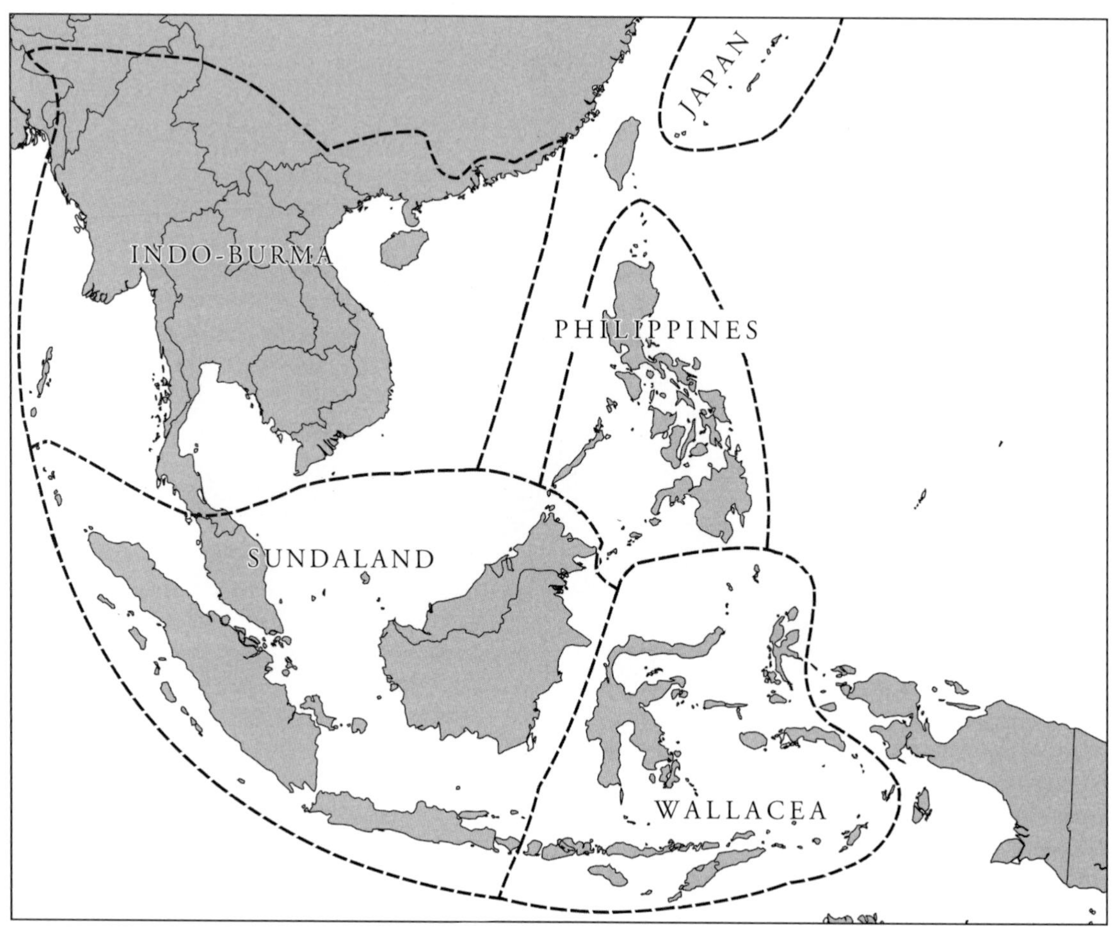

Figure 8.8 Almost all of Tropical East Asia is included in one of Conservation International's Biodiversity Hotspots (Mittermeier et al. 2004). Thus, while they may be useful for attracting conservation funds to the region, they are not useful for setting priorities within it.

region (including New Guinea and excluding China) on the basis of species-richness in 7 plant families. The top 8 eco-regions in TEA were Borneo lowland forest, Luzon rain forest, Southern Annamites montane rain forest, Peninsular Malaysia rain forest, Borneo montane forest, Northern Indochina subtropical forests, Mindanao-Eastern Visayas rain forests, and Kinabalu montane alpine forests. The large differences between these two rankings are partly a reflection of the different taxa used, but also the different weights given to endemism.

In contrast to the Hotspots and Global 200 Eco-regions, BirdLife International's Important Bird Areas (IBAs), which are selected using a standard set of criteria, are small enough for protection in their entirety, making them of direct relevance to fine-scale conservation planning for birds. BirdLife's national partners take responsibility for monitoring and trying to protect IBAs in their country, giving the IBA system a strong local influence. The IBA idea has been generalized from birds to all known biodiversity, resulting in the concept of Key Biodiversity Areas (KBAs), which are sites of global significance for conservation that are large enough to support populations of the species for which they are important (Langhammer et al. 2007; Brooks et al. 2008). As with IBAs, the use of standard criteria and thresholds ensures that the KBA approach is repeatable around the world (Table 8.2).

Within TEA, the Philippines has gone furthest in identifying KBAs, using data for freshwater fishes, amphibians, reptiles, and mammals to build on the IBAs already identified (Anon. 2007). Areas that were suspected to be important, but for which crucial data were lacking, were designated as 'candidate KBAs'. The number of KBAs (128) and candidate KBAs (51) identified so far exceeds the limited resources available for protection in the Philippines and more data are needed for prioritization, but 10 sites, whose loss would lead to the global extinction of one or more species, have been added to a global list of Alliance for Zero Extinction (AZE) sites. Even for well-studied groups, such as mammals, the data for most of TEA are probably still too incomplete to identify a comprehensive regional set of priority areas on this fine scale, but a first approximation would help to focus attention on some of the key areas.

In practice, prioritization of both species and areas will have to take into account non-biological factors, such as cultural importance, public appeal, feasibility, and cost (Miller et al. 2007; Bode et al. 2008; Potts and Vincent 2008). One possible approach is to use cost–benefit analysis to maximize the conservation-gain achieved for the money spent

Table 8.2 Provisional criteria and thresholds for Key Biodiversity Areas (KBAs) (Langhammer et al. 2007).

Criterion	Sub-criteria	Provisional thresholds for triggering KBA status
Vulnerabilty		1 individual of a Critically Endangered (CR) or Endangered (EN) species; 30 individuals or 10 pairs of a Vulnerable (VU) species.
Regular occurrence of a globally threatened species according to the IUCN Red List at the site.		
Irreplaceability	(a) Restricted range species.	Species with a global range of <50,000 km^2; 5% of the global population.
Site holds X% of the global population at any stage of the life-cycle		
	(b) Species with large but clumped distributions.	5% of the global population.
	(c) Globally significant congregations.	1% of the global population.
	(d) Globally significant source populations.	Site is responsible for maintaining 1% of the global population.
	(e) Bioregionally restricted assemblages.	To be defined.

(Naidoo et al. 2006; Underwood et al. 2008). Prioritizations that ignore cost, implicitly assume that this is the same for all species and areas, which is obviously not true. Such an approach would focus conservation resources in TEA on areas with many poorly protected, high-priority species (e.g. the Philippines), and relatively low land and labour costs (e.g. Laos). Other factors, such as the risk of investment failure as a result of corruption or political instability, could also be taken into account. Political boundaries will make this return-on-investment approach to prioritization difficult for national governments, who are expected to spend conservation funds at home, but it should be more practical for donors from outside the region.

It may also be possible to plan protected area networks that simultaneously prioritize the protection of biodiversity and the maintenance of ecosystem services, such as water provision, flood control, carbon storage, and outdoor recreation (Chan et al. 2006; Naidoo et al. 2008). Although there may be some trade-offs between conserving biodiversity and providing services, this approach is likely to increase local support and may attract new sources of funding.

'Prioritization' has become a science in itself that all too often has little, if any, impact on the selection of conservation areas. A recent review concluded that 'two-thirds of conservation assessments published in the peer-reviewed scientific literature do not deliver conservation action, primarily because most researchers never plan for implementation' (Knight et al. 2008). Among the recommendations for avoiding this 'research-implementation gap' is that researchers should formulate their research questions in collaboration with other stakeholders, to ensure that they are meeting real needs.

8.4 Protecting areas

8.4.1 Establishing new protected areas

Protected areas (PAs) are the cornerstone of global biodiversity conservation. Their role is to separate conservation from incompatible land uses, such as agriculture, and to exclude other damaging human activities, such as logging and hunting. Recent reviews have suggested that tropical PAs are, in general, fairly effective, in that the vegetation and fauna are usually in considerably better condition inside parks than outside, but that they are less effective in some parts of TEA than elsewhere in the tropics (Bruner et al. 2001; Curran et al. 2004; Linkie et al. 2004; DeFries et al. 2005; Gaveau et al. 2007; Lee et al. 2007; Wright et al. 2007; Bickford et al. 2008). Within TEA, PAs in Cambodia, Kalimantan, and Sumatra appear to have the biggest problems, from agricultural encroachment, illegal logging, and hunting, with the protective effect of PA designation almost non-existent in the worst cases.

Note, however, that evaluating PA success by comparison with adjacent unprotected areas ignores the fact that many PAs are deliberately located in areas that are at less risk of disturbance (e.g. on steep sites or infertile soils), so the effect of PA designation is confounded with pre-existing differences (Andam et al. 2008). In some cases the effectiveness of the legal protection at withstanding human impacts has simply never been tested. Moreover, an effective PA may simply displace exploitative activities to an unprotected area nearby, reducing the overall benefit (Ewers and Rodrigues 2008).

PAs in populated areas increase the potential for conflict between conservation and human livelihoods, but the available evidence suggests that there is no alternative if more-or-less intact ecosystems, with their full complement of disturbance- and exploitation-sensitive species, are to survive anywhere in TEA. In view of the current explosive expansion of development and exploitation in almost all of TEA, the next decade will probably be the last opportunity to protect additional large areas, particularly in the lowlands (Sodhi et al. 2004; Lee et al. 2007). Although some habitats, such as montane forests in the Sundaland region, may appear relatively unthreatened at present, new threats (such as new crops) can arise quickly and expand too fast for effective mitigation. The precautionary principle (i.e. that one should take action now when the potential damage is great, even if the evidence is inconclusive) suggests, therefore, that the regional PA system should be large and comprehensive enough to preserve all biodiversity, even if all other habitats are lost. In reality, it may already be too late for this, but while extinctions remain few (see 7.5), there is still some hope that disastrous future losses can be avoided.

8.4.2 Enhancing protection for existing protected areas

In much of the region, protecting new areas may be less urgent than enhancing the protection of existing PAs. Doubling the area of 'paper parks' may be less effective than converting the existing ones from paper to reality (although many conservationists view the creation of paper parks as a useful 'foot in the door'). Poverty and corruption are intertwined causes of PA failure, along with poor initial planning that did not consider all stakeholders (Sodhi et al. 2004, 2008; Wright et al. 2007). Increases in funding from national, regional, and international sources can increase protection by providing more, better-trained, and better-motivated staff (Terborgh et al. 2002; Fig. 8.9), but it will also be necessary in many cases to deal with conflicts with local communities and/or influential outsiders.

Even where the PAs themselves are effective, recent satellite-based surveys (and my personal experience on the ground) have shown a rapid increase in human populations and intensification of land-use around many PAs, leaving them increasingly isolated (DeFries et al. 2005). The long-term impacts of this isolation will depend on the size of the PA and the extent to which important habitats (such as water sources and dry season feeding areas) are excluded, as well as the direct impacts of human populations on the reserve biota (Hansen and DeFries 2007). Becker et al. (2007) show that the 'habitat split' that results from the exclusion from protected forest patches of streams and other water bodies needed for breeding is a major problem for forest-associated amphibians in the highly fragmented Brazilian Atlantic Forest, and suggest that this may also be a problem for invertebrates with aquatic larvae and some waterbirds. Mitigating these problems will often require that conservation management extends beyond the boundaries of the PA, increasing the potential for conflicts with human livelihoods. DeFries et al. (2007) suggest that the key management challenge is to identify 'small

Figure 8.9 A Hong Kong-based NGO, Kadoorie Farm and Botanic Garden (KFBG), has put a lot of effort into increasing the capacity and motivation of the field staff at Bawangling National Nature Reserve, Hainan, to conserve the world's rarest gibbon, *Nomascus hainanus* (Fellowes et al. 2008). Almost everything worn or carried by the Reserve's gibbon team in this photograph was supplied by KFBG, including the uniform, with a specially designed logo, and the binoculars, cameras, and GPS. Photograph © KFBG.

loss–big gain' opportunities, which maximize benefits for conservation and minimize negative consequences for human livelihoods.

8.4.3 Community-based conservation

Traditional PAs have often been derided as 'fortress conservation', 'the fences and fines approach', or 'command and control'. They can work, if well-staffed and adequately funded, but at a cost to local people that can become prohibitive in areas with high population densities and/or a long tradition of human use (Sodhi et al. 2008). This has sparked searches for alternative means of achieving biodiversity conservation, in collaboration, rather than competition, with local communities. There is a large and very varied literature on this, with rather few general lessons so far, except that there is no one model that can be adopted everywhere. Key variables include the nature of the traditional uses of the area and the ability of local communities to control incursions by outsiders. Another variable is the effectiveness of government-run PAs, with local communities probably the best hope in those areas where existing PAs are completely ineffective.

Regional experience with community-based conservation has largely involved marine-protected areas (e.g. Alcala and Russ 2006) and it is not clear whether these models can be transferred to land. Community-based conservation is clearly most effective when it involves continuation of traditional practices, such as the annual burning of the Tonle Sap grasslands in Cambodia that provides fodder for domestic cattle and habitat for the endangered Bengal florican (Gray et al. 2007). Many conservationists fear that short-term success with community-based conservation will not translate into long-term protection, while proponents argue that the approach has not yet been given a fair trial.

There should be significant local representation on the management committees of all PAs, but local control is likely to be most effective (and least risky) with small areas that are an addition to the big PAs, rather than an alternative. Remnant forest patches, in otherwise deforested landscapes, are good examples of situations where local management could be more effective than attempts at remote control by urban authorities. Recently deforested regions may lack the community traditions that make social pressures an effective deterrent, but local schools, NGOs, or religious institutions may be effective substitutes. Note also that community conservation need not be area-based, but could act through controls on exploitation (see 8.5).

As with so many new approaches to biodiversity conservation, properly documented experiments are urgently needed in order to substitute evidence for empty argument. Proponents of strict, centrally controlled, well-defended PAs as the only conservation model, tend to forget the diversity of successful models that exist in the developed world, including categories such as Britain's National Parks, which are home to over 300,000 people, many of whom are farmers. The key to these and other successful combinations of conservation and human use is a reasonable, but enforceable, compromise. These compromises do not substitute for strictly protected areas, but they can support many of the more tolerant species and provide a buffer that enhances the capacity of strict PAs. TEA needs both more PAs and more *types* of PAs.

8.4.4 Integrated Conservation and Development Projects (ICDP)

It is also extremely important that lessons are learned from past failures. For more than a decade, a prominent approach to biodiversity conservation in developing countries was the Integrated Conservation and Development Project (ICDP), which linked conservation of biodiversity with the economic development of neighbouring communities. Huge amounts of both conservation and development funding have been sunk into such projects in TEA, despite little evidence for success in either objective, or in reconciling the two (Terborgh et al. 2002; McShane and Wells 2004). The one published evaluation of the effectiveness of an ICDP in TEA—the World Bank funded US$19 million Kerinci Seblat ICDP—showed that participation in the ICDP had no impact on deforestation, although an earlier community questionnaire had shown strong support for conservation in ICDP villages (Linkie et al. 2008). There may be successful examples of ICDPs in TEA, but published evidence is lacking.

8.5 Sustainable exploitation

In theory, the sustainable exploitation of forest resources could encourage the retention of forest cover, rather than its replacement by alternative land-uses that would support far less biodiversity than even the most intensively managed forests. In practice, however, sustainable exploitation is often a much more difficult target to achieve than either protection (i.e. no exploitation) or unsustainable 'mining' of forest resources.

8.5.1 Logging and the collection of non-timber forest products

Most tropical timber at present comes from natural forests, rather than plantations, and, although the total area of timber plantations is growing rapidly in TEA, there is currently no immediate prospect of plantations replacing natural forests as a source of durable, high-grade timbers (Fredericksen and Putz 2003). Management for a sustained yield of high-quality timber is currently the best hope for an economic return from natural forest cover, although payments for environmental services, and both carbon and biodiversity offsets, may soon be competitive alternative—or additional—sources of income. The huge diversity of non-timber forest products (Table 7.3) makes it impossible to generalize, but sustainable management is a potentially attainable target for many of these products.

Logging can be massively damaging to tropical forests, particularly when the density of valuable trees is very high, as it often is in TEA (see 7.4.6). However, the damage caused directly by logging can be reduced considerably by the adoption of a variety of low-impact techniques, collectively known as Reduced Impact Logging (RIL) (Sessions 2007; Putz et al. 2008a). These include the directional felling of trees to minimize damage to other woody plants, including the future tree crop, the careful planning of the logging roads and the skid trails along which logs are extracted, and the avoidance of logging on steep slopes and along watercourses. Whether or not the use of RIL can pay for itself in the short term through increased efficiency is unclear, but wider adoption in legal, logging operations could be ensured by a combination of carrot (certification, carbon credits) and stick (fines for non-compliance) approaches. Carbon credits could be a particularly effective tool for improving forest management, given the huge potential for reducing carbon emissions from intensive logging in the region (Putz et al. 2008b). At the same time, it is essential that the secondary impacts of logging—hunting, fire, clearance—are also controlled. Increased road access is usually the major driver of these secondary impacts, so installing manned barriers during logging operations and permanent closure of roads (with deep ditches or large logs) after logging ceases should reduce them (Sessions 2007). Meijaard et al. (2005) give detailed guidelines for managing production forestry in Borneo in a way that minimizes adverse impacts on biodiversity.

Most logging in TEA currently is illegal (see 7.4.6), so the logger has no financial interest in protecting soils and the future tree crop, while wildlife is a useful source of food for logging crews. The control of illegal logging must clearly be a priority for governments throughout the region, but this will not be achieved without addressing corruption at multiple levels (Ravenel and Granoff 2004; EIA/Telepak 2008). It is not just a matter of poor villagers with chainsaws: illegal logging would not be profitable without the chain of illegal activities that link the forest with national and international markets. Enforcement at the forest level is important, but action is also needed against the other key players in the system. At the far end of the chain, the supply of illegally logged timber and products made from it, should be a criminal offence in major-consumer countries, such as China and the EU, as it has recently become in the USA. The best hope for the future may lie in simple embarrassment at this massive failure of national and international governance, but whether this can outweigh the financial incentives that keep the trade going, is currently not clear. In TEA, control of illegal logging will probably only be achieved if legal logging is expanded and its profitability is not reduced too much by onerous conditions. Conversely, sustainable-forest management will not be competitive, as long as cheap timber from illegal operations is readily available.

During the next decade, all accessible forests in TEA will be either protected or will have had their most valuable timber removed (or both), so attention will need to shift from short-term logging impacts to the long-term management of production forests. If managed natural forests are to be competitive with oil palm and fast-growing timber plantations, the economic yield will have to increase, from some combination of higher prices for certified natural timbers and non-timber products, intensified post-harvest management to promote regeneration and growth of the most valuable species, payments for environmental services, and carbon and biodiversity offsets. Even if protection of biodiversity is not directly targeted, the evidence suggests that, with protection from hunting and clearance, such long-term production-forests could support all but the most disturbance-sensitive of native forest taxa, although their relative abundances may change greatly (Sodhi et al. 2005b; Meijaard and Sheil 2008; see 7.4.6). Conversely, excessive pressure to reduce the direct impacts of logging could promote deforestation by reducing the incentive to maintain forest cover.

8.5.2 Hunting

There is a large literature that addresses the issue of sustainable hunting, but little prospect of this goal being attained in the medium-term in TEA, where there are simply too many people and too little forest (Bennett 2007; Corlett 2007a). In theory, a sustained yield of pigs and, perhaps, deer, could be obtained from managed production-forests, with additional income from the sale of hunting permits. In practice, however, there are few places in the region where this could be effectively policed, so legal hunting of pigs and deer would also increase the chance that other species were killed opportunistically. A big increase in the effectiveness of current enforcement activities is, therefore, needed before licensed hunting can be considered. Moreover, there is no reason to believe that hunting at a level that yields a sustained output of meat or recreation would also maintain services, such as seed dispersal. Enforcement is particularly important in existing protected areas. There is evidence from the literature that enforcement can boost wildlife populations, but also overwhelming evidence that it currently is not working in TEA (Corlett 2007a). Regional and global trade drives the exploitation of rare, high-value, species, so the control of this trade (see below) will have an impact on hunting activity.

8.5.3 Controlling trade

On paper, the control of the international trade in endangered species and their products is a conservation success story. Since 1975, the Convention on International Trade in Endangered Species of Wild Flora and Fauna (CITES) has expanded to include 172 countries as signatory parties. Control of international trade is exerted by parties passing and enforcing laws in their own countries to ensure that the *c.*900 species on Appendix I are not traded at all and the 33,000 species on Appendix II are traded only if this will not risk the survival of the species in the wild. All countries in TEA are parties, although efforts at enforcement vary considerably, both between countries, and between entry and exit points within each country. The regional trade is monitored by TRAFFIC, the international wildlife-trade monitoring network, through its East Asia and South-East Asia offices.

CITES is increasingly being used to control commodities, such as fish and timber, and it is a potential tool to reduce illegal logging of tree species of conservation concern (Chen 2006; Johnson 2007). When a tree species, such as ramin (*Gonystylus bancanus*), is listed on Appendix II, any international trade requires a certificate from the exporting country stating that the exports will not be detrimental to the survival of the species in the wild—a so-called 'non-detriment finding'. New molecular techniques make it possible to trace traded plant and animal parts back to their origins (e.g. Wasser et al. 2007; Smulders et al. 2008).

In practice, CITES has had much less impact in TEA than the above would suggest. Part of the problem is that it aims to control only *international* trade, so domestic trade in the same species may continue to be legal. Other parts of the problem are porous borders, corruption, indifferent officials, reluctance to list regionally endangered species, and insufficient investment in border enforcement. The net result is that trade is still a major threat to endangered species in TEA (Fig. 7.8), despite some local successes, particularly in curbing urban wildlife markets (Bennett 2007; Corlett 2007a).

Like most other conservation interventions in TEA, measures to control the wildlife trade have never been directly evaluated by their impact on wild populations of endangered species. Instead, indirect indicators, such as reports published, workshops held, and people trained, are used as evidence of progress. However, it is clear that almost all trade species are declining in the wild, some precipitously, and that trade is driving much of this decline (Bennett 2007; Corlett 2007a). Overall, therefore, trade interventions are failing. Indeed, a striking feature of many recent reports on the illegal wildlife trade in TEA is the openness with which the trade is carried out (e.g. Ng and Nemora 2007; Shepherd and Nijman 2007). This reflects a combination of genuine ignorance and erratic or non-existent enforcement.

Enforcement may not be enough to completely halt the trade, given the financial incentives, but it would undoubtedly reduce it. The formation of the ASEAN–Wildlife Enforcement Network in 2005, to coordinate enforcement among the ten member states of the Association of South-East Asian Nations (ASEAN), was a big step in the right direction, but the battle is far from won. It is not just a matter of policing the forest and physical markets, as the Internet is becoming an increasingly important forum for the trade (Wu 2007). Action is also needed at the consumer end of the trade chain, in order to reduce demand, particularly in the major East Asian markets (China, Japan, South Korea), but also in urban areas throughout TEA and in Asian expatriate communities worldwide. A recent survey in China found that almost all respondents thought it was important to protect wild tigers, but that 43% had used a product claiming to contain tiger parts, even though most knew this to be illegal (Gratwicke et al. 2008). Education is clearly important, with television the ideal medium for linking consumer actions with conservation impacts, but enforcement and the resulting publicity can help reinforce the message.

8.6 Managing the matrix

The area under formal protection will not be enough. Even if permanent production forests are sustainably managed, much of TEA's biodiversity will survive, if at all, in the human-dominated matrix between the remaining large areas of forest. Very few forest species can adapt to tree-less agricultural and urban habitats, but some can persist in plantations, tree crops, and agroforestry, while others can make use of forest regrowth on abandoned land and in narrow strips of along streams and fences, or survive in tiny patches of primary forest (Corlett 2000; Harvey et al. 2008). As discussed in the previous chapter, the species diversity in recently deforested landscapes is almost certainly a misleading indicator of long-term persistence, but Hong Kong, which was deforested centuries ago and supports few large vertebrates, still retains a rich flora and invertebrate fauna (Hau et al. 2005).

The available evidence suggests that retaining as much tree cover as possible at a landscape level is the key to the persistence of forest species (e.g. Lee et al. 2007). It may be too great an extrapolation from this to assume that increasing tree cover will lead to the re-establishment of lost forest species, or permit their reintroduction, so experiments are needed. In Europe and many other parts of the world, tree planting on private land is encouraged by financial incentives, coupled with the availability of advice and appropriate planting material, and there seems to be no reason why the same methods could not work in TEA. For shade-tolerant crops, the promotion of agroforestry systems with native shade trees can extend tree cover in the matrix, but is likely to require both financial incentives (payment for environmental services, carbon offsets, eco-certification), and education (Bhagwat et al. 2008). A similar combination of incentives and education will also be needed in many areas to reduce the current excessive use of agrochemicals.

In many parts of TEA, matrix habitats are heavily hunted, thus greatly reducing their carrying capacity for native wildlife. The sustainable hunting of pigs, deer, and small mammals in matrix habitats may be viewed as a welcome alternative to hunting in protected areas and managed forests, but enforcement is almost everywhere insufficient to ensure that protected areas and species are avoided. The potential cultural, recreational, and subsistence benefits from hunting need to be weighed against the biodiversity benefits of a blanket ban, except where hunting is needed to protect crops.

The vulnerability of matrix habitats to invasive species is another potential problem. Invasion by the yellow crazy ant (*Anoplolepis gracilipes*) reduces native ant diversity in Indonesian cacao agroforests (Bos et al. 2008), and it is likely that other invasive plants, vertebrates, and invertebrates are having a similar impact, despite the lack of case studies. Direct control of invasives (chemical, physical, or biological) is rarely a practical option, except in the most valuable habitats, but inexpensive changes in management may help in some cases (see 8.7.2).

Linking biodiversity conservation with sustainable agriculture is arguably one of the key global challenges for the next few decades (Scherr and McNeely 2008). A lot of progress has been made in various parts of the world, but very little of this has been in the tropics. The focus of tropical research has so far been more on traditional production systems that are in decline (e.g. homegardens; see 2.5.1), rather than on improvements to the agricultural landscapes that are replacing them (e.g. oil palm plantations). These priorities need to be reversed. A recent study in Borneo, for example, showed that the persistence of biodiversity (birds and butterflies) in oil palm estates is only marginally improved by increasing ground cover and epiphyte abundance within estates, but is significantly enhanced by retaining as much natural forest as possible in the surrounding landscape (Koh 2008a). The birds in turn reduce insect damage to the oil palm, providing a potential incentive (although a minor one) for promoting bird diversity in plantations (Koh 2008b).

8.7 Managing other threats

8.7.1 Controlling fires

Anthropogenic fires are both a conservation and a public health issue in TEA (see 7.4.9). Most fires in the region are probably illegal and the large extent of many is undoubtedly accidental, but active fire-fighting is virtually non-existent in much of TEA. In areas with a prolonged dry season, this is largely because fires are seen as a normal, seasonal phenomenon, with neutral or beneficial impacts. In rainforest areas, in contrast, fires are widely recognized as damaging, but also as the cheapest way of clearing land (Fig. 7.9). As public health problems from fire haze have increased, a lot of money and political pressure has been applied to the problem, but with no visible impact so far (Lohman et al. 2007; Tacconi et al. 2007).

If there was a single simple solution to the fire problem it would have been used already, but the situation need not be as bad as it is now. The proportion of unintentional fires or, at least, unintentional spread of fires away from their starting point, is unclear, but may be large (e.g. Giri and Shreshta 2000). Public education through radio, television, and schools could be as effective in reducing these here as it has been in the developed world. Other fires are deliberate, but unnecessary, and could be reduced by a combination of education and enforcement. The most difficult fires to control will be those lit in seasonal forest by hunters and collectors of non-timber forest products (NTFPs), and those used to clear forest or regrowth for agriculture. Such fires are deliberate and the results are beneficial to those who light them. Enforcement resources will always be limited, so it makes sense to concentrate effort initially on the peat-swamp fires that produce the bulk of the soot and carbon dioxide (Lohman et al. 2007), although even this is a huge target. For the rest, the use of incentives to companies and local communities, perhaps funded by carbon offsets, needs to be explored as a supplement to ineffective law enforcement.

8.7.2 Managing invasive species

Although invasive species are not yet a major threat to intact natural vegetation in TEA, except on oceanic islands, this is a problem that can only get worse as more alien species are introduced (Corlett 2009a) (see 7.4.10). Moreover, aliens already dominate in many anthropogenic habitats, where they may interfere with attempts to enhance the biodiversity value of such areas (see 8.6). Research has shown that introduction is not a simple all-or-nothing process, but that propagule pressure—the number of individuals released—is a major driver for the establishment and invasive spread of exotic species (Reaser et al. 2008). For cultivated ornamental or plantation species, for example, the propagule pressure increases with both the time since first introduction and the number of individuals planted (Kiřvánek et al. 2006; Dehnen-Schmutz et al. 2007). Import numbers may be a reasonable proxy for pet animals, since high numbers increase the risk of both accidental escapes and deliberate releases. For birds, however, origin is more important than numbers, since wild-caught pet species are much better invasives than captive-bred ones (Carrete and Tella 2008).

Prevention is better than cure, but TEA is a region of trade-dependent economies and porous borders, making the strict border-quarantine, practised in countries like Australia, impractical. Improvements to quarantine in TEA need to focus, therefore, on the major pathways by which potential invasives enter the region. Although there is little data from TEA itself, international experience leaves little doubt that the most important of these pathways are the horticultural trade—for the ornamental plants themselves, as well as associated weeds, pests, and pathogens—and the pet trade. In both trades, unfortunately, novelty is an attraction, so 'blacklists' of potential invasives are rapidly outdated. The alternative is a 'whitelist', restricting imports to species considered low risk on the basis of previous experience in TEA or in similar environments elsewhere. For vertebrates, where the ecological arguments are reinforced by concerns for public health and animal welfare, a whitelist, backed up by a blanket ban on the trade in wild-caught animals, would make a lot of sense, but the horticultural trade would undoubtedly see this approach as far too restrictive for plants. An additional problem with plants is the ease with which seeds can be purchased on the Internet and imported by post.

Improvements to border controls will need to be backed up by practices aimed at reducing propagule pressure *after* arrival, in the sometimes prolonged period when an exotic species exists only in cultivation or captivity, and as a few casual escapes. This requires early detection of potential problems followed by immediate action to eliminate or minimize the threat. Neither is happening at present. New invasions are detected by chance, if at all, and attempts at control are *ad hoc* and usually insufficient. Species that have already become invasive in one part of TEA are ignored—or even encouraged—in other parts. What is needed is an early warning system, whereby reports of new invasive threats are encouraged and acted upon. The region's legions of birdwatchers, for example, provide an effective detection system for new bird invasions and other biodiversity NGOs should be encouraged to see the early detection of invasives as part of their function. Concentrations of exotic species, such as zoos and botanical gardens, need to have strategies in place to minimize the risk of escapes and to detect and control those that occur (Dawson et al. 2008). Deliberate releases of captive animals—a widespread practice in TEA, particularly among the region's Buddhists (Fig. 8.10)—must be banned, along with the even more common practice of dumping unwanted pets. As with many other ecologically harmful but socially accepted behaviours, public education is essential, but will usually need to be reinforced by prosecutions of the worst offenders.

The eradication of invasive rodents from small oceanic islands is becoming routine (Howald et al. 2007) and it is likely that an equal effort applied to other taxa would be successful. On the mainland and large islands, however, eradication of well-established aliens is unlikely to be successful and risks collateral damage to native species (Denslow 2007). Instead, the focus should be on reduction of adverse impacts, by managing ecosystems to reduce their invasibility (e.g. encouraging woody canopy closure in forests) and, where necessary, by mechanical, chemical, or biological control. Biological control has had a bad press in recent decades, as the risk of side-effects on non-target organisms have become apparent, but the deliberate release of carefully screened natural enemies (predators, pathogens, parasitoids, and herbivores) is still an indispensable tool for the management of invasive species (Messing and Wright 2006; Van Driesche et al. 2008), and one that is probably underutilized in TEA.

8.7.3 Mitigating climate change impacts

There appear to be few practical options for mitigating the impacts of anthropogenic climate change on biodiversity in TEA. The most obvious ones are also the most difficult to achieve under current conditions: minimizing deforestation; maximizing connectivity of the remaining forests across rainfall and

Figure 8.10 Buddhist groups in Hong Kong and elsewhere in the region release large numbers of birds, including non-native species, and are responsible for many bird invasions. The gloves and face-masks are to guard against infection from avian influenza. Photograph © Chan Sin Wai.

temperature gradients in order to facilitate species movements ('conservation corridors'; e.g. Killeen and Solórzano 2008); minimizing other human impacts, particularly those that tend to open up the canopy; controlling fire use; and enhancing the ability of the matrix between forest patches to support native species. New funding for carbon offsets may provide incentives for all of these actions, but it is more likely that fragmentation, logging, and the degradation of deforested areas will all increase, at least in the medium term, further increasing the vulnerability of the region's biota to climate change.

Even if habitat connectivity could be maintained where still present and restored where lost, the movement speeds needed to keep up with shifting climate envelopes will be too great for many species (Corlett 2009b) (see 7.4.13). Moreover, vegetation connectivity is no help for species confined to immovable geologies or topographies, such as limestone outcrops. The best option in such cases may be human-assisted migration, but this is a highly contentious issue at present (McLachlan et al. 2007). The problem is not so much that of moving a species from where it is threatened by changing climate, but of introducing it to an ecosystem outside its current natural range. There are two opposite concerns in regard to the receptor ecosystem. On the one hand, the introduced species may fail as a result of poor environmental matching and/or interactions with species already present, while on the other it may become invasive as a result of escape from natural enemies and other controls. The experience of invasive-species studies, where biological idiosyncrasies have so far defeated most attempts at useful generalization, suggests that assisted migrations will need species-by-species research, making it unlikely that they will ever be useful for more than a minority of charismatic species. Moreover, assisted migration will be little use if particular current climate types disappear and completely novel climates appear, as will probably be common in TEA (Williams et al. 2007).

The limited options for mitigation on the ground make clear that TEA's major response to climate change must be reducing the region's massive greenhouse-gas emissions, from fossil-fuel use, deforestation, peat-swamp destruction, and agriculture. Not only are TEA's own emissions large enough for reductions to have a significant direct impact on global climate-change, but the active involvement of the region's major emitters, China and Indonesia, in the post-Kyoto negotiations will be essential if a global agreement on emission reductions is to be reached. The region can also contribute by a massive reforestation effort (see 8.2.3, 8.9).

8.7.4 Biofuels

The replacement of fossil fuels by biofuels—combustible materials from plants—could potentially reduce net carbon emissions, because the carbon dioxide released by burning biofuels was withdrawn from the modern atmosphere during growth, while that from fossil fuels was withdrawn tens of millions of years ago. However, initial enthusiasm for this concept has been rapidly replaced by a realization that achieving a net carbon gain is not as easy as was first thought (Fargione et al. 2008; Scharlemann and Laurance 2008; Searchinger et al. 2008).

The biggest problem is that clearance of carbon-rich vegetation, in order to grow biofuels, creates a 'carbon debt' that can take decades to repay. For example, converting lowland rainforest in TEA to oil palm for biofuel, results in a carbon debt that would take an estimated 86 years to repay; while clearing and draining peat-swamp forest could be ten times as bad (Fargione et al. 2008). Growing biofuels on existing cropland is no better, since the continued need for non-biofuel crops ensures the expansion of agriculture elsewhere. There are additional problems with some biofuels, including excessive fossil-fuel consumption during production and the potential for nitrous oxide emissions from crops that require nitrogen fertilizers.

One way out of this dilemma is to use only waste materials, such as crop and logging residues, which would have decayed and released their carbon anyway. This requires efficient conversion of cellulose- and/or lignin-dominated materials into useable (preferably liquid) fuels, and is the subject of much current research. Growing biofuels only on degraded land that has no potential for other crops is another possibility, although the potential for carbon sequestration in spontaneous forest regrowth on abandoned land has to be considered in any carbon accounting. The enormous differences between the costs and benefits of different biofuel-production systems, means that a key issue is going to be reliable certification of the products. Overall, it is clear that a massive expansion in biofuel use has the potential for disastrous impacts in the tropics, which is the best place to grow them, but also for considerable economic and environmental benefits, if done properly. In particular, the use of biofuels with carbon capture and storage technologies in power generation is a potential mechanism for eventually reducing atmospheric carbon dioxide concentrations (Read 2008).

8.7.5 Air pollution

While water pollution is a concern for every manager of a freshwater-, coastal-, or marine-protected area in TEA, the potential impacts of air pollution on biodiversity have been almost completely ignored. To a large extent this reflects a pervasive lack of information on air pollution outside urban areas in the region, plus the fact that the impacts—if indeed there are any—are not immediately visible. Extrapolation from the few relevant studies that have been done suggests that this may be a significant and growing problem (see 7.4.12), but a lot more research will be needed before impacts on biodiversity can be added to the already strong case for a reduction in pollution emissions.

8.8 *Ex situ* conservation

For species and genetically distinct populations that are threatened in the wild, *ex situ* conservation (i.e. conservation in captivity) offers a potential back-up. Zoos, botanical gardens, and other collections of wild organisms in TEA display numerous native species, but this can rarely be considered *ex situ* conservation. *Ex situ* conservation in living collections requires viable, self-sustaining, captive populations, managed so as to conserve genetic diversity, but this is true of very few species in TEA. There are at least 70 Javan gibbons (*Hylobates moloch*) in captivity in Indonesia, for example, but a maximum of five potential breeding pairs and no proof of successful breeding (Nijman 2006). Housing conditions at most zoos in the region are dreadful, there is little collaboration between zoos, and, even in the few species where breeding has been successful, lack of genetic management can result in animals of little or no conservation value. The South China subspecies of the tiger is extinct or near-extinct in the wild and represented in captivity by 73 individuals (Xu et al. 2007), but all of these are descended from six wild founders since 1963 and severe inbreeding contributes to low fertility and high juvenile-mortality. Introducing new genetic diversity from the larger captive population of the genetically very similar Indochinese subspecies is probably the only way that the South China tiger can be saved.

The South East Asian Zoos Association (SEAZA) is attempting to raise standards in regional zoos and participation in *ex situ* breeding programmes is encouraged (SEAZA 2007). However, given the state of the great majority of the region's zoos and the lack of funding, it seems unlikely that captive breeding programmes will make a substantial contribution to biodiversity conservation in the immediate future. An alternative option is to focus on the cryogenic preservation of genetic material, including sperm, embryos, other tissues, and DNA, as already done for some domesticated animals (Andrabi et al. 2007). To do so would be an act of faith in the future development of the reproductive biotechnologies needed to regenerate animals from such material, but the current rapid erosion of genetic diversity in many wild species suggests that waiting until the technologies are mature is not an option. A comprehensive regional strategy for genome resource banking, backed up by focused research, is urgently needed.

Conditions in collections of growing plants are in some ways even worse than for captive animals, since many species are represented by only one or a few genetic individuals, with the wild sources documented poorly, if at all. Seed banks, however, provide a cheap and practical alternative form of *ex situ* conservation for numerous individuals from multiple populations of plant species with desiccation-tolerant seeds: probably a majority of species outside the aseasonal rainforests (see Chapter 4). Long-term seed storage has so far been used largely for crop plants, such as rice and soybeans, in TEA, but storage of wild species is expanding. The most ambitious project is the Southwest China Germplasm Bank of Wild Species, in Yunnan, which has plans to store seeds of thousands of wild species from south-west China (Cyranoski 2003) (Fig. 8.11). This germplasm bank also includes facilities for cryopreservation of animal tissues, as well as an animal DNA bank. Similar facilities for plants and animals are needed in other parts of TEA, with an exchange

Figure 8.11 Long-term seed-banking facilities at the Southwest China Germplasm Bank of Wild Species, based at the Kunming Institute of Botany, Chinese Academy of Sciences, Yunnan. Photograph courtesy of the Kunming Institute of Botany.

system to ensure that all collections are backed-up at multiple sites. Note, however, that seed banks are useless for species with seeds that are intolerant of desiccation, including all members of the important tree families Dipterocarpaceae and Fagaceae.

8.9 Ecological restoration and reintroduction

Forest clearance and, to a lesser extent, forest degradation, are major reasons for local, regional, and global biodiversity losses in TEA, so restoring forest is a major task—arguably, *the* major task—for twenty-first-century conservation. I use the term 'restoration' here in its broadest sense, for any activity that initiates or accelerates the recovery of forest biodiversity after clearance or degradation. Arguably, this is a misuse of a word that normally means returning something to its original state, which is an all but impossible target for most sites, at least on a human time-scale. However, this broader usage has become standard in the ecological literature.

The restoration literature is dominated by the task of actively restoring vegetation structure and species composition, while the fauna is generally assumed to take care of itself (Seddon et al. 2007). In many cases, however, a species that has been lost as a result of habitat destruction and degradation—or direct exploitation—is unable to return without human assistance, as a result of habitat fragmentation and the hostility of the matrix between fragments, or because it survives only in captivity (see above). Reintroduction refers to the human-mediated re-establishment of a wild population of an organism at a site where it previously occurred. The reintroduction literature is dominated by work with animals, although plants too may be reintroduced.

Restoration and reintroduction overlap, since on highly degraded sites the restoration of vegetation necessarily involves the reintroduction of plant populations that were previously extirpated, but the literatures are largely separate (Seddon et al. 2007). The ability to re-establish a self-sustaining population of a previously extirpated plant or animal species could be seen as the ultimate indicator of restoration success. Reintroductions may also be needed, however, at sites where the vegetation is intact, but the fauna impoverished by hunting—the so-called 'empty forests', of which TEA has more than its share (Corlett 2007a).

8.9.1 Restoring forest

You cannot restore a tropical forest in 2–3 years, yet this is the duration of many restoration studies reported in the literature, reflecting the length of a typical PhD project (e.g. Hau and Corlett 2003). There are some longer term studies, particularly in southern China (e.g. Ren et al. 2007), but none that extend over more than one full tree-generation. There is, therefore, a great deal of extrapolation involved in the interpretation of these studies.

Forest cover will usually develop, without any other intervention, if deforested sites are protected from further disturbance. Persistent secondary grasslands in the region are usually maintained by regular burning, so fire control is all that is needed to initiate succession. Cutting of woody plants for fuel and grazing by domestic livestock may also help maintain grasslands. Grasslands persist longer, even in the absence of disturbance, in areas with relatively low rainfall (<1600 mm) and on edaphically extreme sites, but most of these will eventually be invaded by trees, if protected. The problem in TEA, therefore, is not the eventual restoration of forest cover, but rather the slow rate at which this occurs and the low ecological value of the resulting forest.

The rate of development of a woody canopy is a particularly important issue in areas where occasional grassland fires are unavoidable, since a closed canopy will shade out grasses and greatly reduce the flammability of the vegetation. Passive secondary succession is often too slow and patchy to suppress fires, so even a low fire frequency (e.g. once a decade) is sufficient to maintain grassland. Assisted Natural Regeneration (ANR) is a method that tries to accelerate the development of a closed canopy by 'liberating' woody seedlings and saplings that are already present at the site: by preventing disturbances, suppressing competing herbaceous vegetation, and, if necessary, applying fertilizer (Lamb et al. 2005; Shono et al. 2007a). The ANR approach has been widely used in the Philippines and, under different names, elsewhere in the region.

The major advantage of ANR is the low cost in labour and capital in comparison with methods that rely on expensive tree planting. However, the resulting forest is, like spontaneous secondary forests, dominated by a few species of well-dispersed, light-demanding pioneers, and generally supports only a subset of the flora and fauna of the original primary forest. The relative impoverishment of secondary forest floras in the tropics may reflect a variety of filters, from seed dispersal, through seed predation, to seedling establishment and growth in competition with herbaceous vegetation, but the absence of dispersal agents for large-fruited and large-seeded plant species appears to be a universal problem (Hau and Corlett 2003). On degraded hillsides in Hong Kong, for example, almost all seeds dispersed into secondary grasslands pass through the guts of bulbuls (*Pycnonotus* spp.) or other small passerine birds (Au et al. 2006). Natural or artificial perches can locally increase the seed rain into grasslands, but are unlikely to change its composition.

If primary forest remnants still persist in the landscape, the initial 'pioneer desert' will gradually accumulate additional species, but the process can be very slow (Turner et al. 1997). Martínez-Garza and Howe (2003) point out that there is a 'time tax' on the diversification of secondary forests, since primary forest remnants will continue to lose both plant and animal species as long as they are embedded in a low-diversity matrix. Although data are generally lacking, this is likely to be a widespread problem in TEA, suggesting that the restoration of species-rich forest cover should be given a much higher priority than it is at present. 'Saving all the pieces' of the original forest cover will not be enough if the pieces continue to lose species.

Numerous studies throughout the tropics have shown that it is possible to bypass the pioneer desert and, hopefully, beat the time tax, by artificially enriching the forests that develop on abandoned land. The widely used Framework Species Method plants species that are chosen for their ease of propagation in the nursery, their high survival in the field, their spreading, dense, weed-suppressing, crowns, and their attractiveness to seed-dispersing birds and mammals; but it then relies on natural seed dispersal to bring in additional species (Elliott et al. 2003) (Fig. 8.12). This is likely to be less effective at isolated sites, since most seed dispersal in degraded landscapes occurs over distances <1 km (White et al. 2004; Weir and Corlett 2007).

Where money and other resources are available, and the soil not too degraded, high-density planting

Figure 8.12 The Framework Species Method of forest restoration involves planting tree species chosen for their ease of propagation in the nursery, their high survival in the field, their spreading, dense, weed-suppressing, crowns, and their attractiveness to seed-dispersing birds and mammals (Elliott et al. 2003). This restored forest in northern Thailand is nine years old. Photograph © Steve Elliott.

(>2500 trees/ha), with mixtures of early and late successional tree species, can bypass the natural successional sequence altogether and lead to the rapid establishment of a species-rich forest (the Maximum Diversity Method; e.g. Lamb et al. 2005). Some primary forest species will not establish in open sites, but a surprising number will, and the shade-requiring species can be added later (Shono et al. 2007b; Du et al. 2008). Although we do not yet know how many of the planted species will persist over multiple tree-generations, this approach offers at least a medium-term solution to the diversity problem. On sites degraded by prolonged cultivation or repeated burning, growth may be nitrogen-limited (see 6.3.1) and the inclusion of N-fixing legumes in the planting mix may enhance tree growth and speed canopy closure (e.g. Siddique et al. 2008).

Rearing and planting seedlings is expensive, so many studies have looked at the cheaper alternative of direct seeding. Unfortunately, simply broadcasting seeds rarely works and even if seeds are buried individually, only a subset of species do well, often those with larger seeds (Doust et al. 2006). This means that direct seeding can be a lower cost supplement to the high-cost planting of seedlings, but cannot replace it. Another potential alternative to seedlings is planting vegetative stakes (cut a few days before planting and with lateral branches removed) <4 m tall (Zahawi 2008). Again, only a subset of species can be established in this way, but the use of stakes may be advantageous in tall grass and active pastures, and can also provide instant perches for seed-carrying birds.

Yet another source of planting material is 'wildings'—seedlings collected from the wild. Although this practice is apparently widespread in TEA, it is poorly documented. It appears that survival rates are very low if the wildings are planted directly, but that, after a period of recovery in a nursery, they may perform better than seedlings grown from seed (e.g. Ådjers et al.1998). The need for this recovery period means that the use of wildings is unlikely to save much money, but their use may still be advantageous in cases where seedlings are easier to find than seeds, as with mast-fruiting dipterocarps and other species with a persistent 'seedling bank'. This is a topic that deserves more research. For rare species, the impacts of wildling collection on wild populations are a potential concern.

Planting large areas in their entirety with high densities of nursery-grown seedlings will normally be prohibitively expensive, so much recent attention has focused on methods of enriching, rather than replacing, the spontaneous secondary forest. The limitations of both passive succession and the ANR method described above can be mitigated by enrichment with primary-forest species that do not feature in the spontaneous regeneration. Achieving this successfully is likely to require species-specific research, to determine, for example, which species can be sown directly or grown from stakes or wildings, and which need planting as nursery-grown seedlings, and which species should be planted before canopy closure and which afterwards.

It is important that enrichment is planned at the landscape level, so that effort is concentrated where it will do most good, e.g. in buffer zones along streams and around primary forest fragments, and as corridors linking these fragments (Lamb and Erskine 2008). In the Bawangling National Nature Reserve on Hainan Island, for example, forest restoration is focusing on the areas and tree species of most value to the critically endangered Hainan gibbon (Fellowes et al. 2008). It is also important that present and future threats from fires, agricultural expansion, and firewood harvesting, are taken into account. Far too many trees are currently being planted in areas where they have little chance of surviving in the long term. Too often, reforestation projects in TEA become dominated by numerical targets and quotas, with 'success' measured in terms of seedlings planted, rather than forest established.

It has often been suggested that fast-growing, non-invasive, exotic plantations can be used as a tool to 'capture' a site from grasses and other herbaceous species, with subsequent spontaneous invasion or assisted enrichment of the plantation by native species. Although there is little evidence for a general role of exotic monocultures as a 'catalyst' for secondary succession (e.g. Lee et al. 2005, 2008), underplanting with shade-tolerant natives may allow conversion of a plantation into a species-rich native forest. If the exotics can be harvested later without too much damage to the native species, this approach may provide enough cash income to encourage reforestation (McNamara et al. 2006). Moreover, any forest cover—including pure stands of exotics—is likely to support a greater diversity of native species than most other anthropogenic ecosystems, such as degraded pasture or non-tree crops, so exotic plantations can have a role in conservation at the landscape level (Brockerhoff et al. 2008). Research is urgently needed into ways in which the conservation value of these plantations can be enhanced, without excessively compromising their productivity. Timber plantations of mixed native species could potentially combine biodiversity benefits with sufficient financial return to encourage reforestation of large areas (Wardell-Johnson et al. 2008).

Genetic issues have been widely neglected in restoration planting in TEA, with seeds often collected from one or a few convenient mother trees, or seeds and seedlings imported from distant sources. There is evidence for a strong selection against inbreeding in tropical trees (see 4.3.1), suggesting that poor genetic management at the planting stage could lead to low fitness in the next generation. There is also evidence for genetic adaptation to local conditions, suggesting that local seed sources should be preferred, wherever possible (McKay et al. 2005). Nursery managers may tend to view genetic management as one too many additional problems, but this attitude is short-sighted and current practices are probably storing up trouble for the future.

The biggest challenge for the immediate future is scaling this up. We need to be restoring forest at a faster rate than it is cleared, i.e. at least 20,000 km^2 per year in TEA at present. Achieving this will require the initiation of hundreds of thousands of local forest-restoration projects and their continued protection from fire, cutting, and other threats. Carbon and biodiversity offsets, and payments for environment services, are potential sources of funding that are likely to increase in the next decade, and loans, subsidies, free planting material, and regular payments have all been tried somewhere. We now urgently need experimental pilot studies to see how best to convert this funding into species-rich native forests.

8.9.2 Reintroducing species

The generally cautious attitude of conservationists to reintroduction is evident in the mission statement of the IUCN Special Survival Commission (SSC) Re-introduction Specialist Group: 'To combat the ongoing and massive loss of biodiversity by using re-introductions as a *responsible tool* for the management and restoration of biodiversity through actively developing and promoting *sound inter-disciplinary scientific information*, policy, and practice to establish viable wild populations in their natural habitats.' The IUCN/SSC Guidelines for Re-introductions (IUCN 1998) suggest that the principal aim of any reintroduction should be to establish a viable, free-ranging population in the wild, requiring minimal long-term management, with the objective of enhancing the long-term survival of a species, restoring ecological processes, restoring natural biodiversity, and/or promoting conservation awareness. The Guidelines also state that the released organisms should be, as far as possible, from the local subspecies or race, should be free of parasites and diseases acquired in captivity, and should be monitored after release. The need for a good knowledge of the natural history of the species is emphasized, as is the importance of long-term financial and political support.

An increasing number of reintroductions have been carried out worldwide, with widely varying degrees of success. Seddon et al. (2005) list 489 animal species known to be the focus of reintroduction projects, including 172 mammals, 138 birds, 94 reptiles and amphibians, 20 fish, and 65 invertebrates; but this is likely to be a considerable underestimate. In general, translocations of wild-caught animals have had a higher degree of success than releases of individuals born in captivity (e.g. Jule et al. 2008), but there are also examples of success with captive-bred animals. Reintroduction has become part of the conservationist's toolbox in situations where the threat that caused the original extirpation has been removed, a suitable stock for release is available, and adequate unoccupied habitat exists. Well-documented examples from TEA include the rescue, captive breeding, and successful re-establishment in the wild of Romer's tree frog (*Chirixalus romeri*) in Hong Kong (Banks et al. 2008) and the successful reintroduction of Père David's deer (*Elaphurus davidianus*) (Fig. 8.13) to riverine marshland in the lower reaches of the Yangtze River more than a century after the species was extirpated from China (Zeng et al. 2007).

Figure 8.13 Père David's deer, *Elaphurus davidianus*, was extirpated from its marshland habitat in central China, but a herd survived on the Duke of Bedford's estate at Woburn, England, from where it has been reintroduced to China. This photograph, courtesy of Jiang Zhigang, was taken at the Shishou (Tianezhou) Milu National Nature Reserve, on the Yangtze River, where a population was established in 1993–94.

The reality in TEA, however, is usually very different from these examples, with numerous confiscated or abandoned animals simply dumped in the nearest convenient forest, without prior medical examination, pre-release training, or identification of their origins. This is sometimes done simply because the law requires it, but, in other cases, it is seen as the kindest option in the absence of suitable facilities for keeping confiscated wild animals in captivity. The IUCN guidelines for the placement of confiscated animals (IUCN 2002) recommend euthanasia in these circumstances, but this is unlikely to appeal to either wildlife officials or the general public. Such releases are not monitored: it is likely that most animals die, but the establishment of invasive macaque populations in several protected areas in Vietnam (Streicher 2007), and elsewhere in the region, shows that some survive.

Better enforcement of existing laws on the capture, trading, and possession of endangered species would greatly increase the number of animals of potential conservation value in need of placement (Fig. 8.14), so there is an urgent need for a realistic compromise between the IUCN Guidelines and current practices. There are already a variety of projects within the region that illustrate how this could be done, including the release of confiscated Tarictic hornbills in the Philippines (Hembra et al. 2006), orangutans in Kalimantan (Grundmann 2006) and Sumatra (Cocks and Bullo 2008), and langurs (Nadler and Streicher 2003), pygmy lorises (Streicher and Nadler 2003), and a variety of other vertebrates (Streicher 2007) in Vietnam. Captive-raised gibbons from the illegal pet trade have been successfully released in Kalimantan (Cheyne et al. 2008). All the above examples involve animals that were at least born in the wild, but a single zoo-bred Sumatran orangutan has been successfully released in Bukit Tigapuluh National Park, Sumatra, where it joins more than 80 orphaned and pet-trade animals released by the Sumatran Orangutan Conservation Programme in an attempt to re-establish orangutans in an area from which they were extirpated >150 years ago (Cocks and Bullo 2008).

The Elephant Reintroduction Foundation in Thailand is successfully releasing domesticated elephants back to the wild, although in this case the major concern has been the welfare of individual elephants, rather than species conservation. With an estimated 16,000 captive elephants in Asia—<30% of all living Asian elephants—and many of these unwanted as their use in logging declines, the major problem with the Asian elephant is a lack of suitable habitat, rather than a shortage of animals for release

Figure 8.14 Baby orangutans at the Borneo Orangutan Survival Foundation. Photograph © Natalie Behring-Chisholm, Greenpeace.

(Leimgruber et al. 2008). Unfortunately, despite an overall surplus of captive elephants, wild populations are still threatened locally by continued live captures.

The availability of both surplus animals, confiscated from the trade, and fragments of forest, emptied of their wildlife by hunting, provide an ideal opportunity for experimental investigations of reintroduction techniques. These could range from single-sex releases, designed to assess the impact of pre-release conditioning on post-release survival, to studies of the optimum group-size for successful breeding, and assessments of the impact on ecosystem functioning. The need for reintroductions can only grow as declining rural populations and expanding forests, both spontaneous and restored, create new opportunities; so it is important that we learn how to do it successfully.

8.10 Conservation leakage

In conservation, 'leakage' refers to the fact that conservation gains in one area may stimulate losses in another, thus reducing (or negating) the net gain for global conservation (e.g. Gan and McCarl 2007). For example, forest conservation in one area may encourage logging or deforestation in another, as a result of market interactions. Leakage can operate on any scale, from local (e.g. hunters avoiding a well-patrolled section of a protected area may increase pressure in other areas), to regional (e.g. a logging ban in one country, with no decrease in demand, is likely to increase logging in neighbouring countries), to global (e.g. reducing forest conversion to oil palm plantations in TEA may encourage conversion for less productive oil crops elsewhere in the tropics). Conservationists have tended to ignore leakage and most studies of the phenomenon have been in relation to carbon sequestration, where the global linkage is obvious (e.g. Murray et al. 2004). However, it seems likely that many—perhaps most—local conservation successes are at least partly undermined by leakage, so the problem cannot be ignored. In theory, leakage can be reduced or eliminated by increasing the spatial scope of the conservation action, but the increasing globalization of markets means that even regional collaboration may be insufficient for products, such as timber and vegetable oils, that have a global market, while global agreements are notoriously hard to obtain and enforce.

8.11 Education

Hunter-gatherers and forest farmers in TEA possess a wealth of traditional knowledge about species and processes in natural and human-modified ecosystems, but these people form a tiny minority of the region's twenty-first-century population, and most belong to minority ethnic and language groups who have little interaction with the majority. There is also a small but growing minority of well-educated, mostly urban, ecologists and conservationists, both amateur and professional. In between these groups, however, there is a high degree of ignorance about environmental problems in general and biodiversity conservation in particular, encompassing almost everyone from rural smallholders to urban civil servants and politicians. The extent to which this ignorance contributes to behaviour that has harmful impacts on biodiversity is hard to judge, but combating ignorance is a crucial first step towards building the broad public support needed to sustain biodiversity conservation in the region.

Almost all children in TEA attend government-run schools, so this is the obvious way to combat ignorance in the next generation. All countries in the region have central control of at least some aspects of the primary and secondary school syllabus, so a two-pronged approach—getting the syllabus changed, while providing training and teaching materials for schools—is likely to be most effective. Where they exist, independent schools may have more flexibility and can be used to test-run innovative approaches to conservation education.

This is not the place to deal with the details of curriculum reform, but there are three issues that deserve particular mention. First, there is a widespread concern among conservation educators in TEA, as elsewhere, that modern life-styles in general, and high-pressure education systems in particular, are drastically reducing the amount of informal contact between children and nature. Many children—not only in urban areas—no longer play outdoors, except in organized team sports, and nature is something seen on TV, rather than experienced directly. Research elsewhere has suggested that the environmental attitudes and sensitivity of adults is strongly influenced by their participation in nature recreation as children (Pergams and Zaradic 2008). Schools can help compensate for this by encouraging activities that bring children of all ages into contact with wild nature, whether it is in the school grounds, on urban wasteland, or in a local protected area.

Second, there is a widespread conviction among ecologists that an appreciation for biodiversity is best acquired by learning to recognize local wild species one by one. Birdwatching served this purpose for many of the current conservationists in the region, but any taxonomic group for which initial help is available—either in the form of identification manuals or expert tuition—will do. But would this work if it formed part of a formal school syllabus, rather than an enjoyable hobby? This leads to the third point: while there is a vast amount of varied experience in environmental education within TEA, there have been few attempts to evaluate its effectiveness. Educational experiments using before–after and treatment–control designs are difficult, but not impossible, and could have a big impact on the way in we approach environmental education in the future.

Children in schools are not the only possible targets for education. There has been an explosion of biodiversity publishing—books, magazines, and various Internet media—within TEA over the last decade; but the availability of material in local languages varies tremendously (Fig. 8.15). Identification manuals are the basis for serious conservation education, yet most currently available are targeted at the English- (or Chinese-) reading professional, rather than interested amateurs. The need for more educational interpretation in protected areas was mentioned earlier (8.2.5). The region's many zoos and botanical gardens should also be involved, not only through passive signs and other displays—although these are urgently needed—but also, where the expertise is available, through active outreach to local communities and partnerships with protected areas and conservation NGOs (Miller at al. 2004; Mallapur et al. 2008). Indeed, a significant contribution to conservation education should be a requirement for the licensing of all zoos in the region, since there is a risk that zoos without a conservation message may simply encourage an attitude that wild animals are ours to exploit. A commitment to conservation education is central to the stated mission of the South East Asian Zoos Association (SEAZA 2007), but even the 30 zoos that belong to this association are underperforming at present.

Figure 8.15 There has been an explosion of biodiversity publishing in the last few years in several part of the region, but coverage varies greatly by area, language and taxonomic group. This is a selection of recent books from Hong Kong, China. Photograph courtesy of Hugh T.W. Tan.

There is also a need for enhanced tertiary-level training for conservation professionals in the region. Positions that would require a master's degree in conservation biology in the developed world are commonly occupied by school-leavers or people with less relevant degrees in TEA. Improvements to salaries and conditions are probably a pre-requisite for developing a highly educated and motivated conservation profession, but the wider provision of relevant tertiary-level courses and degree programs will be an essential component of this.

8.12 Does conservation work?

Billions of US dollars have been poured into conservation and conservation-related activities in TEA over the last few decades, yet the effectiveness of most interventions is unknown (Ferraro and Pattanayak 2006; Linkie et al. 2008). We have numerous stories and many indirect indicators (numbers of staff trained, changes in conservation awareness etc.), but almost no answers to the obvious question: is biodiversity (or some particular element of it) better off than if the intervention had not occurred? This is a difficult and potentially expensive question to answer, but there can be few other areas of public policy where so much money has been spent with so little empirical evaluation of the results.

The ideal approach is the use of designed experiments, where the intervention is assigned randomly across areas, species, communities, or populations. This will often be impossible or prohibitively expensive, but there are 'quasi-experimental' methods that can be used to identify the impact of intervention, despite a lack of randomization (Ferraro and Pattanayak 2006; Andam et al. 2008). Where even this is impractical, an alternative approach is the questionnaire-based scorecard developed by the Cambridge Conservation Forum to help practitioners to evaluate the success of projects in a systematic and consistent manner (Kapos et al. 2008). The promise of a big increase in conservation funding over the next few decades, as a result of carbon offsets and other initiatives, makes a change in attitudes to rigorous evaluation particularly urgent. Donors may require it, but even if they don't, conservationists should insist that the outcomes of projects they are involved in are evaluated as rigorously as possible. We really need to know what works and what does not.

8.13 The way forward for biodiversity conservation in TEA

This book has been written by an ecologist for other ecologists, so it is not surprising that it ends with a plea for the region's ecologists to take a lead in biodiversity conservation. In an ideal world, academic ecologists could sit back and wait for conservation managers to seek their advice, but we do not live in that world. The situation is critical and we must act now to ensure that as much as possible of the region's biodiversity survives through the next few decades.

While there is a role for confrontation and protest in conservation, this is rarely the most effective role for a scientist. The authority of any science comes from the perception that it is unbiased, and that perception can easily be lost if scientists become too closely identified with one side in a dispute. Achieving the best possible future for biodiversity in TEA will involve compromise, and this will only come from a willingness to talk to all sides and to offer practical alternatives to damaging behaviour. Balancing scientific objectivity with activist passion is a major challenge for the region's ecologists.

Conservation needs both research and action, but the latter can rarely wait for the results of the former. There is a 'time tax' on many of the conservation actions suggested above, with success less likely the longer we wait before starting. The most urgent actions include: the protection of the last large areas of lowland forest in the region; the control of hunting in existing protected areas, and of the wildlife trade that fuels it; the control of illegal logging; protection of the carbon-rich peat-swamp forests from drainage and fire; and a massive expansion in forest restoration. Only slightly less urgent are: regional collaboration for the prioritization of species and areas; the development of national and regional early warning systems for potentially invasive species; and improvements to biodiversity education, in schools, in the media, and in protected areas.

Many priority areas for research will require collaboration between ecologists and social scientists. Key questions include: How can enforcement of conservation laws be made more effective? How can current reforestation efforts be expanded by several orders of magnitude? Can intensive management of small, protected areas make up for their size? Where and how can community-based conservation work? How effective are local NGOs and how can they be strengthened? What is the role for *ex situ* conservation of threatened animal species in TEA and what methods will be most effective? How can confiscated animals of conservation value be reintroduced to the wild? How can funding for carbon offsets best be used to improve biodiversity conservation?

Things are going to get worse for biodiversity in TEA before they start getting better, but how much worse is under our control. There is no 'magic bullet' that will solve all the conservation problems in the region: they will have to be tackled one by one, site by site, species by species, person by person. Success will ultimately depend on our ability to build broad public support for biological conservation in Tropical East Asia among all sectors of society. I hope that this book can contribute to that goal.

References

Aas, W., Shao, M., Jin, L. et al. (2007) Air concentrations and wet deposition of major inorganic ions at five non-urban sites in China, 2001–2003. *Atmospheric Environment*, 41, 1706–1716.

Abe, T. (2006) Threatened pollination systems in native flora of the Ogasawara (Bonin) Islands. *Annals of Botany*, **98**, 317–334.

Abe, T. (2007) Predator or disperser? A test of indigenous fruit preference of alien rats (*Rattus rattus*) on Nishijima (Ogasawara Islands). *Pacific Conservation Biology*, **13**, 213–218.

Abe, H., Matsuki, R., Ueno, S., Nashimoto, M. and Hasegawa, M. (2006) Dispersal of *Camellia japonica* seeds by *Apodemus speciosus* revealed by maternity analysis of plants and behavioral observation of animal vectors. *Ecological Research*, **21**, 732–740.

Abe, T., Makino, S. and Okochi, I. (2008) Why have endemic pollinators declined on the Ogasawara Islands? *Biodiversity and Conservation*, **17**, 1465–1473.

Abram, N.J., Gagan, M.K., Liu, Z.Y., Hantoro, W.S., McCulloch, M.T. and Suwargadi, B.W. (2007) Seasonal characteristics of the Indian Ocean Dipole during the Holocene epoch. *Nature*, **445**, 299–302.

Adeney, J.M., Ginsberg, J.R., Russell, G.J. and Kinnaird, M.F. (2006) Effects of an ENSO-related fire on birds of a lowland tropical forest in Sumatra. *Animal Conservation*, **9**, 292–301.

Ådgers, G., Hadengganan, S., Kuusipalo, J., Otsamo, A. and Vesa, L. (1998) Production of planting stock from wildings of four *Shorea* species. *New Forests*, **16**, 185–197.

Adler, P.B., HilleRisLambers, J. and Levine, J.M. (2007) A niche for neutrality. *Ecology Letters*, **10**, 95–104.

Agetsuma, N. (2007) Ecological function losses caused by monotonous land use induce crop raiding by wildlife on the island of Yakushima, southern Japan. *Ecological Research*, **22**, 390–402.

Aiba, S., Hill, D.A. and Agetsuma, N. (2001) Comparison between old-growth stands and secondary stands regenerating after clear-felling in warm-temperate forests of Yakushima, southern Japan. *Forest Ecology and Management*, **140**, 163–175.

Aiba, S. and Kitayama, K. (1999) Structure, composition and species diversity in an altitude-substrate matrix of rain forest tree communities on Mount Kinabalu, Borneo. *Plant Ecology*, **140**, 139–157.

Aiba, S. and Kitayama, K. (2002) Effects of the 1997–98 El Nino drought on rain forests of Mount Kinabalu, Borneo. *Journal of Tropical Ecology*, **18**, 215–230.

Aiba, M. and Nakashizuka, T. (2007a) Variation in juvenile survival and related physiological traits among dipterocarp species co-existing in a Bornean forest. *Journal of Vegetation Science*, **18**, 379–388.

Aiba, M. and Nakashizuka, T. (2007b) Differences in the dry-mass cost of sapling vertical growth among 56 woody species co-occurring in a Bornean tropical rain forest. *Functional Ecology*, **21**, 41–49.

Aiba, S., Takyu, M. and Kitayama, K. (2005) Dynamics, productivity and species richness of tropical rainforests along elevational and edaphic gradients on Mount Kinabalu, Borneo. *Ecological Research*, **20**, 279–286.

Aide, T.M. (1992) Dry season leaf production: an escape from herbivory. *Biotropica*, **24**, 532–537.

Aitchison, J.C., Ali, J.R. and Davis, A.M. (2007) When and where did India and Asia collide? *Journal of Geophysical Research-Solid Earth*, **112**.

Alcala, A.C. and Russ, G.R. (2006) No-take marine reserves and reef fisheries management in the Philippines: a new people power revolution. *Ambio*, **35**, 245–254.

Aldhous, P. (2004) Land remediation: Borneo is burning. *Nature*, **432**, 144–146.

Aldrian, E. and Susanto, R.D. (2003) Identification of three dominant rainfall regions within Indonesia and their relationship to sea surface temperature. *International Journal of Climatology*, **23**, 1435–1452.

Alexander, I.J. and Lee, S.S. (2005) Mycorrhizas and ecosystem processes in tropical rain forest: implications for diversity. *Biotic interactions in the tropics* (eds D.F.R.P. Burslem, M.A. Pinard and S.E. Hartley), pp. 165–203. Cambridge University Press, Cambridge.

Ali, R. (2004) The effect of introduced herbivores on vegetation in the Andaman Islands. *Current Science*, **86**, 1103–1112.

Ali, R. (2006) Issues relating to invasives in the Andaman Islands. *Journal of the Bombay Natural History Society*, **103**, 349–355.

Ali, J.R. and Aitchison, J.C. (2008) Gondwana to Asia: plate tectonics, paleogeography and the biological connectivity of the Indian sub-continent from the Middle Jurassic through the latest Eocene (166–35 Ma). *Earth Science Reviews*, **88**, 145–166.

Allison, S.D. (2006) Brown ground: a soil carbon analogue for the green world hypothesis? *American Naturalist*, **167**, 619–627.

Amir, H.M.S. and Miller, H.G. (1990) *Shorea leprosula* as an indicator species for site fertility evaluation in dipterocarp forests of Peninsula Malaysia. *Journal of Tropical Forest Science*, **3**, 101–110.

Andam, K., Ferraro, P.J., Pfaff, A., Sanchez-Azofeifa, G.A., and Robalino, J.A. (2008) Measuring the effectiveness of protected area networks in reducing deforestation. *Proceedings of the National Academy of Sciences of the USA*,**105**, 16089–16094.

Anderson, J.A.R. (1983) The tropical peat swamps of western Malesia. *Mires: swamp, bog, fen and moor, regional studies* (ed A.J.P. Gore), pp. 181–199. Elsevier, Amsterdam.

Andrabi, S.M.H. and Maxwell, W.M.C. (2007) A review on reproductive biotechnologies for conservation of endangered mammalian species. *Animal Reproduction Science*, **99**, 223–243.

Anon. (2007) *Priority sites for conservation in the Philippines: Key Biodiversity Areas*. Conservation International Philippines, Department of Environment and Natural Resources, and Haribon Foundation, Quezon City.

Appanah, S. (1993) Mass flowering of dipterocarp forests in the aseasonal tropics. *Journal of Biosciences*, **18**, 457–474.

Appanah, S. and Chan, H.T. (1981) Thrips: the pollinators of some dipterocarps. *Malaysian Forester*, **44**, 234–252.

Aratrakorn, S., Thunhikorn, S. and Donald, P.F. (2006) Changes in bird communities following conversion of lowland forest to oil palm and rubber plantations in southern Thailand. *Bird Conservation International*, **16**, 71–82.

Arnold, A.E. (2008) Endophytic fungi: hidden components of tropical community ecology. *Tropical forest community ecology* (eds W.P. Carson and S.A. Schnitzer), pp. 254–271. Wiley-Blackwell, Oxford.

Arponen, A., Moilanen, A. and Ferrier, S. (2008) A successful community-level strategy for conservation prioritization. *Journal of Applied Ecology*, **45**, 1436–1445.

Ashton, P.S. (2003) Floristic zonation of tree communities on wet tropical mountains revisited. *Perspectives in Plant Ecology Evolution and Systematics*, **6**, 87–104.

Ashton, P.S., Givnish, T.J. and Appanah, S. (1988) Staggered flowering in the Dipterocarpaceae—new insights into floral induction and the evolution of mast fruiting in the aseasonal tropics. *American Naturalist*, **132**, 44–66.

ATBC (2007) *Resolution concerning the 'Heart of Borneo' transboundary conservation initiative*. The Association for Tropical Biology and Conservation.

Au, A.Y.Y., Corlett, R.T. and Hau, B.C.H. (2006) Seed rain into upland plant communities in Hong Kong, China. *Plant Ecology*, **186**, 13–22.

Auffenberg, W. (1988) *Gray's monitor lizard*. University of Florida Press, Florida.

Bacon, A.M., Demeter, F., Rousse, S. et al. (2006) New palaeontological assemblage, sedimentological and chronological data from the Pleistocene Ma U'Oi cave (northern Vietnam). *Palaeogeography Palaeoclimatology Palaeoecology*, **230**, 280–298.

Bailey, R.C., Head, G., Jenike, M., Owen, B., Rechtman, R. and Zechenter, E. (1989) Hunting and gathering in tropical rain forest: is it possible? *American Anthropologist*, **91**, 59–82.

Baillie, I.E. and Ashton, P.S. (1983) Some soil aspects of the nutrient cycle in mixed dipterocarp forests in Sarawak. *Tropical rain forest: ecology and management* (eds S.L. Sutton, T.C. Whitmore and A.C. Chadwick), pp. 347–356. Blackwell, Oxford.

Baillie, I.C., Ashton, P.S., Chin, S.P. et al. (2006) Spatial associations of humus, nutrients and soils in mixed dipterocarp forest at Lambir, Sarawak, Malaysian Borneo. *Journal of Tropical Ecology*, **22**, 543–553.

Baker, D.F. (2007) Reassessing carbon sinks. *Science*, **316**, 1708–1709.

Baker, P.J., Bunyavejchewin, S., Oliver, C.D. and Ashton, P.S. (2005) Disturbance history and historical stand dynamics of a seasonal tropical forest in western Thailand. *Ecological Monographs*, **75**, 317–343.

Bala, G., Caldeira, K., Wickett, M. et al. (2007) Combined climate and carbon-cycle effects of large-scale deforestation. *Proceedings of the National Academy of Sciences of the USA*, **104**, 6550–6555.

Balete, D.S., Rickart, E.A., Rosell-Ambal, R.G.B., Jansa, S. and Heaney, L.R. (2007) Descriptions of two new species of *Rhynchomys* Thomas (Rodentia: Muridae: Murinae) from Luzon Island, Philippines. *Journal of Mammalogy*, **88**, 287–301.

Balmford, A. and Whitten, T. (2003) Who should pay for tropical conservation, and how could the costs be met? *Oryx*, **37**, 238–250.

Baltzer, J.L., Davies, S.J., Bunyavejchewin, S. and Noor, N.S.M. (2008) The role of desiccation tolerance in determining tree species distributions along the Malay-Thai Peninsula. *Functional Ecology*, **22**, 221–231.

Banks, C.B., Lau, M.W.N. and Dudgeon, D. (2008) Captive management and breeding of Romer's tree frog *Chirixalus romeri*. *International Zoo Yearbook*, **42**, 1–10.

Bänziger, H. (1982) Fruit-piercing moths (Lep., Noctuidae) in Thailand: a general survey and some new perspectives. *Mitteilungen der Schweizerischen Entomologischen Gessellschaft*, **55**, 213–240.

Bänziger, H. (1991) Stench and fragrance: unique pollination lure of Thailand's largest flower, *Rafflesia kerrii* Meijer. *Natural History Bulletin of the Siam Society*, **39**, 19–52.

Bänziger, H. (1996a) The mesmerizing wart: the pollination strategy of epiphytic lady slipper orchid *Paphiopedilum villosum* (Lindl) Stein (Orchidaceae). *Botanical Journal of the Linnean Society*, **121**, 59–90.

Bänziger, H. (1996b) Pollination of a flowering oddity: *Rhizanthes zippelii* (Blume) Spach (Rafflesiaceae). *Natural History Bulletin of the Siam Society*, **44**, 113–142.

Bänziger, H. and Pape, T. (2004) Flowers, faeces and cadavers: natural feeding and laying habits of flesh flies in Thailand (Diptera: Sarcophagidae, *Sarcophaga* spp.). *Journal of Natural History*, **38**, 1677–1694.

Baraloto, C., Goldberg, D.E. and Bonal, D. (2005) Performance trade-offs among tropical tree seedlings in contrasting microhabitats. *Ecology*, **86**, 2461–2472.

Barber, C.V., Matthews, E., Brown, D. et al. (2002) *State of the forest: Indonesia*. World Resources Institute, Washington, DC.

Bard, E. (2001) Comparison of alkenone estimates with other paleotemperature proxies. *Geochemistry Geophysics Geosystems*, **2**, 1002.

Barker, G., Barton, H., Bird, M. et al. (2007) The 'human revolution' in lowland tropical Southeast Asia: the antiquity and behavior of anatomically modern humans at Niah Cave (Sarawak, Borneo). *Journal of Human Evolution*, **52**, 243–261.

Barlow, J., Gardner, T.A., Araujo, I.S. et al. (2007) Quantifying the biodiversity value of tropical primary, secondary, and plantation forests. *Proceedings of the National Academy of Sciences of the USA*, **104**, 18555–18560.

Barron, A.R., Wurzburger, N., Bellenger, J.P. et al. (2009) Molybdenum limitation of asymbiotic nitrogen fixation in tropical forest soils. *Nature Geoscience*, **2**, 42–45.

Bartlett, T.O. (2007) The Hylobatidae: small apes of Asia. *Primates in perspective* (eds C.J. Campbell, A. Fuentes, K.C. MacKinnon, M. Panger and S.K. Bearder), pp. 274–289. Oxford University Press, New York.

Barton, H. and Paz, V. (2007) Subterranean diets in the tropical rain forests of Sarawak, Malaysia. *Rethinking agriculture: archaeological and ethnoarchaeological perspectives* (eds T. Denham, J. Iriarte and L. Vrydaghs), pp. 50–77. Left Coast Press, Walnut Creek, California.

Baskin, J.M. and Baskin, C.C. (2004) A classification system for seed dormancy. *Seed Science Research*, **14**, 1–16.

Bautista, A.P. (1991) Recent zooarchaeological researches in the Philippines. *Jurnal Arkeologi Malaysia*, **4**, 45–58.

Beattie, A. (1989) Myrmecotrophy—plants fed by ants. *Trends in Ecology and Evolution*, **4**, 172–176.

Beaufort, L., de Garidel-Thoron, T., Linsley, B., Oppo, D. and Buchet, N. (2003) Biomass burning and oceanic primary production estimates in the Sulu Sea area over the last 380 kyr and the East Asian monsoon dynamics. *Marine Geology*, **201**, 53–65.

Beck, J. and Chey, V.K. (2008) Explaining the elevational diversity pattern of geometrid moths from Borneo: A test of five hypotheses. *Journal of Biogeography*, **35**, 1452–1464.

Beck, J. and Kitching, I.J. (2007) The latitudinal distribution of sphingid species richness in continental southeast Asia: What causes the biodiversity 'hot spot' in northern Thailand? *Raffles Bulletin of Zoology*, **55**, 179–185.

Becker, P., Davies, S.J., Moksin, M., Ismail, M. and Simanjuntak, P.M. (1999) Leaf size distributions of understorey plants in mixed dipterocarp and heath forests of Brunei. *Journal of Tropical Ecology*, **15**, 123–128.

Becker, C.G., Fonseca, C.R., Haddad, C.F.B., Batista, R.F. and Prado, P.I. (2007) Habitat split and the global decline of amphibians. *Science*, **318**, 1775–1777.

Bednarik, R.G. (2001) Replicating the first known sea travel by humans: the Lower Pleistocene crossing of the Lombok Strait. *Human Evolution*, **16**, 229–242.

Begon, M., Townsend, C.R. and Harper, J.L. (2006) *Ecology: from individuals to ecosystems*. Blackwell, Oxford.

Bekken, D.A., Schepartz, L.A., Miller-Antonio, S., Hou, Y.M. and Huang, W. (2004) Taxonomic abundance at Panxian Dadong, a Middle Pleistocene cave in South China. *Asian Perspectives*, **43**, 333–359.

Belcher, B. and Schreckenberg, K. (2007) Commercialisation of non-timber forest products: a reality check. *Development Policy Review*, **25**, 355–377.

Bell, T., Freckleton, R.P. and Lewis, O.T. (2006) Plant pathogens drive density-dependent seedling mortality in a tropical tree. *Ecology Letters*, **9**, 569–574.

Bellwood, P. (1997) *Prehistory of the Indo-Malaysian Archipelago*. University of Hawaii Press, Honolulu.

Bellwood, P. (1999) Archaeology of Southeast Asian hunters and gatherers. *The Cambridge encyclopedia of hunters and gatherers* (eds R.B. Lee and R. Daly), pp. 284–288. Cambridge University Press, Cambridge.

Bellwood, P. (2005) First farmers: origins of agricultural societies. Blackwell Publishing, Oxford.

Benedick, S., Hill, J.K., Mustaffa, N. et al. (2006) Impacts of rain forest fragmentation on butterflies in northern Borneo: species richness, turnover and the value of small fragments. *Journal of Applied Ecology*, **43**, 967–977.

Bennett, E.L. (2007) Hunting, wildlife trade and wildlife consumption patterns in Asia. *Bushmeat and livelihoods:*

wildlife management and poverty reduction (eds G. Davies and D. Brown). Blackwell, Oxford.

Bennett, E.L., Nyaoi, A.J. and Sompud, J. (2000) Saving Borneo's bacon: the sustainability of hunting in Sarawak and Sabah. *Hunting for sustainability in tropical forests* (eds J.G. Robinson and E.L. Bennett), pp. 305–324. Columbia University Press, New York.

Berghoff, S.M., Weissflog, A., Linsenmair, K.E., Hashim, R. and Maschwitz, U. (2002) Foraging of a hypogaeic army ant: a long neglected majority. *Insectes Sociaux*, **49**, 133–141.

Beukema, H., Danielsen, F., Vincent, G., Hardiwinoto, S. and Van Andel, J. (2007) Plant and bird diversity in rubber agroforests in the lowlands of Sumatra, Indonesia. *Agroforestry Systems*, **70**, 217–242.

Bhagwat, S.A., Willis, K.J., Birks, H.J.B. and Whittaker, R.J. (2008) Agroforestry: a refuge for tropical biodiversity? *Trends in Ecology and Evolution*, **23**, 261–267.

Bickford, D., Supriatna, J., Andayani, N. et al. (2008) Indonesia's protected areas need more protection: suggestions from island examples. *Biodiversity and human livelihoods in protected areas: case studies from the Malay Archipelago* (eds N.S. Sodhi, G. Acciaioli, M. Erb and A.K.-J. Tan), pp. 53–77. Cambridge University Press, Cambridge.

Bildstein, K.L. (2006) *Migrating raptors of the world: their ecology and conservation*. Comstock Publishing, Ithaca, New York.

Bird, M.I., Pang, W.C. and Lambeck, K. (2006) The age and origin of the Straits of Singapore. *Palaeogeography Palaeoclimatology Palaeoecology*, **241**, 531–538.

Bird, M.I., Boobyer, E.M., Bryant, C., Lewis, H.A., Paz, V. and Stephens, W.E. (2007) A long record of environmental change from bat guano deposits in Makangit Cave, Palawan, Philippines. *Earth and Environmental Science Transactions of the Royal Society of Edinburgh*, **98**, 59–69.

BirdLife International (2001) *Threatened birds of Asia: the BirdLife International Red Data Book*. BirdLife International, Cambridge.

Bischoff, W., Newbery, D.A., Lingenfelder, M. et al. (2005) Secondary succession and dipterocarp recruitment in Bornean rain forest after logging. *Forest Ecology and Management*, **218**, 174–192.

Blakesley, D., Elliott, S., Kuarak, C., Navakitbumrung, P., Zangkum, S. and Anusarnsunthorn, V. (2002) Propagating framework tree species to restore seasonally dry tropical forest: implications of seasonal seed dispersal and dormancy. *Forest Ecology and Management*, **164**, 31–38.

Blate, G.M., Peart, D.R. and Leighton, M. (1998) Post-dispersal predation on isolated seeds: a comparative study of 40 tree species in a Southeast Asian rainforest. *Oikos*, **82**, 522–538.

Blossey, B. and Hunt-Joshi, T.R. (2003) Belowground herbivory by insects: influence on plants and aboveground herbivores. *Annual Review of Entomology*, **48**, 521–547.

Blundell, A.G. and Burkey, T.V. (2007) A database of schemes that prioritize sites and species based on their conservation value: focusing business on biodiversity. *BMC Ecology*, **7**, 10.

Blüthgen, N. and Fiedler, K. (2004) Competition for composition: lessons from nectar-feeding ant communities. *Ecology*, **85**, 1479–1485.

Blüthgen, N., Mezger, D. and Linsenmair, K.E. (2006) Ant-hemipteran trophobioses in a Bornean rainforest—diversity, specificity and monopolisation. *Insectes Sociaux*, **53**, 194–203.

Bocxlaer, I.V., Roelants, K., Biju, S.D., Nagaraju, J. and Bossuyt, F. (2006) Late Cretaceous vicariance in gondwanan amphibians. *PLoS ONE*, **1**, e74.

Bode, M., Wilson, K.A., Brooks, T.M. et al. (2008) Cost-effective global conservation spending is robust to taxonomic group. *Proceedings of the National Academy of Sciences of the USA*, **105**, 6498–6501.

Boer, E. and Ella, A.B. (2000) *Plant resources of South-East Asia 18. Plants producing exudates*. Backhuys Publishers, Leiden.

Bøgh, A. (1996) Abundance and growth of rattans in Khao Chong National Park, Thailand. *Forest Ecology and Management*, **84**, 71–80.

Boomgaard, P. (2007) *Southeast Asia: an environmental history*. ABC-CLIO, Santa Barbara.

Boonyanuphap, J., Sakurai, K. and Tanaka, S. (2007) Soil nutrient status under upland farming practice in the lower Northern Thailand. *Tropics*, **16**, 215–231.

Borcherding, R., Paarmann, W., Nyawa, S.B. and Bolte, H. (2000) How to be a fig beetle? Observations of ground beetles (Col., Carabidae) associated with fruitfalls in a rain forest of Borneo. *Ecotropica*, **6**, 169–180.

Borchert, R. (1994) Soil and stem water storage determine phenology and distribution of tropical dry forest trees. *Ecology*, **75**, 1437–1449.

Borges, R.M., Bessière, J.M. and Hossaert-McKey, M. (2008) The chemical ecology of seed dispersal in monoecious and dioecious figs. *Functional Ecology*, **22**, 484–493.

Bos, M.M., Tylianakis, J.M., Steffan-Dewenter, I. and Tscharntke, T. (2008) The invasive yellow crazy ant and the decline of forest ant diversity in Indonesian cacao agroforests. *Biological Invasions*, **10**, 1399–1409.

Bowen, M.E., McAlpine, C.A., House, A.P.N. and Smith, G.C. (2007) Regrowth forests on abandoned agricultural land: a review of their habitat values for recovering forest fauna. *Biological Conservation*, **140**, 273–296.

Boyd, R.S. (2004) Ecology of metal hyperaccumulation. *New Phytologist*, **162**, 563–567.

Boyer, S.L., Clouse, R.M., Benavides, L.R. et al. (2007) Biogeography of the world: a case study from cyphophthalmid Opiliones, a globally distributed group of arachnids. *Journal of Biogeography,* **34**, 2070–2085.

Bramwell, D. (2002) How many plant species are there? *Plant Talk,* **28**, 32–34.

Brearley, F.Q. (2005) Nutrient limitation in a Malaysian ultramafic soil. *Journal of Tropical Forest Science,* **17**, 596–609.

Brearley, F.Q., Proctor, J., Suriantata, Nagy, L., Dalrymple, G. and Voysey, B.C. (2007a) Reproductive phenology over a 10-year period in a lowland evergreen rain forest of central Borneo. *Journal of Ecology,* **95**, 828–839.

Brearley, F.Q., Scholes, J.D., Press, M.C. and Palfner, G. (2007b) How does light and phosphorus fertilisation affect the growth and ectomycorrhizal community of two contrasting dipterocarp species? *Plant Ecology,* **192**, 237–249.

Brickle, N.W., Duckworth, J.W., Tordoff, A.W., Poole, C.M., Timmins, R. and McGowan, P.J.K. (2008) The status and conservation of Galliformes in Cambodia, Laos and Vietnam. *Biodiversity and Conservation,* 1–35.

Brink, M. and Escobin, R.P. (2003) *Plant resources of South-East Asia No. 17: fibre plants.* Backhuys Publishers, Leiden.

Brockerhoff, E.G., Jactel, H., Parrotta, J.A., Quine, C.P. and Sayer, J. (2008) Plantation forests and biodiversity: oxymoron or opportunity? *Biodiversity and Conservation,* **17**, 925–951.

Brook, B.W., Sodhi, N.S. and Ng, P.K.L. (2003) Catastrophic extinctions follow deforestation in Singapore. *Nature,* **424**, 420–423.

Brook, B.W., Bradshaw, C.J.A., Koh, L.P. and Sodhi, N.S. (2006) Momentum drives the crash: mass extinction in the tropics. *Biotropica,* **38**, 302–305.

Brook, B.W., Sodhi, N.S. and Bradshaw, C.J.A. (2008) Synergies among extinction drivers under global change. *Trends in Ecology and Evolution,* **23**, 453–460.

Brooke, A.D., Butchart, S.H.M., Garnett, S.T., Crowley, G.M., Mantilla-Beniers, N.B. and Stattersfield, A. (2008) Rates of movement of threatened bird species between IUCN red list categories and towards extinction. *Conservation Biology,* **22**, 417–427.

Brooks, T.M., Pimm, S.L. and Collar, N.J. (1997) Deforestation predicts the number of threatened birds in insular southeast Asia. *Conservation Biology,* **11**, 382–394.

Brooks, T.M., Pimm, S.L. and Oyugi, J.O. (1999) Time lag between deforestation and bird extinction in tropical forest fragments. *Conservation Biology,* **13**, 1140–1150.

Brooks, T.M., De Silva, N., Duya, M.V. et al. (2008) Delineating Key Biodiversity Areas as targets for protecting areas. *Biodiversity and human livelihoods in protected areas: case studies from the Malay Archipelago* (eds N.S. Sodhi, G. Acciaioli, M. Erb and A.K.-J. Tan), pp. 20–35. Cambridge University Press, Cambridge.

Brooks, T.M., Mittermeier, R.A., Mittermeier, C.G. et al. (2002) Habitat loss and extinction in the hotspots of biodiversity. *Conservation Biology,* **16**, 909–923.

Brown, W.C. and Alcala, A.C. (1970) The zoogeography of the herpetofauna of the Philippine islands, a fringing archipelago. *Proceedings of the California Academy of Sciences,* **38**, 105–139.

Brühl, C.A., Mohamed, V. and Linsenmair, K.E. (1999) Altitudinal distribution of leaf litter ants along a transect in primary forests on Mount Kinabalu, Sabah, Malaysia. *Journal of Tropical Ecology,* **15**, 265–277.

Brühl, C.A., Eltz, T. and Linsenmair, K.E. (2003) Size does matter—effects of tropical rainforest fragmentation on the leaf litter ant community in Sabah, Malaysia. *Biodiversity and Conservation,* **12**, 1371–1389.

Bruijnzeel, L.A. (2005) Tropical montane cloud forest: a unique hydrological case. *Forests, water and people in the humid tropics* (eds M. Bonell and L.A. Bruijnzeel). Cambridge University Press, Cambridge.

Bruner, A.G., Gullison, R.E., Rice, R.E. and da Fonseca, G.A.B. (2001) Effectiveness of parks in protecting tropical biodiversity. *Science,* **291**, 125–128.

Buckley, B.M., Palakit, K., Duangsathaporn, K., Sanguantham, P. and Prasomsin, P. (2007) Decadal scale droughts over northwestern Thailand over the past 448 years: links to the tropical Pacific and Indian Ocean sectors. *Climate Dynamics,* **229**, 63–71.

Bungard, R.A., Zipperlen, S.A., Press, M.C. and Scholes, J.D. (2002) The influence of nutrients on growth and photosynthesis of seedlings of two rainforest dipterocarp species. *Functional Plant Biology,* **29**, 505–515.

Bunyavejchewin, S. (1999) Structure and dynamics in seasonal dry evergreen forest in northeastern Thailand. *Journal of Vegetation Science,* **10**, 787–792.

Bunyavejchewin, S., LaFrankie, J.V., Baker, P.J., Kanzaki, M., Ashton, P.S. and Yamakura, T. (2003) Spatial distribution patterns of the dominant canopy dipterocarp species in a seasonal dry evergreen forest in western Thailand. *Forest Ecology and Management,* **175**, 87–101.

Bunyavejchewin, S., Baker, P.J., LaFrankie, J.V. and Ashton, P.S. (2004) Huai Kha Kheng Forest Dynamics Plot, Thailand. *Tropical forest diversity and dynamism: findings from a large-scale plot network* (eds E.C. Losos and E.G. Leigh), pp. 482–491. University of Chicago Press, Chicago.

Burba, G. and Forman, S.L. (2008) Eddy covariance method. *Encyclopedia of earth* (ed C.J. Cleveland). Environmental Information Coalition, National Council for Science and the Environment, Washington, DC.

Burslem, D.F.R.P., Turner, I.M. and Grubb, P.J. (1994) Mineral nutrient status of coastal hill dipterocarp forest and *Adinandra* belukar in Singapore—bioassays of nutrient limitation. *Journal of Tropical Ecology*, **10**, 579–599.

Burslem, D.F.R.P., Grubb, P.J. and Turner, I.M. (1995) Responses to nutrient addition among shade-tolerant tree seedlings of lowland tropical rain-forest in Singapore. *Journal of Ecology*, **83**, 113–122.

Bush, A.B.G. (2007) Extratropical influences on the El Nino-Southern Oscillation through the late Quaternary. *Journal of Climate*, **20**, 788–800.

Butchart, S.H.M., Stattersfield, A.J. and Brooks, T.M. (2006) Going or gone: defining 'possibly extinct' species to give a truer picture of recent extinctions. *Bulletin of the British Ornithologists' Club*, **126A**, 7–24.

Butler, R.A. and Laurance, W.F. (2008) New strategies for conserving tropical forests. *Trends in Ecology and Evolution*, **23**, 469–472.

Cai, Z.Q. and Bongers, F. (2007) Contrasting nitrogen and phosphorus resorption efficiencies in trees and lianas from a tropical montane rain forest in Xishuangbanna, south-west China. *Journal of Tropical Ecology*, **23**, 115–118.

Cai, H.J., Li, Z.S. and You, M.S. (2007) Impact of habitat diversification on arthropod communities: a study in the fields of Chinese cabbage, *Brassica chinensis*. *Insect Science*, **14**, 241–249.

Campbell, C.J., Fuentes, A., MacKinnon, K.C., Panger, M. and Bearder, S.K. (2007) *Primates in perspective*. Oxford University Press, New York.

Campos-Arceiz, A., Lin, T.Z., Htun, W., Takatsuki, S. and Leimgruber, P. (2008a) Working with mahouts to explore the diet of working elephants in Myanmar (Burma). *Ecological Research*, **23**, 1057–1064.

Campos-Arceiz, A., Larrinaga, A.R., Weerasinghe, U.R. et al. (2008b) Behavior rather than diet mediates seasonal differences in seed dispersal by Asian elephants. *Ecology*, **89**, 2684–2691.

Canadell, J.G. and Raupach, M.R. (2008) Managing forests for climate change mitigation. *Science*, **320**, 1456–1457.

Canadell, J.G., Le Quéré, C., Raupach, M.R. et al. (2007) Contributions to accelerating atmospheric CO_2 growth from economic activity, carbon intensity, and efficiency of natural sinks. *Proceedings of the National Academy of Sciences of the USA*, **104**, 18866–18870.

Cannon, C.H. and Leighton, M. (2004) Tree species distributions across five habitats in a Bornean rain forest. *Journal of Vegetation Science*, **15**, 257–266.

Cannon, C.H., Peart, D.R. and Leighton, M. (1998) Tree species diversity in commercially logged Bornean rainforest. *Science*, **281**, 1366–1368.

Carbone, C. and Gittleman, J.L. (2002) A common rule for the scaling of carnivore density. *Science*, **295**, 2273–2276.

Carbone, C., Mace, G.M., Roberts, S.C. and Macdonald, D.W. (1999) Energetic constraints on the diet of terrestrial carnivores. *Nature*, **402**, 286–288.

Carbone, C., Teacher, A. and Rowcliffe, J.M. (2007) The costs of carnivory. *PloS Biology*, **5**, 363–368.

Cardoso, M.F., Nobre, C.A., Lapola, D.M., Oyama, M.D. and Sampaio, G. (2008) Long-term potential for fires in estimates of the occurrence of savannas in the tropics. *Global Ecology and Biogeography*, **17**, 222–235.

Carrete, M. and Tella, J.L. (2008) Wild-bird trade and exotic invasions: a new link of conservation concern? *Frontiers in Ecology and the Environment*, **6**, 207–211.

Cashore, B., Gale, F., Meidinger, E. and Newsom, D. (2006) *Confronting sustainablity: forest certification in developing and transitioning countries*. Yale School of Forestry and Environmental Studies, New Haven.

Catoni, C., Schaefer, H.M. and Peters, A. (2008) Fruit for health: the effects of flavonoids on humoral immune response and food selection in a frugivorous bird. *Functional Ecology*, **22**, 649–654.

Cavaleri, M.A., Oberbauer, S.F. and Ryan, M.G. (2008) Foliar and ecosystem respiration in an old-growth tropical rain forest. *Plant, Cell and Environment*, **31**, 473–483.

Cazetta, E., Schaefer, H.M. and Galetti, M. (2008) Does attraction to frugivores or defense against pathogens shape fruit pulp composition? *Oecologia*, **155**, 277–286.

Chacon, N., Silver, W.L., Dubinsky, E.A. and Cusack, D.F. (2006) Iron reduction and soil phosphorus solubilization in humid tropical forests soils: the roles of labile carbon pools and an electron shuttle compound. *Biogeochemistry*, **78**, 67–84.

Chadwick, O.A., Derry, L.A., Vitousek, P.M., Huebert, B.J. and Hedin, L.O. (1999) Changing sources of nutrients during four million years of ecosystem development. *Nature*, **397**, 491–497.

Chaimanee, Y. (2007) Late Pleistocene of Southeast Asia. *Encyclopedia of Quaternary science* (ed S.A. Elias), pp. 3189–3197. Elsevier, Amsterdam.

Chambers, J.Q., Tribuzy, E.S., Toledo, L.C. et al. (2004) Respiration from a tropical forest ecosystem: Partitioning of sources and low carbon use efficiency. *Ecological Applications*, **14**, S72-S88.

Chambers, J.Q., Asner, G.P., Morton, D.C. et al. (2007) Regional ecosystem structure and function: ecological insights from remote sensing of tropical forests. *Trends in Ecology and Evolution*, **22**, 414–423.

Chan, K.M.A., Shaw, M.R., Cameron, D.R., Underwood, E.C. and Daily, G.C. (2006) Conservation planning for ecosystem services. *PLoS Biology*, **4**, 2138–2152.

Chang, S.C., Yeh, C.F., Wu, M.J., Hsia, Y.J. and Wu, J.T. (2006) Quantifying fog water deposition by in situ exposure experiments in a mountainous coniferous forest in Taiwan. *Forest Ecology and Management*, **224**, 11–18.

Chanthorn, W. and Brockelman, W.Y. (2008) Seed dispersal and seedling recruitment in the light-demanding tree *Choerospondias axillaris* in old-growth forest in Thailand. ScienceAsia, **34**, 129–135.

Chao, K.J., Phillips, O.L., Gloor, E., Monteagudo, A., Torres-Lezama, A. and Martínez, R.V. (2008) Growth and wood density predict tree mortality in Amazon forests. *Journal of Ecology*, **96**, 281–292.

Chapin, M. (2004) A challenge to conservationists. *World Watch Magazine*, **17**, 17–31.

Chapin III, F.S., Matson, P. and Mooney, H.A. (2002) *Principles of terrestrial ecosystem ecology*. Springer, New York.

Chave, J., Condit, R., Muller-Landau, H.C. et al. (2008) Assessing evidence for a pervasive alteration in tropical tree communities. *PloS Biology*, **6**, 455–462.

Chazdon, R.L. (2003) Tropical forest recovery: legacies of human impact and natural disturbances. *Perspectives in Plant Ecology Evolution and Systematics*, **6**, 51–71.

Chen, H.K. (2006) The role of CITES in combating illegal logging—current and potential. TRAFFIC International, Cambridge.

Chen, C.C. and Hsieh, F. (2002) Composition and foraging behaviour of mixed-species flocks led by the grey-cheeked fulvetta in Fushan Experimental Forest, Taiwan. *Ibis*, **144**, 317–330.

Chen, X.Y. and Mulder, J. (2007) Atmospheric deposition of nitrogen at five subtropical forested sites in South China. *Science of the Total Environment*, **378**, 317–330.

Chen, J., Deng, X.B., Bai, Z.L. et al. (2001) Fruit characteristics and *Muntiacus muntijak vaginalis* (Muntjac) visits to individual plants of *Choerospondias axillaris*. *Biotropica*, **33**, 718–722.

Chen, I.C., Shiu, H.J., Benedick, S. et al. (2009) Elevation increases in moth assemblages over 42 years on a tropical mountain. *Proceedings of the National Academy of Sciences of the USA*, **106**, 1479–1483.

Cheng, J.R., Xiao, Z.S. and Zhang, Z.B. (2005) Seed consumption and caching on seeds of three sympatric tree species by four sympatric rodent species in a subtropical forest, China. *Forest Ecology and Management*, **216**, 331–341.

Chesson, P. (2000) Mechanisms of maintenance of species diversity. *Annual Review of Ecology and Systematics*, **31**, 343–366.

Cheyne, S.M., Chivers, D.J. and Sugardjito, J. (2008) Biology and behaviour of reintroduced gibbons. *Biodiversity and Conservation*, **17**, 1741–1751.

Chiba, S. (2007) Morphological and ecological shifts in a land snail caused by the impact of an introduced predator. *Ecological Research*, **22**, 884–891.

Choi, K. and Driwantoro, D. (2007) Shell tool use by early members of *Homo erectus* in Sangiran, central Java, Indonesia: cut mark evidence. *Journal of Archaeological Science*, **34**, 48–58.

Chokkalingam, U., Suyanto, Permana, R.P. et al. (2007) Community fire use, resource change, and livelihood impacts: the downward spiral in the wetlands of southern Sumatra. *Mitigation and Adaptation Strategies for Global Change*, **12**, 75–100.

Choong, M.F. (1996) What makes a leaf tough and how this affects the pattern of *Castanopsis fissa* leaf consumption by caterpillars. *Functional Ecology*, **10**, 668–674.

Chuine, I. and Beaubien, E.G. (2001) Phenology is a major determinant of tree species range. *Ecology Letters*, **4**, 500–510.

Chung, K.P.S. and Corlett, R.T. (2006) Rodent diversity in a highly degraded tropical landscape: Hong Kong, South China. *Biodiversity and Conservation*, **15**, 4521–4532.

Chung, A.Y.C., Eggleton, P., Speight, M.R., Hammond, P.M. and Chey, V.K. (2000) The diversity of beetle assemblages in different habitat types in Sabah, Malaysia. *Bulletin of Entomological Research*, **90**, 475–496.

Ciochon, R.L. and Olsen, J.W. (1991) Paleoanthropological and archaeological discoveries from Lang Trang caves: a new Middle Pleistocene hominid site from northern Vietnam. *Indo-Pacific Prehistory Association Bulletin*, **10**, 59–73.

Ciochon, R.L., Long, V.T., Larick, R. et al. (1996) Dated co-occurrence of *Homo erectus* and *Gigantopithecus* from Tham Khuyen Cave, Vietnam. *Proceedings of the National Academy of Sciences of the USA*, **93**, 3016–3020.

Clark, D.A., Brown, S., Kicklighter, D.W., Chambers, J.Q., Thomlinson, J.R. and Ni, J. (2001) Measuring net primary production in forests: concepts and field methods. *Ecological Applications*, **11**, 356–370.

Clauss, M., Lechner-Doll, M. and Streich, W.J. (2003) Ruminant diversification as an adaptation to the physicomechanical characteristics of forage. A reevaluation of an old debate and a new hypothesis. *Oikos*, **102**, 253–262.

Claussen, M., Berger, A. and Held, H. (2007) A survey of hypotheses for the 100-kyr cycle. *The climate of past interglacials* (eds F. Sirocko, M. Claussen, T. Litt and M.F. Sanchez-Goni), pp. 29–35. Elsevier, Amsterdam.

Cleary, D.F.R. and Mooers, A.O. (2006) Burning and logging differentially affect endemic vs. widely distributed butterfly species in Borneo. *Diversity and Distributions*, **12**, 409–416.

Cleary, D.F.R., Priadjati, A., Suryokusumo, B.K. and Menken, S.B.J. (2006) Butterfly, seedling, sapling and tree diversity and composition in a fire-affected Bornean rainforest. *Austral Ecology*, **31**, 46–57.

Cleary, D.F.R., Boyle, T.J.B., Setyawati, T., Anggraeni, C.D., Van Loon, E.E. and Menken, S.B.J. (2007) Bird species and traits associated with logged and unlogged forest in Borneo. *Ecological Applications*, **17**, 1184–1197.

Cleland, E.E., Chuine, I., Menzel, A., Mooney, H.A. and Schwartz, M.D. (2007) Shifting plant phenology in response to global change. *Trends in Ecology and Evolution*, **22**, 357–365.

Clements, R., Sodhi, N.S., Schilthuizen, M. and Ng, P.K.L. (2006) Limestone karsts of southeast Asia: imperiled arks of biodiversity. *Bioscience*, **56**, 733–742.

Co, L.L., Lagunzad, D.A., LaFrankie, J.V. et al. (2004) Palanan Forest Dynamics Plot, Philippines. *Tropical forest diversity and dynamism: findings from a large-scale plot network* (eds E.C. Losos and E.G. Leigh), pp. 574–584. University of Chicago Press, Chicago.

Cocks, L. and Bullo, K. (2008) The processes for releasing a zoo-bred Sumatran orang-utan *Pongo abelii* at Bukit Tigapuluh National Park, Jambi, Sumatra. *International Zoo Yearbook*, **42**, 183–189.

Codron, D., Lee-Thorp, J.A., Sponheimer, M. and Codron, J. (2007) Nutritional content of savanna plant foods: implications for browser/grazer models of ungulate diversification. *European Journal of Wildlife Research*, **53**, 100–111.

Colling, G., Reckinger, C. and Matthies, D. (2004) Effects of pollen quantity and quality on reproduction and offspring vigor in the rare plant *Scorzonera humilis* (Asteraceae). *American Journal of Botany*, **91**, 1774–1782.

Colwell, R.K., Brehm, G., Cardelús, C.L., Gilman, A.C. and Longino, J.T. (2008) Gllobal warming, elevational arrange shifts, and lowland biotic attrition in the wet tropics. *Science*, **322**, 258–261.

Condit, R., Ashton, P., Bunyavejchewin, S. et al. (2006) The importance of demographic niches to tree diversity. *Science*, **313**, 98–101.

Condit, R., Watts, K., Bohlman, S.A., Perez, R., Foster, R.B. and Hubbell, S.P. (2000) Quantifying the deciduousness of tropical forest canopies under varying climates. *Journal of Vegetation Science*, **11**, 649–658.

Cook, S.C. and Davidson, D.W. (2006) Nutritional and functional biology of exudate-feeding ants. *Entomologia Experimentalis et Applicata*, **118**, 1–10.

Corlett, R.T. (1986) The mangrove understorey: some additional observations. *Journal of Tropical Ecology*, **2**, 93–94.

Corlett, R.T. (1987) The phenology of *Ficus fistulosa* in Singapore. *Biotropica*, **19**, 122–124.

Corlett, R.T. (1990) Flora and reproductive phenology of the rain forest at Bukit Timah, Singapore. *Journal of Tropical Ecology*, **6**, 55–63.

Corlett, R.T. (1991) Plant succession on degraded land in Singapore. *Journal of Tropical Forest Science*, **4**, 151–161.

Corlett, R.T. (1992) The ecological transformation of Singapore, 1819–1990. *Journal of Biogeography*, **19**, 411–420.

Corlett, R.T. (1993) Reproductive phenology of Hong Kong shrubland. *Journal of Tropical Ecology*, **9**, 501–510.

Corlett, R.T. (1994) What is secondary forest? *Journal of Tropical Ecology*, **10**, 445–447.

Corlett, R.T. (1996) Characteristics of vertebrate dispersed fruits in Hong Kong. *Journal of Tropical Ecology*, **12**, 819–833.

Corlett, R.T. (1998) Frugivory and seed dispersal by vertebrates in the Oriental (Indomalayan) Region. *Biological Reviews*, **73**, 413–448.

Corlett, R.T. (2000) Environmental heterogeneity and species survival in degraded tropical landscapes. *The ecological consequences of environmental heterogeneity* (eds M.J. Hutchings, E.A. John and A.J.A. Stewart), pp. 333–355. Blackwell Science, Oxford.

Corlett, R.T. (2001) Pollination in a degraded tropical landscape: a Hong Kong case study. *Journal of Tropical Ecology*, **17**, 155–161.

Corlett, R.T. (2002) Frugivory and seed dispersal in degraded tropical East Asian landscapes. *Seed dispersal and frugivory: ecology, evolution and conservation* (eds D.J. Levey, W.R. Silva and M. Galetti), pp. 451–465. CABI International, Wallingford.

Corlett, R.T. (2004) Flower visitors and pollination in the Oriental (Indomalayan) Region. *Biological Reviews*, **79**, 497–532.

Corlett, R.T. (2005) Vegetation. *The physical geography of Southeast Asia* (ed. A. Gupta), pp. 105–119. Oxford University Press, Oxford.

Corlett, R.T. (2007a) The impact of hunting on the mammalian fauna of tropical Asian forests. *Biotropica*, **39**, 292–303.

Corlett, R.T. (2007b) Pollination or seed dispersal: which should we worry about most? *Seed dispersal: theory and its application in a changing world* (eds A.J. Dennis, E.W. Schupp, R.J. Green and D.A. Westcott), pp. 523–544. CABI, Walllingford.

Corlett, R.T. (2007c) What's so special about Asian tropical forests? *Current Science*, **93**, 1551–1557.

Corlett, R.T. (2009a) Seed dispersal distances and plant migration potential in tropical East Asia. *Biotropica*, **41**, in press.

Corlett, R.T. (2009b) Invasive aliens on tropical East Asian islands. *Biodiversity and Conservation*, in press.

Corlett, R.T. and LaFrankie, J.V. (1998) Potential impacts of climate change on tropical Asian forests through an influence on phenology. *Climatic Change*, **39**, 439–453.

Corlett, R.T. and Lucas, P.W. (1990) Alternative seed-handling strategies in primates—seed-spitting by long-tailed macaques (*Macaca fascicularis*). *Oecologia*, **82**, 166–171.

Corlett, R.T. and Primack, R.B. (2006) Tropical rainforests and the need for cross-continental comparisons. *Trends in Ecology and Evolution*, **21**, 104–110.

Corlett, R.T. and Turner, I.M. (1997) Long-term survival in tropical forest remnants in Singapore and Hong Kong. *Tropical forest remnants: ecology, management and conservation of fragmented communities* (eds W.F. Laurance and R.O. Bierregaard), pp. 333–345. University of Chicago Press, Chicago.

Corre, M.D., Dechert, G. and Veldkamp, E. (2006) Soil nitrogen cycling following montane forest conversion in Central Sulawesi, Indonesia. *Soil Science Society of America Journal*, **70**, 359–366.

Cowling, S.A. (2007) Ecophysiological response of lowland tropical plants to Pleistocene climate. *Tropical rainforest responses to climatic change* (eds M.B. Bush and J.R. Flenley), pp. 333–349. Springer, Berlin.

Cox, C.B. (2001) The biogeographic regions reconsidered. *Journal of Biogeography*, **28**, 511–523.

Cranbrook, Earl of (1988) Report on bones from the Madai and Baturong cave excavations. *Archaeological research in south-eastern Sabah* (ed. P. Bellwood), pp. 142–154. Sabah Museum, Kota Kinabalu.

Cristoffer, C. and Peres, C.A. (2003) Elephants versus butterflies: the ecological role of large herbivores in the evolutionary history of two tropical worlds. *Journal of Biogeography*, **30**, 1357–1380.

Croft, D.A., Heaney, L.R., Flynn, J.J. and Bautista, A.P. (2006) Fossil remains of a new, diminutive *Bubalus* (Artiodactyla: Bovidae: Bovini) from Cebu Island, Philippines. *Journal of Mammalogy*, **87**, 1037–1051.

Crombie, R.I. and Pregill, G.K. (1999) A checklist of the herpetofauna of the Palau Islands (Republic of Belau), Oceania. *Herpetological Monographs*, 29–80.

Croxall, J.P. (1976) Composition and behavior of some mixed species bird flocks in Sarawak. *Ibis*, **118**, 333–346.

Cunningham, A.A., Daszak, P. and Patel, N.G. (2006) Emerging infectious-disease threats to tropical forest ecosystems *Emerging threats to tropical forests* (eds W.F. Laurance and C.A. Peres), pp. 149–164. University of Chicago Press, Chicago.

Curran, L.M. and Leighton, M. (2000) Vertebrate responses to spatiotemporal variation in seed production of mast-fruiting Dipterocarpaceae. *Ecological Monographs*, **70**, 101–128.

Curran, L.M., Trigg, S.N., McDonald, A.K. et al. (2004) Lowland forest loss in protected areas of Indonesian Borneo. *Science*, **303**, 1000–1003.

Curran, T.J., Gersbach, L.N., Edwards, W. and Krockenberger, A.K. (2008) Wood density predicts plant damage and vegetative recovery rates caused by cyclone disturbance in tropical rainforest tree species of North Queensland, Australia. *Austral Ecology*, **33**, 442–450.

Cyranoski, D. (2003) Biodiversity schemes take root in China. *Nature*, **425**, 890–890.

Daehler, C.C. (2006) Invasibility of tropical islands by introduced plants: partitioning the influence of isolation and propagule pressure. *Preslia*, **78**, 389–404.

Daniel, T.F. (2006) Synchronous flowering and monocarpy suggest plietesial life history for Neotropical *Stenostephanus chiapensis* (Acanthaceae). *Proceedings of the California Academy of Sciences*, **57**, 1011–1018.

D'Arrigo, R., Wilson, R., Palmer, J. et al. (2006) Monsoon drought over Java, Indonesia, during the past two centuries. *Geophysical Research Letters*, **33**.

D'Arrigo, R., Wilson, R. and Tudhope, A. (2009) The impact of volcanic forcing on tropical temperatures during the past four centuries. *Nature Geoscience*, **2**, 51–56.

Das, A., Krishnaswamy, J., Bawa, K.S. et al. (2006) Prioritisation of conservation areas in the Western Ghats, India. *Biological Conservation*, **133**, 16–31.

Das, I. (1999) Biogeography of the amphibians and reptiles of the Andaman and Nicobar Islands, India. *Tropical island herpetofauna: origin, current diversity and conservation* (ed. H. Ota), pp. 43–77. Elsevier, Amsterdam.

Davidar, P. (1985) Ecological interactions between mistletoes and their avian pollinators in south India. *Journal of the Bombay Natural History Society*, **82**, 45–60.

Davidar, P., Puyravaud, J.P. and Leigh, E.G. (2005) Changes in rain forest tree diversity, dominance and rarity across a seasonality gradient in the Western Ghats, India. *Journal of Biogeography*, **32**, 493–501.

Davidson, D.W., Cook, S.C., Snelling, R.R. and Chua, T.H. (2003) Explaining the abundance of ants in lowland tropical rainforest canopies. *Science*, **300**, 969–972.

Davidson, E.A., de Carvalho, C.J.R., Vieira, I.C.G. et al. (2004) Nitrogen and phosphorus limitation of biomass growth in a tropical secondary forest. *Ecological Applications*, **14**, S150–S163.

Davidson, E.A., de Carvalho, C.J.R., Figueira, A.M. et al. (2007) Recuperation of nitrogen cycling in Amazonian forests following agricultural abandonment. *Nature*, **447**, 995–998.

Davies, S.J. and Ashton, P.S. (1999) Phenology and fecundity in 11 sympatric pioneer species of *Macaranga* (Euphorbiaceae) in Borneo. *American Journal of Botany*, **86**, 1786–1795.

Davies, S.J. and Becker, P. (1996) Floristic composition and stand structure of mixed dipterocarp and heath forests in Brunei Darussalam. *Journal of Tropical Forest Science*, **8**, 542–569.

Davies, R.G., Eggleton, P., Jones, D.T., Gathorne-Hardy, F.J. and Hernandez, L.M. (2003) Evolution of termite functional diversity: analysis and synthesis of local ecological and regional influences on local species richness. *Journal of Biogeography*, **30**, 847–877.

Davies, S.J., Tan, S., LaFrankie, J.V. and Potts, M.D. (2005) Soil-related floristic variation in the hyperdiverse dipterocarp forest in Lambir Hills, Sarawak. *Pollination ecology and the rain forest* (eds D.W. Roubik, S. Sakai and A.A.H. Karim), pp. 22–34. Springer, New York.

Davis, A.J. (2000) Species richness of dung-feeding beetles (Coleoptera: Aphodiidae, Scarabaeidae, Hybosoridae) in tropical rainforest at Danum Valley, Sabah, Malaysia. *The Coleopterists' Bulletin*, **54**, 221–231.

Daws, M.I., Garwood, N.C. and Pritchard, H.W. (2006) Prediction of desiccation sensitivity in seeds of woody species: a probabilistic model based on two seed traits and 104 species. *Annals of Botany*, **97**, 667–674.

Dawson, W., Mndolwa, A.S., Burslem, D.F.R.P. and Hulme, P.E. (2008) Assessing the risks of plant invasions arising from collections in tropical botanical gardens. *Biodiversity and Conservation*, **17**, 1979–1995.

de Bruyn, M., Nugroho, E., Hossain, M.M., Wilson, J.C. and Mather, P.B. (2005) Phylogeographic evidence for the existence of an ancient biogeographic barrier: the Isthmus of Kra Seaway. *Heredity*, **94**, 370–378.

DeFries, R., Hansen, A., Newton, A.C. and Hansen, M.C. (2005) Increasing isolation of protected areas in tropical forests over the past twenty years. *Ecological Applications*, **15**, 19–26.

DeFries, R., Hansen, A., Turner, B.L., Reid, R. and Liu, J.G. (2007) Land use change around protected areas: management to balance human needs and ecological function. *Ecological Applications*, **17**, 1031–1038.

Dehnen-Schmutz, K., Touza, J., Perrings, C. and Williamson, M. (2007) A century of the ornamental plant trade and its impact on invasion success. *Diversity and Distributions*, **13**, 527–534.

de Lang, R. and Vogel, G. (2005) *The snakes of Sulawesi: a field guide to the land snakes of Sulawesi with identification keys*. Edition Chimaira, Frankfurt am Main.

Delang, C.O. and Wong, T. (2006) The livelihood-based forest classification system of the Pwo Karen in western Thailand. *Mountain Research and Development*, **26**, 138–145.

Delgado, R.A. and Van Schaik, C.P. (2000) The behavioral ecology and conservation of the orangutan (*Pongo pygmaeus*): a tale of two islands. *Evolutionary Anthropology*, **9**, 201–218.

DeLucia, E.H., Drake, J.E., Thomas, R.B. and Gonzalez-Meler, M. (2007) Forest carbon use efficiency: is respiration a constant fraction of gross primary production? *Global Change Biology*, **13**, 1157–1167.

Denslow, J.S. (2003) Weeds in paradise: thoughts on the invasibility of tropical islands. *Annals of the Missouri Botanical Garden*, **90**, 119–127.

Denslow, J.S. (2007) Managing dominance of invasive plants in wildlands. *Current Science*, **93**, 1579–1586.

Dent, D.H., Bagchi, R., Robinson, D., Majalap-Lee, N. and Burslem, D.F.R.P. (2006) Nutrient fluxes via litterfall and leaf litter decomposition vary across a gradient of soil nutrient supply in a lowland tropical rain forest. *Plant and Soil*, **288**, 197–215.

Deutsch, C.A., Tewksbury, J.J., Huey, R.B. et al. (2008) Impacts of climate warming on terrestrial ectotherms across latitude. *Proceedings of the National Academy of Sciences of the USA*, **105**, 6668–6672.

DeWalt, S.J., Denslow, J.S. and Ickes, K. (2004) Natural-enemy release facilitates habitat expansion of the invasive tropical shrub *Clidemia hirta*. *Ecology*, **85**, 471–483.

Dick, C.W., Bermingham, E., Lemes, M.R. and Gribel, R. (2007) Extreme long-distance dispersal of the lowland tropical rainforest tree *Ceiba pentandra* L. (Malvaceae) in Africa and the Neotropics. *Molecular Ecology*, **16**, 3039–3049.

Ding, T.S., Yuan, H.W., Geng, S., Koh, C.N. and Lee, P.F. (2006) Macro-scale bird species richness patterns of the East Asian mainland and islands: energy, area and isolation. *Journal of Biogeography*, **33**, 683–693.

Dodson, J.R., Hickson, S., Khoo, R., Li, X.Q., Toia, J. and Zhou, W.J. (2006) Vegetation and environment history for the past 14 000 yr BP from Dingnan, Jiangxi Province, South China. *Journal of Integrative Plant Biology*, **48**, 1018–1027.

Dominy, N.J., Grubb, P.J., Jackson, R.V. et al. (2008) In tropical lowland rain forests monocots have tougher leaves than dicots, and include a new kind of tough leaf. *Annals of Botany*, **101**, 1363–1377.

Donald, P.F. (2004) Biodiversity impacts of some agricultural commodity production systems. *Conservation Biology*, **18**, 17–37.

Donnegan, J.A., Butler, S.L., Kuegler, O., Stroud, B.J., Hiserote, B.A. and Rengulbai, K. (2007) *Palau's forest resources, 2003*. USDA, Portland.

Donoghue, M.J. (2008) A phylogenetic perspective on the distribution of plant diversity. *Proceedings of the National Academy of Sciences of the USA*,**105**, 11549–11555.

Donovan, S.E., Eggleton, P. and Bignell, D.E. (2001) Gut content analysis and a new feeding group classification of termites. *Ecological Entomology*, **26**, 356–366.

Donovan, S.E., Griffiths, G.J.K., Homathevi, R. and Winder, L. (2007) The spatial pattern of soil-dwelling termites in primary and logged forest in Sabah, Malaysia. *Ecological Entomology*, **32**, 1–10.

Douglas, A.E. (2006) Phloem-sap feeding by animals: problems and solutions. *Journal of Experimental Botany*, **57**, 747–754.

Doust, S.J., Erskine, P.D. and Lamb, D. (2006) Direct seeding to restore rainforest species: microsite effects on the early establishment and growth of rainforest tree seedlings on degraded land in the wet tropics of Australia. *Forest Ecology and Management*, **234**, 333–343.

Dowie, M. (2006) The hidden cost of paradise: indigenous people are being displaced to create wilderness areas, to the detriment of all. *Stanford Social Innovation Review*, **2006**, 31–38.

Dransfield, J. and Manokaran, N. (1993) *Plant resources of South-East Asia. No. 6. Rattans*. Pudoc, Wageningen.

Dransfield, S. and Widjaja, E.A. (1995) *Plant resources of South-East Asia No. 7. Bamboos*. Backhuys, Leiden.

Du, X.J., Guo, Q.F., Gao, X.M. and Ma, K.P. (2007) Seed rain, soil seed bank, seed loss and regeneration of *Castanopsis fargesii* (Fagaceae) in a subtropical evergreen broad-leaved forest. *Forest Ecology and Management*, **238**, 212–219.

Du, X.J., Liu, C., Yu, X. and Ma, K. (2008) Effects of shading on early growth of *Cyclobalanopsis glauca* (Fagaceae) in subtropical abandoned fields: implications for vegetation restoration. *Acta Oecologica*, **33**, 154–161.

Duckworth, J.W. and Nettelback, A.R. (2007) Observations of small-toothed palm civets *Arctogalidia trivirgata* in Khao Yai National Park, Thailand, with notes on feeding techniques. *Natural History Bulletin of the Siam Society*, **55**, 187–192.

Dudal, R. (2005) Soils of Southeast Asia. *The physical geography of Southeast Asia* (ed. A. Gupta), pp. 94–104. Oxford University Press, Oxford.

Dudgeon, D. (2000) Riverine wetlands and biodiversity conservation in tropical Asia. *Biodiversity in wetlands: assessment, function and conservation* (eds B. Gopal, W.J. Junk and J.A. Davis), pp. 35–60. Backhuys Publishers, The Hague.

Dudgeon, D. and Corlett, R.T. (2004) *The ecology and biodiversity of Hong Kong*. Joint Publishing, Hong Kong.

Dunn, R.R. (2004) Recovery of faunal communities during tropical forest regeneration. *Conservation Biology*, **18**, 302–309.

Dunn, R.R., Gove, A.D., Barraclough, T.G., Givnish, T.J. and Majer, J.D. (2007) Convergent evolution of an ant-plant mutualism across plant families, continents, and time. *Evolutionary Ecology Research*, **9**, 1349–1362.

Dyer, L.A., Singer, M.S., Lill, J.T. et al. (2007) Host specificity of Lepidoptera in tropical and temperate forests. *Nature*, **448**, 696–699.

Ebeling, J. and Yasué, M. (2008) Generating carbon finance through avoided deforestation and its potential to create climatic, conservation and human development benefits. *Philosophical Transactions of the Royal Society B: Biological Sciences*, **363**, 1917–1924.

Edwards, I.D., MacDonald, A.A. and Proctor, J. (1993) *The natural history of Seram*. Intercept, Andover.

Ehrlich, P.R. and Goulder, L.H. (2007) Is current consumption excessive? A general framework and some indications for the United States. *Conservation Biology*, **21**, 1145–1154.

EIA/Telepak (2008) *Borderlines: Vietnam's booming furniture industry and timber smuggling in the Mekong region*. Environmental Investigation Agency, London.

Eichhorn, M.P., Fagan, K.C., Compton, S.G., Dent, D.H. and Hartley, S.E. (2007) Explaining leaf herbivory rates on tree seedlings in a Malaysian rain forest. *Biotropica*, **39**, 416–421.

Elliott, S., Navakitbumrung, P., Kuarak, C., Zangkum, S., Anusarnsunthorn, V. and Blakesley, D. (2003) Selecting framework tree species for restoring seasonally dry tropical forests in northern Thailand based on field performance. *Forest Ecology and Management*, **184**, 177–191.

Elliott, S., Promkutkaew, S. and Maxwell, J.F. (1994) Flowering and seed production phenology of dry tropical forest trees in northern Thailand. *Proceedings of the International Symposium on Genetic Conservation and Production of Tropical Forest Tree Seed* (eds R.M. Drysdale, S.E.T. John and A.C. Yapa), pp. 52–61. ASEAN—Canada Forest Tree Seed Centre, Saraburi, Thailand.

Elliott, S., Baker, P.J. and Borchert, R. (2006) Leaf flushing during the dry season: the paradox of Asian monsoon forests. *Global Ecology and Biogeography*, **15**, 248–257.

Ellwood, M.D.F. and Foster, W.A. (2004) Doubling the estimate of invertebrate biomass in a rainforest canopy. *Nature*, **429**, 549–551.

Ellwood, M.D.F., Jones, D.T. and Foster, W.A. (2002) Canopy ferns in lowland dipterocarp forest support a prolific abundance of ants, termites, and other invertebrates. *Biotropica*, **34**, 575–583.

Elser, J.J., Fagan, W.F., Denno, R.F. et al. (2000) Nutritional constraints in terrestrial and freshwater food webs. *Nature*, **408**, 578–580.

Elvidge, C.D., Tuttle, B.T., Sutton, P.S., Baugh, K.E., Howard, A.T., Milesi, C., Bhaduri, B.L. and Nemani, R. (2007) Global distribution and density of constructed impervious surfaces. *Sensors*, **7**, 1962–1979.

Elvin, M. (2004) *The retreat of the elephant*. Island Press, Washington, DC.

Emmons, L.H. (2000) *Tupai: a field study of Bornean treeshrews*. University of California Press, Berkeley.

Endicott, K. (1999a) Introduction: Southeast Asia. *The Cambridge encyclopedia of hunters and gatherers* (eds R.B. Lee and R. Daly), pp. 275–283. Cambridge University Press, Cambridge.

Endicott, K. (1999b) The Batek of Peninsula Malaysia. *The Cambridge encyclopedia of hunters and gatherers* (eds R.B. Lee and R. Daly), pp. 298–306. Cambridge University Press, Cambridge.

Engel, V.L. and Parrotta, J.A. (2001) An evaluation of direct seeding for reforestation of degraded lands in central São Paulo state, Brazil. *Forest Ecology and Management*, **152**, 169–181.

Engelbrecht, B.M.J., Comita, L.S., Condit, R. et al. (2007) Drought sensitivity shapes species distribution patterns in tropical forests. *Nature*, **447**, 80-U2.

Enoki, T. (2003) Microtopography and distribution of canopy trees in a subtropical evergreen broad-leaved forest in the northern part of Okinawa Island, Japan. *Ecological Research*, **18**, 103–113.

Ewers, R.M. and Didham, R.K. (2006) Confounding factors in the detection of species responses to habitat fragmentation. *Biological Reviews*, **81**, 117–142.

Ewers, R.M. and Rodrigues, A.S.L. (2008) Estimates of reserve effectiveness are confounded by leakage. *Trends in Ecology and Evolution*, **23**, 113–116.

Fa, J.E. and Funk, S.M. (2007) Global endemicity centres for terrestrial vertebrates: an ecoregions approach. *Endangered Species Research*, **3**, 31–42.

Fang, H., Gundersen, P., Mo, J.M. and Zhu, W.X. (2008) Input and output of dissolved organic and inorganic nitrogen in subtropical forests of South China under high air pollution. *Biogeosciences*, **5**, 339–352.

FAO (2006) *The state of food insecurity in the world 2006*. Food and Agriculture Organization of the United Nations, Rome.

FAO (2007) *World Reference Base for soil resources 2006. A framework for international classification, correlation and communication. First update 2007*. Food and Agriculture Organization of the United Nations, Rome.

Fargione, J., Hill, J., Tilman, D., Polasky, S. and Hawthorne, P. (2008) Land clearing and the biofuel carbon debt. *Science*, **319**, 1235–1238.

Fearnside, P.M. (2006) Mitigation of climatic change in the Amazon. *Emerging threats to tropical forests* (eds W.F. Laurance and C.A. Peres), pp. 353–375. University of Chicago Press, Chicago.

Feeley, K.J., Davies, S.J., Ashton, P.S. et al. (2007a) The role of gap phase processes in the biomass dynamics of tropical forests. *Proceedings of the Royal Society B: Biological Sciences*, **274**, 2857–2864.

Feeley, K.J., Wright, S.J., Supardi, M.N.N., Kassim, A.R. and Davies, S.J. (2007b) Decelerating growth in tropical forest trees. *Ecology Letters*, **10**, 461–469.

Fellowes, J.R. (2006) Ant (Hymenoptera: Formicidae) genera in southern China: observations on the Oriental-Palearctic boundary. *Myrmecologische Nachrichten*, **8**, 239–249.

Fellowes, J.R., Chan, B.P.L., Zhou, J., Chen, S., Yang, S. and Ng, S.-C. (2008) Current status of the Hainan gibbon (*Nomascus hainanus*): progress of population monitoring and other priority actions. *Asian Primates Journal*, **1**, 2–9.

Ferraro, P.J. (2008) Assymetric information and contract design for payments for environmental services. *Ecological Economics*, **65**, 810–821.

Ferraro, P.J. and Pattanayak, S.K. (2006) Money for nothing? A call for empirical evaluation of biodiversity conservation investments. *PLoS Biology*, **4**, 482–488.

Ferraz, G., Russell, G.J., Stouffer, P.C., Bierregaard, R.O., Pimm, S.L. and Lovejoy, T.E. (2003) Rates of species loss from Amazonian forest fragments. *Proceedings of the National Academy of Sciences of the USA*, **100**, 14069–14073.

Fine, P.V.A. (2001) An evaluation of the geographic area hypothesis using the latitudinal gradient in North American tree diversity. *Evolutionary Ecology Research*, **3**, 413–428.

Fitzherbert, E.B., Struebig, M.J., Morel, A. et al. (2008) How will oil palm expansion affect biodiversity? *Trends in Ecology and Evolution*, **23**, 538–545.

Flach, M. and Rumawas, F. (1996) *Plant resources of South-East Asia no. 9. Plants yielding non-seed carbohydrates*. Backhuys Publishers, Leiden.

Flannery, T.F. (1995) *Mammals of the South West Pacific and Moluccan Islands*. Reed, Sydney.

Fleming, T.H. and Muchhala, N. (2008) Nectar-feeding bird and bat niches in two worlds: Pantropical comparisons of vertebrate pollination systems. *Journal of Biogeography*, **35**, 764–780.

Flenley, J.R. (2007) Ultraviolet insolation and the tropical rainforest: altitudinal variations, Quaternary and recent change, extinctions, and biodiversity. *Tropical rainforest responses to climatic change* (eds M.B. Bush and J.R. Flenley), pp. 219–235. Springer, New York.

Flint, E.P. (1994) Changes in land use in South and Southeast Asia from 1880 to 1980: a data base prepared as part of a coordinated research program on carbon fluxes in the tropics. *Chemosphere*, **29**, 1015–1062.

Fontaine, B., Gargominy, O. and Neubert, E. (2007) Priority sites for conservation of land snails in Gabon: testing the umbrella species concept. *Diversity and Distributions*, **13**, 725–734.

Forget, P.-M., Lambert, J.E., Hulme, P.E. and Vander Wall, S.B. (2005) *Seed fate: predation, dispersal and seedling establishment*. CABI Publishing, Wallingford.

FORRU (2006) *How to plant a forest: the principals and practice of restoring tropical forests*. Chiang Mai University, Chiang Mai.

Fox, E.A., van Schaik, C.P., Sitompul, A. and Wright, D.N. (2004) Intra- and interpopulational differences in orangutan (*Pongo pygmaeus*) activity and diet: implications for the invention of tool use. *American Journal of Physical Anthropology*, **125**, 162–174.

Francis, C.M. and Wells, D.R. (2003) The bird community at Pasoh: composition and population dynamics. *Ecology of a lowland rain forest in Southeast Asia* (eds T. Okuda, N. Manokaran, Y. Matsumoto, K. Niiyama, S.C. Thomas and P.S. Ashton), pp. 375–393. Springer, Tokyo.

Freckleton, R.P. and Lewis, O.T. (2006) Pathogens, density dependence and the coexistence of tropical trees.

Proceedings of the Royal Society B: Biological Sciences, **273**, 2909–2916.

Fredericksen, T.S. and Putz, F.E. (2003) Silvicultural intensification for tropical forest conservation. *Biodiversity and Conservation,* **12**, 1445–1453.

Freiberg, M. and Turton, S.M. (2007) Importance of drought on the distribution of the birds nest fern, *Asplenium nidus*, in the canopy of a lowland tropical rainforest in north-eastern Australia. *Austral Ecology,* **32**, 70–76.

FSC (2004) *FSC principles and criteria for forest stewardship*. Forest Stewardship Council, Bonn.

Fuchs, J., Ohlson, J.I., Ericson, P.G.P. and Pasquet, E. (2007) Synchronous intercontinental splits between assemblages of woodpeckers suggested by molecular data. *Zoologica Scripta,* **36**, 11–25.

Fuentes, M. (2008) Biological conservation and global poverty. *Biotropica,* **40**, 139–140.

Fuentes, A., Kalchik, S., Gettler, L., Kwiatt, A. and Konecki, M. (2008) Characterizing human-macaque interactions in Singapore. *American Journal of Primatology,* **70**, 1–5.

Fukushima, M., Kanzaki, M., Hara, M., Ohkubo, T., Preechapanya, P. and Choocharoen, C. (2008) Secondary forest succession after the cessation of swidden cultivation in the montane forest area in Northern Thailand. *Forest Ecology and Management,* **255**, 1994–2006.

Funakoshi, K., Watanabe, H. and Kunisaki, T. (1993) Feeding ecology of the Northern Ryukyu fruit bat, *Pteropus dasymallus dasymallus*, in a warm-temperate region. *Journal of Zoology,* **230**, 221–230.

Gagan, M.K., Hendy, E.J., Haberle, S.G. and Hantoro, W.S. (2004) Post-glacial evolution of the Indo-Pacific Warm Pool and El Nino-Southern Oscillation. *Quaternary International,* **118**, 127–143.

Gallery, R.E., Dalling, J.W. and Arnold, A.E. (2007) Diversity, host affinity, and distribution of seed-infecting fungi: a case study with *Cecropia*. *Ecology,* **88**, 582–588.

Galloway, J.N., Dentener, F.J., Capone, D.G. et al. (2004) Nitrogen cycles: past, present, and future. *Biogeochemistry,* **70**, 153–226.

Gan, J.B. and McCarl, B.A. (2007) Measuring transnational leakage of forest conservation. *Ecological Economics,* **64**, 423–432.

García, C., Jordano, P. and Godoy, J.A. (2007) Contemporary pollen and seed dispersal in a *Prunus mahaleb* population: patterns in distance and direction. *Molecular Ecology,* **16**, 1947–1955.

Gardner, T.A., Barlow, J., Parry, L.W. and Peres, C.A. (2007) Predicting the uncertain future of tropical forest species in a data vacuum. *Biotropica,* **39**, 25–30.

Garrity, D.P., Soekardi, M., Van Noordwijk, M. et al. (1996) The *Imperata* grasslands of tropical Asia: area, distribution, and typology. *Agroforestry Systems,* **36**, 3–29.

Gathorne-Hardy, F.J. and Harcourt-Smith, W.E.H. (2003) The super-eruption of Toba, did it cause a human bottleneck? *Journal of Human Evolution,* **45**, 227–230.

Gathorne-Hardy, F.J., Syaukani, Davies, R.G., Eggleton, P. and Jones, D.T. (2002) Quaternary rainforest refugia in south-east Asia: using termites (Isoptera) as indicators. *Biological Journal of the Linnean Society,* **75**, 453–466.

Gaveau, D.L.A., Wandono, H. and Setiabudi, F. (2007) Three decades of deforestation in southwest Sumatra: Have protected areas halted forest loss and logging, and promoted re-growth? *Biological Conservation,* **134**, 495–504.

Gehring, C., Denich, M., Kanashiro, M. and Vlek, P.L.G. (1999) Response of secondary vegetation in Eastern Amazonia to relaxed nutrient availability constraints. *Biogeochemistry,* **45**, 223–241.

George, W. (1981) Wallace and his line. *Wallace's Line and plate tectonics* (ed. T.C. Whitmore), pp. 3–8. Clarendon Press, Oxford.

Giannini, N.P. and Simmons, N.B. (2003) A phylogeny of megachiropteran bats (Mammalia: Chiroptera: Pteropodidae) based on direct optimization analysis of one nuclear and four mitochondrial genes. *Cladistics—the International Journal of the Willi Hennig Society,* **19**, 496–511.

Gibbons, P. and Lindenmayer, D.B. (2007) Offsets for land clearing: no net loss or the tail wagging the dog. *Ecological Management and Restoration,* **8**, 26–31.

Gibbs, D., Barnes, E. and Cox, J. (2001) *Pigeons and doves: a guide to the pigeons and doves of the world*. Yale University Press, New Haven.

Gilbert, B., Wright, S.J., Muller-Landau, H.C., Kitajima, K. and Hernandez, A. (2006) Life history trade-offs in tropical trees and lianas. *Ecology,* **87**, 1281–1288.

Giri, C. and Shrestha, S. (2000) Forest fire mapping in Huay Kha Khaeng Wildlife Sanctuary, Thailand. *International Journal of Remote Sensing,* **21**, 2023–2030.

Glass, B.P. and Koeberl, C. (2006) Australasian microtektites and associated impact ejecta in the South China Sea and the Middle Pleistocene supereruption of Toba. *Meteoritics and Planetary Science,* **41**, 305–326.

Global Witness (2007) *Cambodia's family tree: illegal logging and the stripping of public assets by Cambodia's elite*. Global Witness.

Goldammer, J.G. (2007) History of equatorial vegetation fires and fire research in Southeast Asia before the 1997–98 episode: a reconstruction of creeping environmental changes. *Mitigation and Adaptation Strategies for Global Change,* **12**, 13–22.

Goldman, R.L., Tallis, H., Kareiva, P. and Daily, G.C. (2008) Field evidence that ecosystem service projects support biodiversity and diversify options. *Proceedings of the National Academy of Sciences of the USA,* **105**, 9445–9448.

Gorog, A.J., Sinaga, M.H. and Engstrom, M.D. (2004) Vicariance or dispersal? Historical biogeography of three Sunda shelf murine rodents (*Maxomys surifer, Leopoldamys sabanus* and *Maxomys whiteheadi*). *Biological Journal of the Linnean Society,* **81**, 91–109.

Graham, E.A., Mulkey, S.S., Kitajima, K., Phillips, N.G. and Wright, S.J. (2003) Cloud cover limits net CO_2 uptake and growth of a rainforest tree during tropical rainy seasons. *Proceedings of the National Academy of Sciences of the USA,* **100**, 572–576.

Gratwicke, B., Mills, J., Dutton, A. et al. (2008) Attitudes towards consumption and conservation of tigers in China. *PLoS ONE,* **3**, e2544.

Gray, T.N.E., Chamnan, H., Borey, R., Collar, N.J. and Dolman, P.M. (2007) Habitat preferences of a globally threatened bustard provide support for community-based conservation in Cambodia. *Biological Conservation,* **138**, 341–350.

Green, W.A. and Hickey, L.J. (2005) Leaf architectural profiles of angiosperm floras across the Cretaceous/Tertiary boundary. *American Journal of Science,* **305**, 983–1013.

Green, J.J., Dawson, L.A., Proctor, J., Duff, E.I. and Elston, D.A. (2005) Fine root dynamics in a tropical rain forest is influenced by rainfall. *Plant and Soil,* **276**, 23–32.

Greenberg, R. (1995) Insectivorous migratory birds in tropical ecosystems—the breeding currency hypothesis. *Journal of Avian Biology,* **26**, 260–264.

Greene, H.W. (1997) *Snakes: the evolution of mystery in nature*. University of California Press, Berkeley.

Grenyer, R., Orme, C.D.L., Jackson, S.F. et al. (2006) Global distribution and conservation of rare and threatened vertebrates. *Nature,* **444**, 93–96.

Grove, R. (1998) Global impact of the 1789–1793 El Nino. *Nature,* **393**, 318–319.

Grundmann, E. (2006) Back to the wild: will reintroduction and rehabilitation help the long-term conservation of orang-utans in Indonesia? *Social Science Information,* **45**, 265–284.

Grytnes, J.A. and Beaman, J.H. (2006) Elevational species richness patterns for vascular plants on Mount Kinabalu, Borneo. *Journal of Biogeography,* **33**, 1838–1849.

Guy, C.L. (2003) Freezing tolerance of plants: current understanding and selected emerging concepts. *Canadian Journal of Botany,* **81**, 1216–1223.

Haines, P.W., Howard, K.T., Ali, J.R., Burrett, C.F. and Bunopas, S. (2004) Flood deposits penecontemporaneous with ~0.8Ma tektite fall in NE Thailand: impact-induced environmental effects? *Earth and Planetary Science Letters,* **225**, 19–28.

Hajra, P.K., Rao, P.S.N. and Mudgal, V. (1999) *Flora of Andaman and Nicobar Islands*. Botanical Survey of India, Calcutta.

Hall, R. (2002) Cenozoic geological and plate tectonic evolution of SE Asia and the SW Pacific: computer-based reconstructions, model and animations. *Journal of Asian Earth Sciences,* **20**, 353–431.

Halpern, B.S., Pyke, C.R., Fox, H.E., Haney, C., Schlaepfer, M.A. and Zaradic, P. (2006) Gaps and mismatches between global conservation priorities and spending. *Conservation Biology,* **20**, 56–64.

Halpin, K., Hyatt, A.D., Plowright, R.K. et al. (2007) Emerging viruses: coming in on a wrinkled wing and a prayer. *Clinical Infectious Diseases,* **44**, 711–717.

Hamann, A. (2004) Flowering and fruiting phenology of a Philippine submontane rain forest: climatic factors as proximate and ultimate causes. *Journal of Ecology,* **92**, 24–31.

Hamann, A., Barbon, E.B., Curio, E. and Madulid, D.A. (1999) A botanical inventory of a submontane tropical rainforest on Negros Island, Philippines. *Biodiversity and Conservation,* **8**, 1017–1031.

Hamer, K.C., Hill, J.K., Benedick, S., Mustaffa, N., Chey, V.K. and Maryati, M. (2006) Diversity and ecology of carrion- and fruit-feeding butterflies in Bornean rain forest. *Journal of Tropical Ecology,* **22**, 25–33.

Hampe, A. (2008) Fruit tracking, frugivore satiation, and their consequences for seed dispersal. *Oecologia,* **156**, 137–145.

Hansen, J.E. (2007) Scientific reticence and sea level rise. *Environmental Research Letters,* **2**, 024002.

Hansen, A.J. and DeFries, R. (2007) Ecological mechanisms linking protected areas to surrounding lands. *Ecological Applications,* **17**, 974–988.

Hansen, J., Sato, M., Kharecha, P., Russell, G., Lea, D.W. and Siddall, M. (2007) Climate change and trace gases. *Philosophical Transactions of the Royal Society A: Mathematical, Physical and Engineering Sciences,* **365**, 1925–1954.

Hansen, M.C., Stehman, S.V., Potapov, P.V. et al. (2008) Humid tropical forest clearing from 2000 to 2005 quantified by using multitemporal and multiresolution remotely sensed data. *Proceedings of the National Academy of Sciences of the USA,***105**, 9439–9444.

Hanski, I. and Krikken, J. (1991) Dung beetles in tropical forests in South-East Asia. *Dung beetle ecology* (eds I. Hanski and Y. Cambefort), pp. 179–197. Princeton University Press, Princeton.

Hanya, G. (2004) Diet of a Japanese macaque troop in the coniferous forest of Yakushima. *International Journal of Primatology,* **25**, 55–71.

Hanya, G., Matsubara, M., Sugiura, H. et al. (2004) Mass mortality of Japanese macaques in a western coastal forest of Yakushima. *Ecological Research,* **19**, 179–188.

Hanya, G., Kiyono, M., Takafumi, H., Tsujino, R. and Agetsuma, N. (2007) Mature leaf selection of Japanese macaques: effects of availability and chemical content. *Journal of Zoology*, **273**, 140–147.

Harcourt, A.H. (1999) Biogeographic relationships of primates on South-East Asian islands. *Global Ecology and Biogeography*, **8**, 55–61.

Hardesty, B.D., Hubbell, S.P. and Bermingham, E. (2006) Genetic evidence of frequent long-distance recruitment in a vertebrate-dispersed tree. *Ecology Letters*, **9**, 516–525.

Hardy, O.J., Maggia, L., Bandou, E. et al. (2006) Fine-scale genetic structure and gene dispersal inferences in 10 Neotropical tree species. *Molecular Ecology*, **15**, 559–571.

Harrison, R.D. (2005) Figs and the diversity of tropical rainforests. *Bioscience*, **55**, 1053–1064.

Harrison, R.D. (2006) Mortality and recruitment of hemi-epiphytic figs in the canopy of a Bornean rain forest. *Journal of Tropical Ecology*, **22**, 477–480.

Harrison, M.E. and Chivers, D.J. (2007) The orang-utan mating system and the unflanged male: a product of increased food stress during the late Miocene and Pliocene? *Journal of Human Evolution*, **52**, 275–293.

Harrison, R.D., Yamamura, N. and Inoue, T. (2000) Phenology of a common roadside fig in Sarawak. *Ecological Research*, **15**, 47–61.

Harrison, T., Ji, X. and Su, D. (2002) On the systematic status of the late Neogene hominoids from Yunnan Province, China. *Journal of Human Evolution*, **43**, 207–227.

Harrison, R.D., Hamid, A.A., Kenta, T. et al. (2003) The diversity of hemi-epiphytic figs (*Ficus*; Moraceae) in a Bornean lowland rain forest. *Biological Journal of the Linnean Society*, **78**, 439–455.

Harrison, T., Krigbaum, J. and Manser, J. (2006) Primate biogeography and ecology on the Sunda Shelf Islands: a paleontological and zooarchaeological perspective. *Primate Biogeography: Progress and Prospects*, (eds J.G. Fleagle and S. Lehman), pp. 331–372. Springer, New York.

Harteveld, M., Hertle, D., Wiens, M. and Leuschner, C. (2005) Spatial and temporal variability of fine root abundance and growth in tropical moist forests and agroforestry systems (Sulawesi, Indonesia). *Ecotropica*, **13**, 111–120.

Harvey, C.A., Komar, O., Chazdon, R. et al. (2008) Integrating agricultural landscapes with biodiversity conservation in the Mesoamerican hotspot. *Conservation Biology*, **22**, 8–15.

Hasegawa, M., Ito, M.T. and Kitayama, K. (2006) Community structure of oribatid mites in relation to elevation and geology on the slope of Mount Kinabalu, Sabah, Malaysia. *European Journal of Soil Biology*, **42**, S191-S196.

Hau, B.C.H. and Corlett, R.T. (2003) Factors affecting the early survival and growth of native tree seedlings planted on a degraded hillside grassland in Hong Kong, China. *Restoration Ecology*, **11**, 483–488.

Hau, B.C.H., Dudgeon, D. and Corlett, R.T. (2005) Beyond Singapore: Hong Kong and Asian biodiversity. *Trends in Ecology and Evolution*, **20**, 281–282.

Hawkins, B.A., Diniz-Filho, J.A.F., Jaramillo, C.A. and Soeller, S.A. (2007) Climate, niche conservatism, and the global bird diversity gradient. *American Naturalist*, **170**, S16-S27.

Hayaishi, S. and Kawamoto, Y. (2006) Low genetic diversity and biased distribution of mitochondrial DNA haplotypes in the Japanese macaque (*Macaca fuscata yakui*) on Yakushima Island. *Primates*, **47**, 158–164.

Heaney, L.R. (1984) Mammalian species richness on islands on the Sunda Shelf, Southeast Asia. *Oecologia*, **61**, 11–17.

Heaney, L.R. (1991) A synopsis of climatic and vegetational change in Southeast Asia. *Climatic Change*, **19**, 53–61.

Heaney, L.R. (2001) Small mammal diversity along elevational gradients in the Philippines: an assessment of patterns and hypotheses. *Global Ecology and Biogeography*, **10**, 15–39.

Heaney, L.R., Balete, D., Dolar, L. et al. (1998) A synopsis of the mammalian fauna of the Philippine Islands. *Fieldiana: Zoology, n.s.*, **88**, 1–61.

Hedin, L.O., Vitousek, P.M. and Matson, P.A. (2003) Nutrient losses over four million years of tropical forest development. *Ecology*, **84**, 2231–2255.

Heil, M. (2008) Indirect defence via tritrophic interactions. *New Phytologist*, **178**, 41–61.

Heil, M., Fiala, B., Maschwitz, U. and Linsenmair, K.E. (2001) On benefits of indirect defence: short- and long-term studies of antiherbivore protection via mutualistic ants. *Oecologia*, **126**, 395–403.

Hembra, S.S., Tacud, B., Geronimo, E. et al. (2006) Saving Philippine hornbills on Panay Island, Philippines. *Re-introduction News*, **25**, 45–46.

Herre, E.A., Van Bael, S.A., Maynard, Z. et al. (2005) Tropical plants as chimera: implications of foliar endophytic fungi for the study of host-plant defence, physiology and genetics. *Biotic interactions in the tropics* (eds D.F.R.P. Burslem, M.A. Pinard and S.E. Hartley), pp. 226–240. Cambridge University Press, Cambridge.

Higham, C.F.W. (1989) The archaeology of mainland Southeast Asia: from 10,000 BC to the fall of Angkor. Cambridge University Press, Cambridge.

Higham, C.F.W. (2002) *Early cultures of mainland Southeast Asia*. Thames and Hudson, London.

Hill, C., Soares, P., Mormina, M. et al. (2007a) A mitochondrial stratigraphy for island southeast Asia. *American Journal of Human Genetics*, **80**, 29–43.

Hill, J., Woodland, W. and Gough, G. (2007b) Can visitor satisfaction and knowledge about tropical rainforests be enhanced through biodiversity interpretation, and does this promote a positive attitude towards ecosystem conservation? *Journal of Ecotourism*, **6**, 75–85.

Hirano, T., Segah, H., Harada, T. et al. (2007) Carbon dioxide balance of a tropical peat swamp forest in Kalimantan, Indonesia. *Global Change Biology*, **13**, 412–425.

Hirata, R., Saigusa, N., Yamamoto, S. et al. (2008) Spatial distribution of carbon balance in forest ecosystems across East Asia. *Agricultural and Forest Meteorology*, **148**, 761–775.

Hiratsuka, M., Toma, T., Mindawati, N., Heriansyah, I. and Morikawa, Y. (2005) Biomass of a man-made forest of timber tree species in the humid tropics of West Java, Indonesia. *Journal of Forest Research*, **10**, 487–491.

Hirosawa, H., Higashi, S. and Mohamed, M. (2000) Food habits of *Aenictus* army ants and their effects on the ant community in a rain forest of Borneo. *Insectes Sociaux*, **47**, 42–49.

Ho, P.-T. (1955) The introduction of American food plants into China. *American Anthropologist*, **57**, 191–201.

Hodgkison, R., Ayasse, M., Kalko, E.K.V. et al. (2007) Chemical ecology of fruit bat foraging behavior in relation to the fruit odors of two species of paleotropical bat-dispersed figs (*Ficus hispida* and *Ficus scortechinii*). *Journal of Chemical Ecology*, **33**, 2097–2110.

Hodkinson, D.J. and Thompson, K. (1997) Plant dispersal: the role of man. *Journal of Applied Ecology*, **34**, 1484–1496.

Hooijer, A., Silvius, M., Wösten, H. and Page, S.E. (2006) *PEAT-CO2, Assessment of CO_2 emissions from drained peatlands in SE Asia*. Delft Hydraulics, Delft.

Hope, G.S., Kershaw, A.P., Van der Kaars, S. et al. (2004) History of vegetation and habitat change in the Austral-Asian region. *Quaternary International*, **118–119**, 103–126.

Hou, Y.M., Potts, R., Yuan, B.Y. et al. (2000) Mid-Pleistocene Acheulean-like stone technology of the Bose basin, South China. *Science*, **287**, 1622–1626.

Houlton, B.Z., Sigman, D.M., Schuur, E.A.G. and Hedin, L.O. (2007) A climate-driven switch in plant nitrogen acquisition within tropical forest communities. *Proceedings of the National Academy of Sciences of the USA*, **104**, 8902–8906.

Houlton, B.Z., Wang, Y.-P., Vitousek, P.M. and Field, C.B. (2008) A unifying framework for dinitrogen fixation in the terrestrial biosphere. *Nature*, **454**, 327–330.

Howald, G., Donlan, C.J., Galvan, J.P. et al. (2007) Invasive rodent eradication on islands. *Conservation Biology*, **21**, 1258–1268.

Hsu, C.C., Horng, F.W. and Kuo, C.M. (2002) Epiphyte biomass and nutrient capital of a moist subtropical forest in north-eastern Taiwan. *Journal of Tropical Ecology*, **18**, 659–670.

Hu, H., Liu, W. and Cao, M. (2008) Impact of land use and land cover changes on ecosystem services in Menglun, Xishuangbanna, Southwest China. *Environmental Monitoring and Assessment*, **146**, 146–156.

Huang, Z. (2000) The interactions of population dynamics of *Thalassodes quadraria* and the plant community structure and climatic factors in Dinghushan. *Chinese Journal of Ecology*, **19**, 24–27.

Huang, C.Y. and Hou, P.C.L. (2004) Density and diversity of litter amphibians in a monsoon forest of southern Taiwan. *Zoological Studies*, **43**, 795–802.

Huang, C.Y., Zhao, M.X., Wang, C.C. and Wei, G.J. (2001) Cooling of the South China Sea by the Toba eruption and correlation with other climate proxies similar to 71,000 years ago. *Geophysical Research Letters*, **28**, 3915–3918.

Huang, C., Wu, H., Zhou, Q., Li, Y. and Cai, X. (2008) Feeding strategy of Francois langur and white-headed langur at Fusui, China. *Americam Journal of Primatology*, **70**, 320–326.

Hubbell, S.P. (1980) Seed predation and the coexistence of tree species in tropical forests. *Oikos*, **35**, 214–229.

Hubbell, S.P. (2001) *The unified neutral theory of biodiversity and biogeography*. Princeton University Press, Princeton.

Hubbell, S.P. (2006) Neutral theory and the evolution of ecological equivalence. *Ecology*, **87**, 1387–1398.

Huchon, D., Chevret, P., Jordan, U. et al. (2007) Multiple molecular evidences for a living mammalian fossil. *Proceedings of the National Academy of Sciences of the USA*, **104**, 7495–7499.

Huete, A.R., Restrepo-Coupe, N., Ratana, P. et al. (2008) Multiple site tower flux and remote sensing comparisons of tropical forest dynamics in Monsoon Asia. *Agricultural and Forest Meteorology*, **148**, 748–760.

Hughes, J.B., Round, P.D. and Woodruff, D.S. (2003) The Indochinese-Sundaic faunal transition at the Isthmus of Kra: an analysis of resident forest bird species distributions. *Journal of Biogeography*, **30**, 569–580.

Hulcr, J., Mogia, M., Isua, B. and Novotny, V. (2007) Host specificity of ambrosia and bark beetles (Col., Curculionidae: Scolytinae and Platypodinae) in a New Guinea rainforest. *Ecological Entomology*, **32**, 762–772.

Human Rights Watch (2006) Too high a price: the human costs of the Indonesian military's economic activities. *Human Rights Watch*, **18**, 1–140.

Hunt, C.O., Gilbertson, D.D. and Rushworth, G. (2007) Modern humans in Sarawak, Malaysian Borneo, during Oxygen Isotope Stage 3: palaeoenvironmental evidence from the Great Cave of Niah. *Journal of Archaeological Science*, **34**, 1953–1969.

Husa, K. and Wohlschlägl, H. (2007) From 'baby boom' to 'grey boom'? *Geographische Rundschau International Edition*, **4**, 20–27.

Hyatt, L.A., Rosenberg, M.S., Howard, T.G. et al. (2003) The distance dependence prediction of the Janzen-Connell hypothesis: a meta-analysis. *Oikos*, **103**, 590–602.

Ichie, T., Hiromi, T., Yoneda, R. et al. (2004) Short-term drought causes synchronous leaf shedding and flushing in a lowland mixed dipterocarp forest, Sarawak, Malaysia. *Journal of Tropical Ecology*, **20**, 697–700.

Ichie, T., Kenzo, T., Kitahashi, Y., Koike, T. and Nakashizuka, T. (2005) How does *Dryobalanops aromatica* supply carbohydrate resources for reproduction in a masting year? *Trees—Structure and Function*, **19**, 703–710.

Ickes, K., Dewalt, S.J. and Thomas, S.C. (2003) Resprouting of woody saplings following stem snap by wild pigs in a Malaysian rain forest. *Journal of Ecology*, **91**, 222–233.

ICS (2008) *International stratigraphic chart*. International Commission on Stratigraphy.

Inger, R.F. (1980) Densities of floor-dwelling frogs and lizards in lowland forests of Southeast Asia and Central America. *American Naturalist*, **115**, 761–770.

Inger, R.F. and Voris, H.K. (2001) The biogeographical relations of the frogs and snakes of Sundaland. *Journal of Biogeography*, **28**, 863–891.

IPCC (2007) *Climate change 2007: the physical science basis*. Cambridge University Press, New York.

Irwin, R.E., Brody, A.K. and Waser, N.M. (2001) The impact of floral larceny on individuals, populations, and communities. *Oecologia*, **129**, 161–168.

Isaac, N.J.B., Turvey, S.T., Collen, B., Waterman, C. and Baillie, J.E.M. (2007) Mammals on the EDGE: conservation priorities based on threat and phylogeny. *PLoS ONE*, **3**, e296.

Ishii, S., Bell, J.N.B. and Marshall, F.M. (2007) Phytotoxic risk assessment of ambient air pollution on agricultural crops in Selangor State, Malaysia. *Environmental Pollution*, **150**, 267–279.

Itioka, T., Inoue, T., Kaliang, H. et al. (2001) Six-year population fluctuation of the giant honey bee *Apis dorsata* (Hymenoptera: Apidae) in a tropical lowland dipterocarp forest in Sarawak. *Annals of the Entomological Society of America*, **94**, 545–549.

Itioka, T. and Yamauti, M. (2004) Severe drought, leafing phenology, leaf damage and lepidopteran abundance in the canopy of a Bornean aseasonal tropical rain forest. *Journal of Tropical Ecology*, **20**, 479–482.

Ito, E., Araki, M., Tani, A. et al. (2008) Leaf-shedding phenology in tropical seasonal forests of Cambodia estimated from NOAA satellite images. *Geoscience and Remote Sensing Symposium, 2007. IGARSS 2007. IEEE International*, pp. 4331–4335.

Itoh, A., Yamakura, T., Ohkubo, T. et al. (2003) Importance of topography and soil texture in the spatial distribution of two sympatric dipterocarp trees in a Bornean rainforest. *Ecological Research*, **18**, 307–320.

ITTO (2006) *Annual review and assessment of the world timber situation: 2006*. International Tropical Timber Organization, Yokohama.

IUCN (1998) *IUCN guidelines for re-introductions*. IUCN, Gland.

IUCN (2001) *IUCN Red List Categories and Criteria version 3.1*. IUCN Species Survival Commission, Gland.

IUCN (2002) *IUCN guidelines for the placement of confiscated animals*. IUCN, Gland.

IUCN (2003) *Guidelines for application of IUCN red list criteria at regional level*. IUCN Species Survival Commission, Gland.

IUCN (2004) *2004 IUCN red list of threatened species: a global species assessment*. IUCN Species Survival Commission, Gland.

IUCN (2005) *Guidelines for using the IUCN red list categories and criteria*. IUCN Species Survival Commission, Gland.

Jaeger, J.-J. (2003) Mammalian evolution: isolationist tendencies. *Nature*, **426**, 509–511.

Jankowska-Błaszczuk, M. and Grubb, P.J. (2006) Changing perspectives on the role of the soil seed bank in northern temperate deciduous forests and in tropical lowland rain forests: parallels and contrasts. *Perspectives in Plant Ecology Evolution and Systematics*, **8**, 3–21.

Jansen, S., Broadley, M.R., Robbrecht, E. and Smets, E. (2002) Aluminum hyperaccumulation in angiosperms: a review of its phylogenetic significance. *Botanical Review*, **68**, 235–269.

Janzen, D.H. (1970) Herbivores and the number of species in tropical forests. *American Naturalist*, **104**, 501–528.

Janzen, D.H. (1974) Tropical blackwater rivers, animals, and mast fruiting by the Dipterocarpaceae. *Biotropica*, **6**, 69–103.

Janzen, D.H. (1976) Why bamboos wait so long to flower. *Annual Review of Ecology and Systematics*, **7**, 347–391.

Janzen, D.H. (2001) Latent extinctions: the living dead. *Encyclopedia of biodiversity* (ed. S. A. Levin), pp. 689–699. Academic Press, New York.

Jing, Y. and Flad, R.K. (2002) Pig domestication in ancient China. *Antiquity*, **76**, 724–732.

Jog, M.M., Marathe, R.R., Goel, S.S., Ranade, S.P., Kunte, K.K. and Watve, M.G. (2005) Sarcocystosis of chital (*Axis axis*) and dhole (*Cuon alpinus*): ecology of a mammalian prey-predator-parasite system in Peninsular India. *Journal of Tropical Ecology*, **21**, 479–482.

John, R., Dalling, J.W., Harms, K.E. et al. (2007) Soil nutrients influence spatial distributions of tropical tree species. *Proceedings of the National Academy of Sciences of the USA*, **104**, 864–869.

Johns, A.G. (1997) *Timber production and biodiversity conservation in tropical rainforests*. Cambridge University Press, Cambridge.

Johnson, S. (2007) CITES branches out. *Tropical Forest Update*, **17**, 1–2.

Johnson, A., Vongkhamheng, C., Hedemark, M. and Saithongdam, T. (2006) Effects of human-carnivore conflict on tiger (*Panthera tigris*) and prey populations in Lao PDR. *Animal Conservation*, **9**, 421–430.

Jones, G.S., Gregory, J.M., Stott, P.A., Tett, S.F.B. and Thorpe, R.B. (2005) An AOGCM simulation of the climate response to a volcanic super-eruption. *Climate Dynamics*, **25**, 725–738.

Jones, K.E., Patel, N.G., Levy, M.A. et al. (2008) Global trends in emerging infectious diseases. *Nature*, **451**, 990–993.

Jongwutiwes, S., Putaporntip, C., Iwasaki, T., Sata, T. and Kanbara, H. (2004) Naturally acquired *Plasmodium knowlesi* malaria in human, Thailand. *Emerging Infectious Diseases*, **10**, 2211–2213.

Jønsson, K.A., Irestedt, M., Fuchs, J. et al. (2008) Explosive avian radiations and multi-directional dispersal across Wallacea: Evidence from the Campephagidae and other Crown Corvida (Aves). *Molecular Phylogenetics and Evolution*, **47**, 221–236.

Jule, K.R., Leaver, L.A. and Lea, S.E.G. (2008) The effects of captive experience on reintroduction survival in carnivores: a review and analysis. *Biological Conservation*, **141**, 355–363.

Kaimowitz, D. and Sheil, D. (2007) Conserving what and for whom? Why conservation should help meet basic human needs in the tropics. *Biotropica*, **39**, 567–574.

Kalka, M.B., Smith, A.R. and Kalko, E.K.V. (2008) Bats limit arthropods and herbivory in a tropical forest. *Science*, **320**, 71.

Kalkman, V.J., Clausnitzer, V., Dijkstra, K.-B., Orr, A.G., Paulson, D.R. and Van Tol, J. (2008) Global diversity of dragonflies (Odonata) in freshwater. *Hydrobiologia*, **595**, 351–363.

Kanzaki, M., Yap, S.K., Kimura, K., Okauchi, T. and Yamakura, T. (1997) Survival and germination of buried seeds of non-dipterocarp species in a tropical rain forest, Pasoh, West Malaysia. *Tropics*, **7**, 9–20.

Kanzaki, M., Hara, M., Yamakura, T. et al. (2004) Doi Inthanon Forest Dynamics Plot, Thailand. *Tropical forest diversity and dynamism: findings from a large-scale plot network* (eds E.C. Losos and E.G. Leigh), pp. 474–481. University of Chicago Press, Chicago.

Kapos, V., Balmford, A., Aveling, R. et al. (2008) Calibrating conservation: new tools for measuring success. *Conservation Letters*, **1**, 155–164.

Karasov, W.H. and Martínez del Rio, C. (2007) *Physiological ecology: how animals process energy, nutrients, and toxins*. Princeton University Press, Princeton.

Kartawinata, K. (1990) A review of natural vegetation studies in Malesia, with special reference to Indonesia. *The plant diversity of Malesia* (eds P. Baas, K. Kalkman and R. Geesink), pp. 121–132. Kluwer Academic, Dordrecht.

Karube, H. and Suda, S. (2004) A preliminary report on influence of an introduced lizard *Anolis carolinensis* on the native insect fauna of Ogasawara Islands. *Research Reports of the Kanagawa Prefectural Museum of Natural History*, **12**, 21–30.

Kaspari, M., Garcia, M.N., Harms, K.E., Santana, M., Wright, S.J. and Yavitt, J.B. (2008) Multiple nutrients limit litterfall and decomposition in a tropical forest. *Ecology Letters*, **11**, 35–43.

Kato, M., Kosaka, Y., Kawakita, A. et al. (2008) Plant-pollinator interactions in tropical monsoon forests in Southeast Asia. *American Journal of Botany*, **95**, 1375–1394.

Kaufmann, E. and Maschwitz, U. (2006) Ant-gardens of tropical Asian rainforests. *Naturwissenschaften*, **93**, 216–227.

Kawakami, K. and Higuchi, H. (2002) Predation by domestic cats on birds of Hahajima Island of the Bonin Islands, southern Japan. *Ornithological Science*, **1**, 143–144.

Kawakita, A. and Kato, M. (2006) Assessment of the diversity and species specificity of the mutualistic association between *Epicephala* moths and *Glochidion* trees. *Molecular Ecology*, **15**, 3567–3581.

Kawanishi, K. and Sunquist, M.E. (2004) Conservation status of tigers in a primary rainforest of Peninsular Malaysia. *Biological Conservation*, **120**, 329–344.

Kaya, M., Kammesheidt, L. and Weidelt, H.J. (2002) The forest garden system of Saparua island, Central Maluku, Indonesia, and its role in maintaining tree species diversity. *Agroforestry Systems*, **54**, 225–234.

Keeley, J.E. and Bond, W.J. (1999) Mast flowering and semelparity in bamboos: the bamboo fire cycle hypothesis. *American Naturalist*, **154**, 383–391.

Keeling, H.C. and Phillips, O.L. (2007) The global relationship between forest productivity and biomass. *Global Ecology and Biogeography*, **16**, 618–631.

Keogh, J.S., Barker, D.G. and Shine, R. (2001) Heavily exploited but poorly known: systematics and biogeography of commercially harvested pythons (*Python curtus* group) in Southeast Asia. *Biological Journal of the Linnean Society*, **73**, 113–129.

Kershaw, A.P., van der Kaars, S., Moss, P. et al. (2006) Environmental change and the arrival of people in the Australian region. *Before Farming*, **1**, article 2.

Kessler, M., Keßler, P.J.A., Gradstein, S.R., Bach, K., Schmull, M. and Pitopang, R. (2005) Tree diversity in primary forest and different land use systems in Central Sulawesi, Indonesia. *Biodiversity and Conservation*, **14**, 547–560.

Khiem, N.T., Cuong, L.Q. and Chien, H.V. (2003) Market study of meat from field rats in the Mekong delta. *Rats, mice and people: rodent biology and management* (eds G. R. Singleton, L. A. Hinds, C. J. Krebs and D. M. Spratt), pp. 543–547. Australian Centre for International Agricultural Research, Canberra.

Khurana, E. and Singh, J.S. (2001) Ecology of tree seed and seedlings: implications for tropical forest conservation and restoration. *Current Science*, **80**, 748–757.

Kikuta, T., Gunsalam, G., Kon, M. and Ochi, T. (1997) Altitudinal change of fauna, diversity and food preference of dung and carrion beetles on Mt. Kinabalu, Borneo. *Tropics*, **7**, 123–132.

Killeen, T.J. and Solórzano, L.A. (2008) Conservation strategies to mitigate impacts from climate change in Amazonia. *Philosophical Transactions of the Royal Society B: Biological Sciences*, **363**, 1881–1888.

Kimura, K., Yumoto, T. and Kikuzawa, K. (2001) Fruiting phenology of fleshy-fruited plants and seasonal dynamics of frugivorous birds in four vegetation zones on Mt. Kinabalu, Borneo. *Journal of Tropical Ecology*, **17**, 833–858.

King, D.A., Davies, S.J., Supardi, M.N.N. and Tan, S. (2005) Tree growth is related to light interception and wood density in two mixed dipterocarp forests of Malaysia. *Functional Ecology*, **19**, 445–453.

King, D.A., Davies, S.J. and Noor, N.S.M. (2006a) Growth and mortality are related to adult tree size in a Malaysian mixed dipterocarp forest. *Forest Ecology and Management*, **223**, 152–158.

King, D.A., Davies, S.J., Tan, S. and Noor, N.S.M. (2006b) The role of wood density and stem support costs in the growth and mortality of tropical trees. *Journal of Ecology*, **94**, 670–680.

King, D.A., Wright, S.J. and Connell, J.H. (2006c) The contribution of interspecific variation in maximum tree height to tropical and temperate diversity. *Journal of Tropical Ecology*, **22**, 11–24.

Kingdon-Ward, F. (1945) A sketch of the botany and geography of north Burma. *Journal of the Bombay Natural History Society*, **45**, 16–30, 133–148.

Kingston, T., Francis, C.M., Akbar, Z. and Kunz, T.H. (2003) Species richness in an insectivorous bat assemblage from Malaysia. *Journal of Tropical Ecology*, **19**, 67–79.

Kinnaird, M.F. and O'Brien, T.G. (2005) Fast foods of the forest: the influence of figs on primates and hornbills across Wallace's line. *Tropical fruits and frugivores: the search for strong interactors* (eds J. L. Dew and J. P. Boubli), pp. 155–184. Kluwer, Dordrecht.

Kinnaird, M.F. and O'Brien, T.G. (2007) *The ecology and conservation of Asian hornbills: farmers of the forest*. University of Chicago Press, Chicago.

Kinnaird, M.F., Obrien, T.G. and Suryadi, S. (1996) Population fluctuation in Sulawesi red-knobbed hornbills: tracking figs in space and time. *Auk*, **113**, 431–440.

Kirkendall, L.R. and Ødegaard, F. (2007) Ongoing invasions of old-growth tropical forests: establishment of three incestuous beetle species in southern Central America (Curculionidae: Scolytinae). *Zootaxa*, **1588**, 53–62.

Kirkpatrick, R.C. (2007) The Asian colobines: diversity among leaf-eating monkeys. *Primates in perspective* (eds C. J. Campbell, A. Fuentes, K. C. MacKinnon, M. Panger and S. Bearder), pp. 186–200. Oxford University Press, Oxford.

Kishimoto-Yamada, K. and Itioka, T. (2008a) Survival of flower-visiting chrysomelids during non general-flowering periods in Bornean dipterocarp forests. *Biotropica*, **40**, 600–606.

Kishimoto-Yamada, K. and Itioka, T. (2008b) Consequences of a severe drought associated with an El Nino-Southern Oscillation on a light-attracted leaf-beetle (Coleoptera, Chrysomelidae) assemblage in Borneo. *Journal of Tropical Ecology*, **24**, 229–233.

Kitamura, S., Suzuki, S., Yumoto, T. et al. (2004) Dispersal of *Aglaia spectabilis*, a large-seeded tree species in a moist evergreen forest in Thailand. *Journal of Tropical Ecology*, **20**, 421–427.

Kitamura, S., Suzuki, S., Yumoto, T. et al. (2005) A botanical inventory of a tropical seasonal forest in Khao Yai National Park, Thailand: implications for fruit-frugivore interactions. *Biodiversity and Conservation*, **14**, 1241–1262.

Kitamura, S., Yumoto, T., Poonswad, P. and Wohandee, P. (2007) Frugivory and seed dispersal by Asian elephants, *Elephas maximus*, in a moist evergreen forest of Thailand. *Journal of Tropical Ecology*, **23**, 373–376.

Kitamura, S., Yumoto, T., Poonswad, P., Suzuki, S. and Wohandee, P. (2008) Rare seed-predating mammals determine seed fate of *Canarium euphyllum*, a large-seeded tree species in a moist evergreen forest, Thailand. *Ecological Research*, **23**, 169–177.

Kitayama, K. (1992) An altitudinal transect study of the vegetation on Mount Kinabalu, Borneo. *Vegetatio*, **102**, 149–171.

Kitayama, K. (2005) Comment on 'Ecosystem properties and forest decline in contrasting long-term chronosequences'. *Science*, **308**, 63b.

Kitayama, K. and Aiba, S.I. (2002a) Ecosystem structure and productivity of tropical rain forests along altitudinal gradients with contrasting soil phosphorus pools on Mount Kinabalu, Borneo *Journal of Ecology*, **90**, 37–51.

Kitayama, K. and Aiba, S.I. (2002b) Control of organic carbon density in vegetation and soil of tropical rain forest ecosystems on Mount Kinabalu. *Sabah Parks Nature Journal*, **5**, 71–90.

Kitayama, K., Aiba, S.I., Takyu, M., Majalap, N. and Wagai, R. (2004) Soil phosphorus fractionation and phosphorus-use efficiency of a Bornean tropical montane rain forest during soil aging with podozolization. *Ecosystems*, **7**, 259–274.

Klaassen, M. and Nolet, B.A. (2008) Stoichiometry of endothermy: shifting the quest from nitrogen to carbon. *Ecology Letters*, **11**, 785–792.

Klaus, G., Klaus-Hugi, C. and Schmid, B. (1998) Geophagy by large mammals at natural licks in the rain forest of the Dzanga National Park, Central African Republic. *Journal of Tropical Ecology*, **14**, 829–839.

Knight, A.T., Cowling, R.M., Rouget, M., Balmford, A., Lombard, A.T. and Campbell, B.M. (2008) Knowing but not doing: selecting priority conservation areas and the research-implementation gap. *Conservation Biology*, **22**, 610–617.

Knight, T.M., Steets, J.A., Vamosi, J.C. et al. (2005) Pollen limitation of plant reproduction: pattern and process. *Annual Review of Ecology Evolution and Systematics*, **36**, 467–497.

Knott, C.D. (1998) Changes in orangutan caloric intake, energy balance, and ketones in response to fluctuating fruit availability. *International Journal of Primatology*, **19**, 1061–1079.

Ko, I.W.P., Corlett, R.T. and Xu, R.J. (1998) Sugar composition of wild fruits in Hong Kong, China. *Journal of Tropical Ecology*, **14**, 381–387.

Kochian, L.V., Hoekenga, O.A. and Pineros, M.A. (2004) How do crop plants tolerate acid soils?—Mechanisms of aluminum tolerance and phosphorous efficiency. *Annual Review of Plant Biology*, **55**, 459–493.

Koh, L.P. (2008a) Can oil palm plantations be made more hospitable for forest butterflies and birds? *Journal of Applied Ecology*, **45**, 1002–1009.

Koh, L.P. (2008b) Birds defend oil palms from herbivorous insects. *Ecological Applications*, **18**, 821–825.

Koh, L.P. and Wilcove, D.S. (2007) Cashing in palm oil for conservation. *Nature*, **448**, 993–994.

Körner, C. and Paulsen, J. (2004) A world-wide study of high altitude treeline temperatures. *Journal of Biogeography*, **31**, 713–732.

Kosugi, Y., Takanashi, S., Ohkubo, S. et al. (2008) CO_2 exchange of a tropical rainforest at Pasoh in Peninsular Malaysia. *Agricultural and Forest Meteorology*, **148**, 439–452.

Kraft, N.J.B., Cornwell, W.K., Webb, C.O. and Ackerly, D.D. (2007) Trait evolution, community assembly, and the phylogenetic structure of ecological communities. *American Naturalist*, **170**, 271–283.

Kreft, H. and Jetz, W. (2007) Global patterns and determinants of vascular plant diversity. *Proceedings of the National Academy of Sciences of the USA*, **104**, 5925–5930.

Kremen, C., Cameron, A., Moilanen, A. et al. (2008) Aligning conservation priorities across taxa in Madagascar with high-resolution planning tools. *Science*, **320**, 222–226.

Krigbaum, J. (2005) Reconstructing human subsistence in the West Mouth (Niah Cave, Sarawak) burial series using stable isotopes of carbon. *Asian Perspectives*, **44**, 73–89.

Kring, D.A. (2007) The Chicxulub impact event and its environmental consequences at the Cretaceous-Tertiary boundary. *Palaeogeography Palaeoclimatology Palaeoecology*, **255**, 4–21.

Kriticos, D.J., Yonow, T. and McFadyen, R.E. (2005) The potential distribution of *Chromolaena odorata* (Siam weed) in relation to climate. *Weed Research*, **45**, 246–254.

Křivánek, M., Pyšek, P. and Jarošík, V. (2006) Planting history and propagule pressure as predictors of invasion by woody species in a temperate region. *Conservation Biology*, **20**, 1487–1498.

Krupnick, G.A. and Kress, W.J. (2003) Hotspots and ecoregions: a test of conservation priorities using taxonomic data. *Biodiversity and Conservation*, **12**, 2237–2253.

Kuang, Y.W., Sun, F.F., Wen, D.Z., Zhou, G.Y. and Zhao, P. (2008) Tree-ring growth patterns of Masson pine (*Pinus massoniana* L.) during the recent decades in the acidification Pearl River Delta of China. *Forest Ecology and Management*, **255**, 3534–3540.

Kuchikura, Y. (1988) Efficiency and focus of blowpipe hunting among Semaq Beri hunter-gatherers of Peninsula Malaysia. *Human Ecology*, **16**, 271–305.

Kuhnt, W., Holbourn, A., Hall, R., Zuvela, M. and Kase, R. (2004) Neogene history of the Indonesian throughflow. *Continent-ocean interactions within East Asian marginal seas* (eds P. Clift, P. Wang, W. Kuhnt and D.E. Hayes), pp. 299–320. AGU Geophysical Monograph.

Kulju, K.K.M., Sierra, S.E.C., Draisma, S.G.A., Samuel, R. and van Welzen, P.C. (2007) Molecular phylogeny of *Macaranga*, *Mallotus*, and related genera (Euphorbiaceae s.s.): insights from plastid and nuclear DNA sequence data. *American Journal of Botany*, **94**, 1726–1743.

Kumar, B.M. and Nair, P.K.R. (2004) The enigma of tropical homegardens. *Agroforestry Systems*, **61–62**, 135–152.

Kurokawa, H., Yoshida, T., Nakamura, T., Lai, J.H. and Nakashizuka, T. (2003) The age of tropical rain-forest canopy species, Borneo ironwood (*Eusideroxylon zwageri*), determined by C-14 dating. *Journal of Tropical Ecology*, **19**, 1–7.

Kurokawa, H., Kitahashi, Y., Koike, T., Lai, J. and Nakashizuka, T. (2004) Allocation to defense or growth in dipterocarp forest seedlings in Borneo. *Oecologia*, **140**, 261–270.

Kurzel, B.P., Schnitzer, S.A. and Carson, W.P. (2006) Predicting liana crown location from stem diameter in three Panamanian lowland forests. *Biotropica*, **38**, 262–266.

Kusters, K., Achdiawan, R., Belcher, B. and Perez, M.R. (2006) Balancing development and conservation? An assessment of livelihood and environmental outcomes of nontimber forest product trade in Asia, Africa, and Latin America. *Ecology and Society*, **11**.

Kusumoto, B. and Enoki, T. (2008) Contribution of a liana species, *Mucuna macrocarpa* Wall., to litterfall production and nitrogen input in a subtropical evergreen broad-leaved forest. *Journal of Forest Research*, **13**, 35–42.

Kuzmin, Y.V. (2006) Chronology of the earliest pottery in East Asia: progress and pitfalls. *Antiquity*, **80**, 362–371.

Kwok, H.K. and Corlett, R.T. (1999) Seasonality of a forest bird community in Hong Kong, South China. *Ibis*, **141**, 70–79.

Kwok, H.K. and Corlett, R.T. (2002) Seasonality of forest invertebrates in Hong Kong, South China. *Journal of Tropical Ecology*, **18**, 637–644.

LaFrankie, J.V. and Chan, H.T. (1991) Confirmation of sequential flowering in *Shorea* (Dipterocarpaceae). *Biotropica*, **23**, 200–203.

LaFrankie, J.V., Ashton, P.S., Chuyong, G.B. et al. (2006) Contrasting structure and composition of the understory in species-rich tropical rain forests. *Ecology*, **87**, 2298–2305.

Lai, J., Zhang, M. and Xie, Z. (2006) Characteristics of the evergreen broad-leaved forest in Shiping Forest Park, Three Gorges Reservoir Area. *Biodiversity Science*, **14**, 435–443.

Lam, S.K.Y., Lee, S.K. and LaFrankie, J.V. (2004) Bukit Timah Forest Dynamics Plot, Singapore. *Tropical forest diversity and dynamism: findings from a large-scale plot network* (eds E.C. Losos and E.G. Leigh), pp. 464–473. University of Chicago Press, Chicago.

Laman, T.G. (1995) *Ficus stupenda* germination and seedling establishment in a Bornean rain-forest canopy. *Ecology*, **76**, 2617–2626.

Laman, T.G. (1996a) *Ficus* seed shadows in a Bornean rain forest. *Oecologia*, **107**, 347–355.

Laman, T.G. (1996b) The impact of seed harvesting ants (*Pheidole* sp nov) on *Ficus* establishment in the canopy. *Biotropica*, **28**, 777–781.

Lamb, D. and Erskine, P. (2008) Forest restoration at a landscape scale. *Living in a dynamic tropical forest landscape* (eds N. E. Stork and S. M. Turton), pp. 469–484. Blackwell, Malden.

Lamb, D., Erskine, P.D. and Parrotta, J.A. (2005) Restoration of degraded tropical forest landscapes. *Science*, **310**, 1628–1632.

Lan, G.-Y., Hu, Y.-H., Cao, M. et al. (2008) Establishment of Xishuangbanna Tropical Forest Dynamics Plot: species composition and spatial distribution patterns. *Journal of Plant Ecology*, **32**, 287–298.

Langhammer, P.F., Bakarr, M.I., Bennun, L.A. et al. (2007) Identification and gap analysis of key biodiversity areas: targets for comprehensive protected area systems. IUCN, Gland.

Langner, A., Miettinen, J. and Siegert, F. (2007) Land cover change 2002–2005 in Borneo and the role of fire derived from MODIS imagery. *Global Change Biology*, **13**, 2329–2340.

Larson, G., Cucchi, T., Fujita, M. et al. (2007) Phylogeny and ancient DNA of *Sus* provides insights into neolithic expansion in island southeast Asia and Oceania. *Proceedings of the National Academy of Sciences of the USA*, **104**, 4834–4839.

Lasco, R.D., MacDicken, K.G., Pulhin, F.B., Guillermo, I.Q., Sales, R.F. and Cruz, R.V.O. (2006) Carbon stocks assessment of a selectively logged dipterocarp forest and wood processing mill in the Philippines. *Journal of Tropical Forest Science*, **18**, 212–221.

Latimer, W. and Hill, D. (2007) Mitigation banking: securing no net loss to bodiversity? A UK perspective. *Planning Practice and Research*, **22**, 155–175.

Laumonier, Y. (1997) *The vegetation and physiography of Sumatra*. Kluwer, Dordrecht.

Laurance, W.F. (2005) When bigger is better: the need for Amazonian megareserves. *Trends in Ecology and Evolution*, **20**, 645–648.

Laurance, W.F. (2007a) Forest destruction in tropical Asia. *Current Science*, **93**, 1544–1550.

Laurance, W.F. (2007b) A new initiative to use carbon trading for tropical forest conservation. *Biotropica*, **39**, 20–24.

Laurance, W.F. (2008a) Can carbon trading save vanishing forests? *Bioscience*, **58**, 286–287.

Laurance, W.F. (2008b) The need to cut China's illegal timber imports. *Science*, **319**, 1184–1184.

Laurance, W.F. (2008c) Theory meets reality: how habitat fragmentation research has transcended island biogeographic theory. *Biological Conservation*, **141**, 1731–1744.

Laurance, W.F. (2008d) The real cost of minerals. *New Scientist*, **199** (2669), 16.

Laurance, W.F., Bierregaard, R.O., Gascon, C. et al. (1997) Tropical forest fragmentation: synthesis of a diverse and dynamic discipline. *Tropical forest remnants: ecology, management, and conservation of fragmented communities*, pp. 502–514. University of Chicago Press, Chicago.

Laurance, W.F., Lovejoy, T.E., Vasconcelos, H.L. et al. (2002) Ecosystem decay of Amazonian forest fragments: a 22-year investigation. *Conservation Biology,* **16**, 605–618.

Laurance, W.F., Peres, C.A., Jansen, P.A. and D'Croz, L. (2006) Emerging threats to tropical forests: what we know and what we don't know. *Emerging threats to tropical forests* (eds W.F. Laurance and C.A. Peres). University of Chicago Press, Chicago.

Leake, J.R. (2005) Plants parasitic on fungi: unearthing the fungi in myco-heterotrophs and debunking the 'saprophytic' plant myth. *Mycologist,* **19**, 113–122.

Lee, T.M. and Jetz, W. (2008) Future battlegrounds for conservation under global change. *Proceedings of the Royal Society B: Biological Sciences,* **275**, 1261–1270.

Lee, E.W.S., Hau, B.C.H. and Corlett, R.T. (2005) Natural regeneration in exotic tree plantations in Hong Kong, China. *Forest Ecology and Management,* **212**, 358–366.

Lee, E.W.S., Hau, B.C.H. and Corlett, R.T. (2008) Seed rain and natural regeneration in *Lophostemon confertus* plantations in Hong Kong, China. *New Forests,* **35**, 119–130.

Lee, H.S., Davies, S.J., Lafrankie, J.V. et al. (2002) Floristic and structural diversity of mixed dipterocarp forest in Lambir Hills National Park, Sarawak, Malaysia. *Journal of Tropical Forest Science,* **14**, 379–400.

Lee, T.M., Sodhi, N.S. and Prawiradilaga, D.M. (2007) The importance of protected areas for the forest and endemic avifauna of Sulawesi (Indonesia). *Ecological Applications,* **17**, 1727–1741.

Leigh, E.G. (2007) Neutral theory: a historical perspective. *Journal of Evolutionary Biology,* **20**, 2075–2091.

Leighton, M. (1993) Modeling dietary selectivity by Bornean orangutans—evidence for integration of multiple criteria in fruit selection. *International Journal of Primatology,* **14**, 257–313.

Leighton, M. and Leighton, D.R. (1983) Vertebrate responses to fruiting seasonality within a Bornean rain forest. *Tropical rain forest: ecology and management* (eds P.S. Sutton, T.C. Whitmore and A.C. Chadwick), pp. 181–196. Blackwell Scientific, Oxford.

Leimgruber, P., Senior, B., Uga et al. (2008) Modeling population viability of captive elephants in Myanmar (Burma): implications for wild populations. *Animal Conservation,* **11**, 198–205.

Leishman, M.R., Masters, G.J., Clarke, I.P. and Brown, V.K. (2000) Seed bank dynamics: the role of fungal pathogens and climate change. *Functional Ecology,* **14**, 293–299.

Lenton, T.M., Held, H., Kriegler, E. et al. (2008) Tipping elements in the Earth's climate system. *Proceedings of the National Academy of Sciences of the USA,* **105**, 1786–1793.

Leung, G.P.C., Hau, B.C.H. and Corlett, R.T. (2009) Exotic plant invasion in the highly degraded upland landscape of Hong Kong, China. *Biodiversity and Conservation,* **18**, 191–202.

Leuschner, C., Moser, G., Bertsch, C., Roderstein, M. and Hertel, D. (2007) Large altitudinal increase in tree root/shoot ratio in tropical mountain forests of Ecuador. *Basic and Applied Ecology,* **8**, 219–230.

Lever, C. (2003) *Naturalized reptiles and amphibians of the* world. Oxford University Press, Oxford.

Levey, D.J., Tewksbury, J.J., Izhaki, I., Tsahar, E. and Haak, D.C. (2007) Evolutionary ecology of secondary compounds in ripe fruit: case studies with capsaicin and emodin. *Seed dispersal: theory and its application in a changing world* (eds A.J. Dennis, E.W. Schupp, R.J. Green and D.A. Westcott), pp. 37–58. CAB International, Wallingford.

Levin, D.A. (2000) *The origin, expansion and demise of plant species*. Oxford University Press, Oxford.

Levy, H., Schwarzkopf, M.D., Horowitz, L., Ramaswamy, V. and Findell, K.L. (2008) Strong sensitivity of late 21st century climate to projected changes in short-lived air pollutants. *Journal of Geophysical Research-Atmospheres,* **113**, D06102.

Lewinsohn, T. and Roslin, T. (2008) Four ways toward tropical herbivore megadiversity. *Ecology Letters,* **11**, 398–416.

Lewis, O.T., Memmott, J., Lasalle, J., Lyal, C.H.C., Whitefoord, C. and Godfray, H.C.J. (2002) Structure of a diverse tropical forest insect-parasitoid community. *Journal of Animal Ecology,* **71**, 855–873.

Lewis, H., Paz, V., Lara, M. et al. (2008) Terminal Pleistocene to mid-Holocene occupation and an early cremation burial at Ille Cave, Palawan, Philippines. *Antiquity,* **82**, 318–335.

Li, M.J. and Wang, Z.H. (1984) The phenology of common plants on Mt. Dinghushan in Guangdong. *Tropical and Subtropical Forest Ecosystems,* **2**, 1–11.

Li, Y.D., Comiskey, J.A. and Dallmeier, F. (1998) Structure and composition of tropical mountain rain forest at the Jianfengling Natural Reserve, Hainan Island, PR China. *Forest Biodiversity Research, Monitoring and Modeling,* **20**, 551–562.

Li, X., Wilson, S.D. and Song, Y. (1999) Secondary succession in two subtropical forests. *Plant Ecology,* **143**, 13–21.

Li, B.G., Pan, R.L. and Oxnard, C.E. (2002) Extinction of snub-nosed monkeys in China during the past 400 years. *International Journal of Primatology,* **23**, 1227–1244.

Li, Z.A., Zou, B., Xia, H.P., Ren, H., Mo, J.M. and Weng, H. (2005) Litterfall dynamics of an evergreen broadleaf forest and a pine forest in the subtropical region of China. *Forest Science,* **51**, 608–615.

Li, Z., Saito, Y., Matsumoto, E., Wang, Y.J., Tanabe, S. and Vu, Q.L. (2006) Climate change and human impact on the Song Hong (Red River) Delta, Vietnam, during the Holocene. *Quaternary International,* **144**, 4–28.

Li, H.M., Aide, T.M., Ma, Y.X., Liu, W.J. and Cao, M. (2007) Demand for rubber is causing the loss of high diversity rain forest in SW China. *Biodiversity and Conservation*, **16**, 1731–1745.

Lieth, H., Berlekamp, J., Fuest, S. and Riediger, S. (1999) *Climate diagram world atlas* (CD-ROM). Backhuys Publishers, Leiden, Netherlands.

Liew, P.M., Lee, C.Y. and Kuo, C.M. (2006) Holocene thermal optimal and climate variability of East Asian monsoon inferred from forest reconstruction of a subalpine pollen sequence, Taiwan. *Earth and Planetary Science Letters*, **250**, 596–605.

Lin, K.C., Hamburg, S.P., Tang, S., Hsia, Y.J. and Lin, T.C. (2003) Typhoon effects on litterfall in a subtropical forest. *Canadian Journal of Forest Research*, **33**, 2184–2192.

Linkie, M., Smith, R.J. and Leader-Williams, N. (2004) Mapping and predicting deforestation patterns in the lowlands of Sumatra. *Biodiversity and Conservation*, **13**, 1809–1818.

Linkie, M., Smith, R.J., Zhu, Y. et al. (2008) Evaluating biodiversity conservation around a large Sumatran protected area. *Conservation Biology*, **22**, 683–690.

Litton, C.M., Raich, J.W. and Ryan, M.G. (2007) Carbon allocation in forest ecosystems. *Global Change Biology*, **13**, 2089–2109.

Liu, H. (1998) The change of geographical distribution of two Asian species of rhinoceros in Holocene. *Journal of Chinese Geography*, **8**, 83–88.

Liu, W., Fox, J.E.D. and Xu, Z. (2002) Litterfall and nutrient dynamics in a montane moist evergreen broad-leaved forest in Ailao Mountains, SW China. *Plant Ecology*, **164**, 157–170.

Liu, F., Chen, J., Chai, J. et al. (2007a) Adaptive functions of defensive plant phenolics and a non-linear bee response to nectar components. *Functional Ecology*, **21**, 96–100.

Liu, Y., Zhang, Y., He, D., Cao, M. and Zhu, H. (2007b) Climatic control of plant species richness along elevation gradients in the longitudinal range-gorge region. *Chinese Science Bulletin*, **52** (suppl. II), 50–58.

Liu, J., Li, S., Ouyang, Z., Tam, C. and Chen, X. (2008a) Ecological and socioeconomic effects of China's policies for ecosystem services. *Proceedings of the National Academy of Sciences of the USA*, **105**, 9477–9482.

Liu, W.J., Wang, P.Y., Li, J.T., Li, P.J. and Liu, W.Y. (2008b) The importance of radiation fog in the tropical seasonal rain forest of Xishuangbanna, south-west China. *Hydrology Research*, **39**, 79–87.

Lloyd, J. and Farquhar, G.D. (2008) Effects of rising temperatures and [CO2] on the physiology of tropical forest trees. *Philosophical Transactions of the Royal Society B: Biological Sciences*, **363**, 1811–1817.

Lohman, D.J., Bickford, D. and Sodhi, N.S. (2007) The burning issue. *Science*, **316**, 376–376.

Londo, J.P., Chiang, Y.C., Hung, K.H., Chiang, T.Y. and Schaal, B.A. (2006) Phylogeography of Asian wild rice, *Oryza rufipogon*, reveals multiple independent domestications of cultivated rice, *Oryza sativa*. *Proceedings of the National Academy of Sciences of the USA*, **103**, 9578–9583.

Losos, E.C. and Leigh, E.G. (2004) *Tropical forest diversity and dynamism: findings from a large-scale plot network*. University of Chicago Press, Chicago.

Lotz, C.N. and Schondube, J.E. (2006) Sugar preferences in nectar- and fruit-eating birds: Behavioral patterns and physiological causes. *Biotropica*, **38**, 3–15.

Louys, J. (2007) Limited effect of the Quaternary's largest super-eruption (Toba) on land mammals from Southeast Asia. *Quaternary Science Reviews*, **26**, 3108–3117.

Louys, J. (2008) Quaternary extinctions in southeast Asia. *Mass extinction* (ed. A.M.T. Elewa), pp. 159–190. Springer, Berlin.

Louys, J., Curnoe, D. and Tong, H.W. (2007) Characteristics of Pleistocene megafauna extinctions in Southeast Asia. *Palaeogeography Palaeoclimatology Palaeoecology*, **243**, 152–173.

Lovelock, C.E., Andersen, K. and Morton, J.B. (2003) Arbuscular mycorrhizal communities in tropical forests are affected by host tree species and environment. *Oecologia*, **135**, 268–279.

Lovelock, C.E., Feller, I.C., Ball, M.C., Ellis, J. and Sorrell, B. (2007) Testing the growth rate vs. geochemical hypothesis for latitudinal variation in plant nutrients. *Ecology Letters*, **10**, 1154–1163.

Lucas, P.W. and Corlett, R.T. (1991) Relationship between the diet of *Macaca fascicularis* and forest phenology. *Folia Primatologica*, **57**, 201–215.

Lucas, P.W., Turner, I.M., Dominy, N.J. and Yamashita, N. (2000) Mechanical defenses to herbivory. *Annals of Botany*, **86**, 913–920.

Lugo, A.E. (2008) Visible and invisible effects of hurricanes on forest ecosystems: an international review. *Austral Ecology*, **33**, 368–398.

Luiselli, L. (2006) Resource partitioning and interspecific competition in snakes: the search for general geographical and guild patterns. *Oikos*, **114**, 193–211.

Luizão, F.J., Luizão, R.C.C. and Proctor, J. (2007) Soil acidity and nutrient deficiency in central Amazonian heath forest soils. *Plant Ecology*, **192**, 209–224.

Luo, T.X., Li, W.H. and Zhu, H.Z. (2002) Estimated biomass and productivity of natural vegetation on the Tibetan Plateau. *Ecological Applications*, **12**, 980–997.

Luo, T.X., Pan, Y.D., Ouyang, H. et al. (2004) Leaf area index and net primary productivity along subtropical to alpine gradients in the Tibetan Plateau. *Global Ecology and Biogeography*, **13**, 345–358.

Luo, L.P., Guo, X.G., Qian, T.J., Wu, D., Men, X.Y. and Dong, W.G. (2007) Distribution of gamasid mites on small mammals in Yunnan Province, China. *Insect Science*, **14**, 71–78.

Lynam, A.J., Round, P.D. and Brockelman, W.Y. (2006) *Status of birds and large mammals in Thailand's Dong Phayayen-Khao Yai forest complex*. Wildlife Conservation Society, Bangkok.

Macaulay, V., Hill, C., Achilli, A. et al. (2005) Single, rapid coastal settlement of Asia revealed by analysis of complete mitochondrial genomes. *Science*, **308**, 1034–1036.

Machida, H. and Sugiyama, S. (2002) The impact of the Kikai-Akahoya explosive eruptions on human societies. *Natural disasters and cultural change* (eds R. Torrence and J. Gratten), pp. 313–325. Routledge, London.

Maeto, K. and Fukuyama, K. (2003) Vertical stratification of ambrosia beetle assemblage in a lowland rain forest at Pasoh, Peninsular Malaysia. *Pasoh: ecology of a lowland rain forest in Southeast Asia* (eds Okuda T, N. Monokaran, Y. Matsumoto, Niiyama K., S.C. Thomas and P.S. Ashton), pp. 325–336. Springer-Verlag, Tokyo.

Majolo, B. and Ventura, R. (2004) Apparent feeding association between Japanese macaques (*Macaca fuscata yakui*) and sika deer (*Cervus nippon*) living on Yakushima Island, Japan. *Ethology Ecology and Evolution*, **16**, 33–40.

Mallapur, A., Waran, N. and Sinha, A. (2008) The captive audience: the educative influence of zoos on their visitors in India. *International Zoo Yearbook*, **42**, 1–11.

Manokaran, N., Seng, Q.E., Ashton, P.S. et al. (2004) Pasoh Forest Dynamics Plot, Peninsular Malaysia. *Tropical forest diversity and dynamism: findings from a large-scale plot network* (eds E.C. Losos and E.G. Leigh), pp. 585–598. University of Chicago Press, Chicago.

Marchant, R., Mumbi, C., Behera, S. and Yamagata, T. (2007) The Indian Ocean dipole—the unsung driver of climatic variability in East Africa. *African Journal of Ecology*, **45**, 4–16.

Markesteijn, L., Poorter, L. and Bongers, F. (2007) Light-dependent leaf trait variation in 43 tropical dry forest tree species. *American Journal of Botany*, **94**, 515–525.

Marod, D., Kutintara, U., Yarwudhi, C., Tanaka, H. and Nakashisuka, T. (1999) Structural dynamics of a natural mixed deciduous forest in western Thailand. *Journal of Vegetation Science*, **10**, 777–786.

Marod, D., Kutintara, U., Tanaka, H. and Nakashizuka, T. (2002) The effects of drought and fire on seed and seedling dynamics in a tropical seasonal forest in Thailand. *Plant Ecology*, **161**, 41–57.

Marod, D., Kutintara, U., Tanaka, H. and Nakashizuka, T. (2004) Effects of drought and fire on seedling survival and growth under contrasting light conditions in a seasonal tropical forest. *Journal of Vegetation Science*, **15**, 691–700.

Martínez-Garza, C. and Howe, H.F. (2003) Restoring tropical diversity: beating the time tax on species loss. *Journal of Applied Ecology*, **40**, 423–429.

Maschwitz, U. and Hänel, H. (1985) The migrating herdsman *Dolichoderus* (*Diabolus*) *cuspidatus*—an ant with a novel mode of life. *Behavioral Ecology and Sociobiology*, **17**, 171–184.

Maschwitz, U., Steghauskovac, S., Gaube, R. and Hänel, H. (1989) A South East Asian ponerine ant of the genus *Leptogenys* (Hym, Form) with army ant life habits. *Behavioral Ecology and Sociobiology*, **24**, 305–316.

Mason, B.G., Pyle, D.M. and Oppenheimer, C. (2004) The size and frequency of the largest explosive eruptions on Earth. *Bulletin of Volcanology*, **66**, 735–748.

Massey, F.P., Press, M.C. and Hartley, S.E. (2005) Have the impacts of insect herbivores on the growth of tropical tree seedlings been underestimated? *Biotic interactions in the tropics* (eds D.F.R.P. Burslem, M.A. Pinard and S.E. Hartley), pp. 347–365. Cambridge University Press, Cambridge.

Masuko, K. (1984) Studies on the predatory biology of oriental dacetine ants (Hymenoptera: Formicidae) I. Some Japanese species of *Strumigenys*, *Pentastruma*, and *Epitritus*, and a Malaysian *Labidogenys*, with special reference to hunting tactics in short-mandibulate forms. *Insectes Sociaux*, **31**, 429–451.

Masuko, K. (2008) Larval stenocephaly related to specialized feeding in the ant genera *Amblyopone*, *Leptanilla* and *Myrmecina* (Hymenoptera: Formicidae). *Arthropod Structure and Development*, **37**, 109–117.

Matsubayashi, H., Lagan, P. and Sukor, J.R. (2006) Utilization of *Macaranga* trees by the Asian elephants (*Elephas maximus*) in Borneo. *Mammal Study*, **31**, 115–118.

Matsubayashi, H., Lagan, P., Majalap, N., Tangah, J., Sukor, J.R.A. and Kitayama, K. (2007) Importance of natural licks for the mammals in Bornean inland tropical rain forests. *Ecological Research*, **22**, 742–748.

McCall, A.C. and Irwin, R.E. (2006) Florivory: the intersection of pollination and herbivory. *Ecology Letters*, **9**, 1351–1365.

McCanny, S.J. (1985) Alternatives in parent-offspring relationships in plants. *Oikos*, **45**, 148–149.

McClure, H.E. (1967) Composition of mixed species flocks in lowland and sub-montane forests of Malaya. *Wilson Bulletin*, **79**, 131–154.

McConkey, K.R. (2005) Influence of faeces on seed removal from gibbon droppings in a dipterocarp forest in Central Borneo. *Journal of Tropical Ecology*, **21**, 117–120.

McConkey, K.R. and Chivers, D.J. (2007) Influence of gibbon ranging patterns on seed dispersal distance and deposition site in a Bornean forest. *Journal of Tropical Ecology*, **23**, 269–275.

McConkey, K.R., Aldy, F., Ario, A. and Chivers, D.J. (2002) Selection of fruit by gibbons (*Hylobates muelleri* x *agilis*) in the rain forests of Central Borneo. *International Journal of Primatology,* **23**, 123–145.

McConkey, K.R., Ario, A., Aldy, F. and Chivers, D.J. (2003) Influence of forest seasonality on gibbon food choice in the rain forests of Barito Ulu, Central Kalimantan. *International Journal of Primatology,* **24**, 19–32.

McGregor, G.R. and Nieuwolt, S. (1998) *Tropical climatology*. John Wiley, Chichester.

McGroddy, M.E., Daufresne, T. and Hedin, L.O. (2004) Scaling of C: N: P stoichiometry in forests worldwide: implications of terrestrial redfield-type ratios. *Ecology,* **85**, 2390–2401.

McGuire, K.L. (2007) Common ectomycorrhizal networks may maintain monodominance in a tropical rain forest. *Ecology,* **88**, 567–574.

McJannet, D., Wallace, J., Fitch, P., Disher, M. and Reddell, P. (2007) Water balance of tropical rainforest canopies in north Queensland, Australia. *Hydrological Processes,* **21**, 3473–3484.

McKay, J.K., Christian, C.E., Harrison, S. and Rice, K.J. (2005) 'How local is local?'—A review of practical and conceptual issues in the genetics of restoration. *Restoration Ecology,* **13**, 432–440.

McLachlan, J.S., Hellmann, J.J. and Schwartz, M.W. (2007) A framework for debate of assisted migration in an era of climate change. *Conservation Biology,* **21**, 297–302.

McLoughlin, S., Carpenter, R.J., Jordan, G.J. and Hill, R.S. (2008) Seed ferns survived the end-Cretaceous mass extinction in Tasmania. *American Journal of Botany,* **95**, 465–471.

McMahon, G., Subdibjo, E.R., Aden, J., Bouzaher, A., Dore, G. and Kunanayagam, R. (2000) *Mining and the environment in Indonesia: long-term trends and repercussions of the Asian economic crisis*. Environment and Social Development Unit (EASES), East Asia and Pacific Region, World Bank.

McNamara, S., Tinh, D.V., Erskine, P.D., Lamb, D., Yates, D. and Brown, S. (2006) Rehabilitating degraded forest land in central Vietnam with mixed native species plantings. *Forest Ecology and Management,* **233**, 358–365.

McNeely, J.A. (2007) A zoological perspective on payments for ecosystem services. *Integrative Zoology,* **2**, 68–78.

McPhaden, M.J., Zebiak, S.E. and Glantz, M.H. (2006) ENSO as an integrating concept in Earth science. *Science,* **314**, 1740–1745.

McShane, T.O. and Wells, M.P. (2004) *Getting biodiversity projects to work: towards more effective conservation and development*. Columbia University Press, New York.

Medway, Lord (1972) The Quaternary mammals of Malesia: a review. *The Quaternary era in Malesia* (eds P.S. Ashton and H.M. Ashton), pp. 63–98. University of Hull, Hull.

Medway, Lord (1979) The Niah excavations and an assessment of the impact of early man on mammals in Borneo. *Asian Perspectives,* **20**, 51–69.

Meijaard, E. (2003) Mammals of south-east Asian islands and their Late Pleistocene environments. *Journal of Biogeography,* **30**, 1245–1257.

Meijaard, E. and Sheil, D. (2007) A logged forest in Borneo is better than none at all. *Nature,* **446**, 974–974.

Meijaard, E. and Sheil, D. (2008) The persistence and conservation of Borneo's mammals in lowland rain forests managed for timber: observations, overviews and opportunities. *Ecological Research,* **23**, 21–34.

Meijaard, E., Sheil, D., Augeri, D. et al. (2005) Life after logging: reconciling wildlife conservation and production forestry in Indonesian Borneo. CIFOR, Jakarta.

Meijaard, E., Sheil, D., Nasi, R. and Stanley, S.A. (2006) Wildlife conservation in Bornean timber concessions. *Ecology and Society,* **11**.

Meiri, S., Meijaard, E., Wich, S.A., Groves, C.P. and Helgen, K.M. (2008) Mammals of Borneo—Small size on a large island. *Journal of Biogeography,* **35**, 1087–1094.

Mellars, P. (2006) Going east: new genetic and archaeological perspectives on the modern human colonization of Eurasia. *Science,* **313**, 796–800.

Meng, K., Li, S. and Murphy, R.W. (2008) Biogeographical patterns of Chinese spiders (Arachnida: Araneae) based on a parsimony analysis of endemicity. *Journal of Biogeography* **35**, 1241–1249.

Mercer, J.M. and Roth, V.L. (2003) The effects of Cenozoic global change on squirrel phylogeny. *Science,* **299**, 1568–1572.

Messing, R.H. and Wright, M.G. (2006) Biological control of invasive species: solution or pollution. *Frontiers in Ecology and Environment,* **4**, 132–140.

Metcalfe, I. (2005) Asia: South-east. *Encyclopedia of geology. Volume 1*. (eds R.C. Selley, L. Robin, M. Cocks and I.R. Plimer), pp. 169–198. Elsevier, Amsterdam.

Metcalfe, D.J. and Turner, I.M. (1998) Soil seed bank from lowland rain forest in Singapore: canopy-gap and litter-gap demanders. *Journal of Tropical Ecology,* **14**, 103–108.

Metcalfe, D.J., Grubb, P.J. and Turner, I.M. (1998) The ecology of very small-seeded shade-tolerant trees and shrubs in lowland rain forest in Singapore. *Plant Ecology,* **134**, 131–149.

Metz, M.R., Comita, L.S., Chen, Y.Y. et al. (2008) Temporal and spatial variability in seedling dynamics: a cross-site comparison in four lowland tropical forests. *Journal of Tropical Ecology,* **24**, 9–18.

Meyfroidt, P. and Lambin, E.F. (2008) Forest transition in Vietnam and its environmental impacts. *Global Change Biology,* **14**, 1319–1336.

Miller, B., Conway, W., Reading, R.P. et al. (2004) Evaluating the conservation mission of zoos, aquariums, botanical gardens, and natural history museums. *Conservation Biology*, **18**, 86–93.

Miller, K.G., Kominz, M.A., Browning, J.V. et al. (2005) The phanerozoic record of global sea-level change. *Science*, **310**, 1293–1298.

Miller, R.M., Rodriguez, J.P., Aniskowicz-Fowler, T. et al. (2007) National threatened species listing based on IUCN criteria and regional guidelines: current status and future perspectives. *Conservation Biology*, **21**, 684–696.

Mingram, J., Schettler, G., Nowaczyk, N. et al. (2004) The Huguang maar lake—a high-resolution record of palaeoenvironmental and palaeoclimatic changes over the last 78,000 years from South China. *Quaternary International*, **122**, 85–107.

Mittermeier, R.A., Gill, P.R., Hoffman, M. et al. (2004) *Hotspots revisited: Earth's biologically richest and most endangered ecoregions*. CEMEX, Mexico City.

Miyamoto, K., Suzuki, E., Kohyama, T., Seino, T., Mirmanto, E. and Simbolon, H. (2003) Habitat differentiation among tree species with small-scale variation of humus depth and topography in a tropical heath forest of Central Kalimantan, Indonesia. *Journal of Tropical Ecology*, **19**, 43–54.

Miyamoto, K., Rahajoe, J.S., Kohyama, T. and Mirmanto, E. (2007) Forest structure and primary productivity in a Bornean heath forest. *Biotropica*, **39**, 35–42.

Mizoguchi, Y., Miyata, A., Ohatani, Y., Hirata, R. and Yuta, S. (2009) A review of tower flux observation sites in Asia. *Journal of Forest Research*, **14**, 1–9.

Moe, S.R. (1993) Mineral content and wildlife use of soil licks in southwestern Nepal. *Canadian Journal of Zoology*, **71**, 933–936.

Moffett, M.W. (1987) Division-of-labor and diet in the extremely polymorphic ant *Pheidologeton diversus*. *National Geographic Research*, **3**, 282–304.

Moffett, M.W. (1988) Foraging behavior in the Malayan swarm-raiding ant *Pheidologeton silenus* (Hymenoptera Formicidae Myrmicinae). *Annals of the Entomological Society of America*, **81**, 356–361.

Mohd Azlan, J. (2006) Mammal diversity and conservation in a secondary forest in peninsular Malaysia. *Biodiversity and Conservation*, **15**, 1013–1025.

Mokany, K., Raison, R.J. and Prokushkin, A.S. (2006) Critical analysis of root: shoot ratios in terrestrial biomes. *Global Change Biology*, **12**, 84–96.

Moles, A.T. and Westoby, M. (2004) Seedling survival and seed size: a synthesis of the literature. *Journal of Ecology*, **92**, 372–383.

Moles, A.T., Ackerly, D.D., Tweddle, J.C. et al. (2007) Global patterns in seed size. *Global Ecology and Biogeography*, **16**, 109–116.

Mollman, S. (2008) Birders flock east. *The Wall Street Journal*, **September 12**, W1.

Momose, K., Yumoto, T., Nagamitsu, T. et al. (1998) Pollination biology in a lowland dipterocarp forest in Sarawak, Malaysia. I. Characteristics of the plant-pollinator community in a lowland dipterocarp forest. *American Journal of Botany*, **85**, 1477–1501.

Monk, K.A., De Fretes, Y. and Reksodiharjo-Lilley, G. (1997) *The ecology of Nusa Tenggara and Maluku*. Oxford University Press, Oxford.

Moog, J., Saw, L.G., Hashim, R. and Maschwitz, U. (2005) The triple alliance: how a plant-ant, living in an ant-plant, acquires the third partner, a scale insect. *Insectes Sociaux*, **52**, 169–176.

Morley, R.J. (1998) Palynological evidence for Tertiary plant dispersals in the SE Asia region in relation to plate tectonics and climate. *Biogeography and geological evolution of SE Asia* (eds R. Hall and J.D. Holloway), pp. 177–200. Backhuys, Leiden.

Morley, R.J. (2003) Interplate dispersal paths for megathermal angiosperms. *Perspectives in Plant Ecology Evolution and Systematics*, **6**, 5–20.

Morley, R.J. (2007) Cretaceous and Tertiary climate change and the past distribution of megathermal rainforests. *Tropical rainforest responses to climatic change* (eds J.R. Flenley and M.B. Bush), pp. 1–31. Springer, New York.

Morrison, K.D. and Junker, L.L. (2002) *Forager-traders in South and Southeast Asia: long-term histories*. Cambridge University Press, Cambridge.

Morrison, J.C., Sechrest, W., Dinerstein, E., Wilcove, D.S. and Lamoreux, J.F. (2007) Persistence of large mammal faunas as indicators of global human impacts. *Journal of Mammalogy*, **88**, 1363–1380.

Morrissey, T., Ashmore, M.R., Emberson, L.D., Cinderby, S. and Büker, P. (2007) The impacts of ozone on nature conservation: a review and recommendations for research and policy. JNCC, Peterborough.

Morwood, M.J., O'Sullivan, P.B., Aziz, F. and Raza, A. (1998) Fission-track ages of stone tools and fossils on the east Indonesian island of Flores. *Nature*, **392**, 173–176.

Morwood, M.J., Brown, P., Jatmiko et al. (2005) Further evidence for small-bodied hominins from the Late Pleistocene of Flores, Indonesia. *Nature*, **437**, 1012–1017.

Morwood, M.J., Sutikna, T., Saptomo, E.W. et al. (2008) Climate, people and faunal succession on Java, Indonesia: evidence from Song Gupuh. *Journal of Archaeological Science*, **35**, 1776–1789.

Moss, S.J. and Wilson, E.J. (1998) Biogeographic implications of. the Tertiary palaeogeographic evolution of Sulawesi and. Borneo. *Biogeography and geological evolution of SE Asia* (eds R. Hall and J.D. Holloway), pp. 133–163. Backhuys Publishers, Leiden.

Mouissie, A.M., Lengkeek, W. and van Diggelen, R. (2005) Estimating adhesive seed-dispersal distances: field experiments and correlated random walks. *Functional Ecology*, **19**, 478–486.

Moyle, R.G. (2004) Phylogenetics of barbets (Aves: Piciformes) based on nuclear and mitochondrial DNA sequence data. *Molecular Phylogenetics and Evolution*, **30**, 187–200.

Moyle, R.G. and Marks, B.D. (2006) Phylogenetic relationships of the bulbuls (Aves: Pycnonotidae) based on mitochondrial and nuclear DNA sequence data. *Molecular Phylogenetics and Evolution*, **40**, 687–695.

Müller, A., Diener, S., Schnyder, S., Stutz, K., Sedivy, C. and Dorn, S. (2006) Quantitative pollen requirements of solitary bees: implications for bee conservation and the evolution of bee-flower relationships. *Biological Conservation*, **130**, 604–615.

Muñoz-Piña, C., Guevara, A., Torres, J.M. and Braña, J. (2008) Paying for the hydrological services of Mexico's forests: analysis, negotiations and results. *Ecological Economics*, **65**, 725–736.

Murdiyarso, D. and Adiningsih, E.S. (2007) Climate anomalies, Indonesian vegetation fires and terrestrial carbon emissions. *Mitigation and Adaptation Strategies for Global Change*, **12**, 101–112.

Murray, B.C., McCarl, B.A. and Lee, H.C. (2004) Estimating leakage from forest carbon sequestration programs. *Land Economics*, **80**, 109–124.

Nabuurs, G.J., Masera, O., Andrasko, K. et al. (2007) Forestry. *Climate change 2007: mitigation* (eds B. Metz, O.R. Davidson, P.R. Bosch, R. Dave and L.A. Meyer), pp. 543–578. Cambridge University Press, Cambridge.

Nadler, T. and Streicher, U. (2003) Re-introduction possibilities for endangered langurs in Vietnam. *Re-introduction News*, **23**, 35–37.

Naidoo, R., Balmford, A., Ferraro, P.J., Polasky, S., Ricketts, T.H. and Rouget, M. (2006) Integrating economic costs into conservation planning. *Trends in Ecology and Evolution*, **21**, 681–687.

Naidoo, R., Balmford, A., Costanza, R. et al. (2008) Global mapping of ecosystem services and conservation priorities. *Proceedings of the National Academy of Sciences of the USA*, **105**, 9495–9500.

Naito, Y., Kanzaki, M., Numata, S. et al. (2008a) Size-related flowering and fecundity in the tropical canopy tree species, *Shorea acuminata* (Dipterocarpaceae) during two consecutive general flowerings. *Journal of Plant Research*, **121**, 33–42.

Naito, Y., Kanzaki, M., Iwata, H. et al. (2008b) Density-dependent selfing and its effects on seed performance in a tropical canopy tree species, *Shorea acuminata* (Dipterocarpaceae). *Forest Ecology and Management*, **256**, 375–383.

Nakagawa, M., Tanaka, K., Nakashizuka, T. et al. (2000) Impact of severe drought associated with the 1997–1998 El Nino in a tropical forest in Sarawak. *Journal of Tropical Ecology*, **16**, 355–367.

Nakagawa, M., Takeuchi, Y., Kenta, T. and Nakashizuka, T. (2005) Predispersal seed predation by insects vs. vertebrates in six dipterocarp species in Sarawak, Malaysia. *Biotropica*, **37**, 389–396.

Nakagawa, M., Miguchi, H., Sato, K., Shoko, S. and Nakashizuka, T. (2007) Population dynamics of arboreal and terrestrial small mammals in a tropical rainforest, Sarawak, Malaysia. *Raffles Bulletin of Zoology*, **55**, 389–395.

Nakamoto, A., Kinjo, K. and Izawa, M. (2007) Food habits of Orii's flying-fox, *Pteropus dasymallus inopinatus*, in relation to food availability in an urban area of Okinawa-jima Island, the Ryukyu Archipelago, Japan. *Acta Chiropterologica*, **9**, 237–249.

Nanami, S., Kawaguchi, H., Tateno, R., Li, C.H. and Katagiri, S. (2004) Sprouting traits and population structure of co-occurring *Castanopsis* species in an evergreen broad-leaved forest in southern China. *Ecological Research*, **19**, 341–348.

Nascimento, H.E.M., Laurance, W.F., Condit, R., Laurance, S.G., D'Angelo, S. and Andrade, A.C. (2005) Demographic and life-history correlates for Amazonian trees. *Journal of Vegetation Science*, **16**, 625–634.

Nathan, R. and Casagrandi, R. (2004) A simple mechanistic model of seed dispersal, predation and plant establishment: Janzen-Connell and beyond. *Journal of Ecology*, **92**, 733–746.

Nelson, S.L., Kunz, T.H. and Humphrey, S.R. (2005) Folivory in fruit bats: leaves provide a natural source of calcium. *Journal of Chemical Ecology*, **31**, 1683–1691.

Nepstad, D.C., Stickler, C.M. and Almeida, O.T. (2006) Globalization of the Amazon soy and beef industries: opportunities for conservation. *Conservation Biology*, **20**, 1595–1603.

Newbery, D.M., Campbell, E.J.F., Proctor, J. and Still, M.J. (1996) Primary lowland dipterocarp forest at Danum Valley, Sabah, Malaysia. Species composition and patterns in the understorey. *Vegetatio*, **122**, 193–220.

Newstrom, L.E., Frankie, G.W. and Baker, H.G. (1994) A new classification for plant phenology based on flowering patterns in lowland tropical rain-forest trees at La Selva, Costa Rica. *Biotropica*, **26**, 141–159.

Ng, F.S.P. (1978) Strategies of establishment in Malayan forest trees. *Tropical trees as living systems* (eds P.B. Tomlinson and M.H. Zimmermann), pp. 129–162. Cambridge University Press, Cambridge.

Ng, J. and Nemora (2007) *Tiger trade revisited in Sumatra, Indonesia*. TRAFFIC Southeast Asia, Petaling Jaya.

Nicholson, D.I. (1965) A study of virgin forest near Sandakan North Borneo. *Proceedings of the symposium on ecological research in humid tropics vegetation*, pp. 67–87. UNESCO, Kuching.

Nieh, J.C. (2004) Recruitment communication in stingless bees (Hymenoptera, Apidae, Meliponini). *Apidologie*, **35**, 159–182.

Nijman, V. (2006) In-situ and ex-situ status of the Javan Gibbon and the role of zoos in conservation of the species. *Contributions to Zoology*, **75**, 161–168.

Nilus, N. (2004) Effect of edaphic variation of forest structure, dynamics, diversity and regeneration in a lowland tropical rain forest in Borneo. PhD, Aberdeen University.

Noguchi, H., Itoh, A., Mizuno, T. et al. (2007) Habitat divergence in sympatric Fagaceae tree species of a tropical montane forest in northern Thailand. *Journal of Tropical Ecology*, **23**, 549–558.

Noma, N. (1997) Annual fluctuations of sapfruits production and synchronization within and inter species in a warm temperate forest on Yakushima Island. *Tropics*, **6**, 441–449.

Noma, N. and Yumoto, T. (1997) Fruiting phenology of animal-dispersed plants in response to winter migration of frugivores in a warm temperate forest on Yakushima Island, Japan. *Ecological Research*, **12**, 119–129.

Nor, S.M. (2001) Elevational diversity patterns of small mammals on Mount Kinabalu, Sabah, Malaysia. *Global Ecology and Biogeography*, **10**, 41–62.

Norden, N., Chave, J., Caubere, A. et al. (2007) Is temporal variation of seedling communities determined by environment or by seed arrival? A test in a neotropical forest. *Journal of Ecology*, **95**, 507–516.

Novotny, V. and Basset, Y. (2005) Review—Host specificity of insect herbivores in tropical forests. *Proceedings of the Royal Society B: Biological Sciences*, **272**, 1083–1090.

Novotny, V. and Wilson, M.R. (1997) Why are there no small species among xylem-sucking insects? *Evolutionary Ecology*, **11**, 419–437.

Novotny, V., Basset, Y., Miller, S.E. et al. (2004) Local species richness of leaf-chewing insects feeding on woody plants from one hectare of a lowland rainforest. *Conservation Biology*, **18**, 227–237.

Novotny, V., Drozd, P., Miller, S.E. et al. (2006) Why are there so many species of herbivorous insects in tropical rainforests? *Science*, **313**, 1115–1118.

Nunn, C.L. and Altizer, S. (2006) *Infectious diseases in primates*. Oxford University Press, Oxford.

Obendorf, P.J., Oxnard, C.E. and Kefford, B.J. (2008) Are the small human-like fossils found on Flores human endemic cretins? *Proceedings of the Royal Society B: Biological Sciences*, **275**, 1287–1296.

O'Hanlon-Manners, D.L. and Kotanen, P.M. (2006) Losses of seeds of temperate trees to soil fungi: effects of habitat and host ecology. *Plant Ecology*, **187**, 49–58.

Ohkubo, T., Tani, M., Akojima, I. et al. (2007) Spatial pattern of landslides due to heavy rains in a mixed dipterocarp forest, north-western Borneo. *Tropics*, **16**, 59–69.

Okamoto, T., Kawakita, A. and Kato, M. (2007) Interspecific variation of floral scent composition in *Glochidion* and its association with host-specific pollinating seed parasite (*Epicephala*). *Journal of Chemical Ecology*, **33**, 1065–1081.

Olson, D.M. and Dinerstein, E. (2002) The Global 200: priority ecoregions for global conservation. *Annals of the Missouri Botanical Garden*, **89**, 199–224.

Olson, D.M., Dinerstein, E., Wikramanayake, E.D. et al. (2001) Terrestrial ecoregions of the worlds: a new map of life on Earth. *Bioscience*, **51**, 933–938.

Oota, H., Pakendorf, B., Weiss, G. et al. (2005) Recent origin and cultural reversion of a hunter-gatherer group. *PLoS Biology*, **3**, 536–542.

Oren, R., Hseih, C.I., Stoy, P. et al. (2006) Estimating the uncertainty in annual net ecosystem carbon exchange: spatial variation in turbulent fluxes and sampling errors in eddy-covariance measurements. *Global Change Biology*, **12**, 883–896.

Oshiro, I. and Nohara, T. (2000) Distribution of Pleistocene terrestrial vertebrates and their migration to the Ryukyus. *Tropics*, **10**, 41–50.

Otani, T. (2004) Effects of macaque ingestion on seed destruction and germination of a fleshy-fruited tree, *Eurya emarginata*. *Ecological Research*, **19**, 495–501.

Otani, T. and Shibata, E. (2000) Seed dispersal and predation by Yakushima macaques, *Macaca fuscata yakui*, in a warm temperate forest of Yakushima Island, southern Japan. *Ecological Research*, **15**, 133–144.

Ouyang, Z.Y., Xu W.H., Wang, X.Z. et al. (2008) Impact assessment of Wenchuan Earthquake on ecoystems. *Acta Ecologica Sincica*, **28**, 5801–5809.

Page, S.E., Siegert, F., Rieley, J.O., Boehm, H.D.V., Jaya, A. and Limin, S. (2002) The amount of carbon released from peat and forest fires in Indonesia during 1997. *Nature*, **420**, 61–65.

Page, S.E., Wust, R.A.J., Weiss, D., Rieley, J.O., Shotyk, W. and Limin, S.H. (2004) A record of Late Pleistocene and Holocene carbon accumulation and climate change from an equatorial peat bog (Kalimantan, Indonesia): implications for past, present and future carbon dynamics. *Journal of Quaternary Science*, **19**, 625–635.

Paine, C.E., Harms, K.E., Schnitzer, S.A. and Carson, W.P. (2008) Weak competition among tropical tree seedlings: implications for species coexistence. *Biotropica*, **40**, 432–440.

Palm, C., Sanchez, P., Ahamed, S. and Awiti, A. (2007) Soils: a contemporary perspective. *Annual Review of Environment and Resources*, **32**, 99–129.

Palmiotto, P.A., Davies, S.J., Vogt, K.A., Ashton, M.S., Vogt, D.J. and Ashton, P.S. (2004) Soil-related habitat specialization in dipterocarp rain forest tree species in Borneo. *Journal of Ecology*, **92**, 609–623.

Paoli, G.D. and Curran, L.M. (2007) Soil nutrients limit fine litter production and tree growth in mature lowland forest of Southwestern Borneo. *Ecosystems*, **10**, 503–518.

Paoli, G.D., Curran, L.M. and Zak, D.R. (2006) Soil nutrients and beta diversity in the Bornean Dipterocarpaceae: evidence for niche partitioning by tropical rain forest trees. *Journal of Ecology*, **94**, 157–170.

Paoli, G.D., Curran, L.M. and Slik, J.W.F. (2008) Soil nutrients affect spatial patterns of aboveground biomass and emergent tree density in southwestern Borneo. *Oecologia*, **155**, 287–299.

Paperna, I., Soh, M.C.K., Yap, C.A.M. et al. (2005) Blood parasite prevalence and abundance in the bird communities of several forested locations in Southeast Asia. *Ornithological Science*, **4**, 129–138.

Partin, J.W., Cobb, K.M., Adkins, J.F., Clark, B. and Fernandez, D.P. (2007) Millennial-scale trends in west Pacific warm pool hydrology since the Last Glacial Maximum. *Nature*, **449**, 452–455.

Patiño, S., Grace, J. and Bänziger, H. (2000) Endothermy by flowers of *Rhizanthes lowii* (Rafflesiaceae). *Oecologia*, **124**, 149–155.

Paz, V. (2005) Rock shelters, caves, and archaeobotany in island Southeast Asia. *Asian Perspectives*, **44**, 107–118.

Pearce, P.L. (2008) The nature of rainforest tourism: insights from a tourism social science research programme. *Living in a dynamic tropical forest landscape* (eds N.E. Stork and S.M. Turton), pp. 94–106. Blackwell, Malden.

Pearson, D.L. and Cassola, F. (1992) World-wide species richness patterns of tiger beetles (Coleoptera: Cicindelidae): indicator taxon for biodiversity and conservation studies *Conservation Biology*, **6**, 376–391.

Peeters, P.J., Sanson, G. and Read, J. (2007) Leaf biomechanical properties and the densities of herbivorous insect guilds. *Functional Ecology*, **21**, 246–255.

Pennington, R.T., Richardson, J.E. and Lavin, M. (2006) Insights into the historical construction of species-rich biomes from dated plant phylogenies, neutral ecological theory and phylogenetic community structure. *New Phytologist*, **172**, 605–616.

Pergams, O.R.W. and Zaradic, P.A. (2008) Reply to Jacobs and Manfredo: more support for a pervasive decline in nature-based recreation. *Proceedings of the National Academy of Sciences of the USA*, **105**, E41-E42.

Perkins, D.H. (1969) *Agricultural development in China 1368–1968*. Aldine Publishing, Chicago.

Peters, H.A. (2001) *Clidemia hirta* invasion at the Pasoh Forest Reserve: an unexpected plant invasion in an undisturbed tropical forest. *Biotropica*, **33**, 60–68.

Peters, H.A. (2003) Neighbour-regulated mortality: the influence of positive and negative density dependence on tree populations in species-rich tropical forests. *Ecology Letters*, **6**, 757–765.

Petraglia, M., Korisettar, R., Boivin, N. et al. (2007) Middle paleolithic assemblages from the Indian subcontinent before and after the Toba super-eruption. *Science*, **317**, 114–116.

Pfeiffer, M., Nais, J. and Linsenmair, K.E. (2004) Myrmecochory in the Zingiberaceae: seed removal of *Globba franciscii* and *G. propinpua* by ants (Hymenoptera-Formicidae) in rain forests on Borneo. *Journal of Tropical Ecology*, **20**, 705–708.

Pfeiffer, M., Nais, J. and Linsenmair, K.E. (2006) Worker size and seed size selection in 'seed'-collecting ant ensembles (Hymenoptera: Formicidae) in primary rain forests on Borneo. *Journal of Tropical Ecology*, **22**, 685–693.

Pfeiffer, M., Tuck, H.C. and Lay, T.C. (2008) Exploring arboreal ant community composition and co-occurrence patterns in plantations of oil palm *Elaeis guineensis* in Borneo and Peninsular Malaysia. *Ecography*, **31**, 21–32.

Phillips, O.L., Lewis, S.L., Baker, T.R., Chao, K.J. and Higuchi, N. (2008) The changing Amazon forest. *Philosophical Transactions of the Royal Society B: Biological Sciences*, **363**, 1819–1827.

Pielke, R.A., Adegoke, J., Beltran-Przekurat et al. (2007) An overview of regional land-use and land-cover impacts on rainfall. *Tellus Series B-Chemical and Physical Meteorology*, **59**, 587–601.

Pipoly, J.J. and Madulid, D.A. (1998) Composition, structure and species richness of a submontane moist forest on Mt Kinasalapi, Mindanao, Philippines. *Forest Biodiversity Research, Monitoring and Modeling*, **20**, 591–600.

PlantLife International (2004) *Identifying and protecting the world's most important plant areas*. PlantLife International, Salisbury.

Pokon, R., Novotny, V. and Samuelson, G.A. (2005) Host specialization and species richness of root-feeding chrysomelid larvae (Chrysomelidae, Coleoptera) in a New Guinea rain forest. *Journal of Tropical Ecology*, **21**, 595–604.

Polasky, S. (2008) Why conservation planning needs socioeconomic data. *Proceedings of the National Academy of Sciences of the USA*, **105**, 6505–6506.

Poorter, L. and Kitajima, K. (2007) Carbohydrate storage and light requirements of tropical moist and dry forest tree species. *Ecology*, **88**, 1000–1011.

Poorter, L. and Markesteijn, L. (2008) Seedling traits determine drought tolerance of tropical tree species. *Biotropica*, **40**, 321–331.

Pope, K.O. and Terrell, J.E. (2008) Environmental setting of human migrations in the circum-Pacific region. *Journal of Biogeography*, **35**, 1–21.

Porder, S., Vitousek, P.M., Chadwick, O.A., Chamberlain, C.P. and Hilley, G.E. (2007) Uplift, erosion, and phosphorus limitation in terrestrial ecosystems. *Ecosystems*, **10**, 158–170.

Posa, M.R.C., Diesmos, A.C., Sodhi, N.S. and Brooks, T.M. (2008) Hope for threatened tropical biodiversity: Lessons from the Philippines. *BioScience*, **58**, 231–240.

Potts, M.D. and Vincent, J.R. (2008) Spatial distribution of species populations, relative economic values, and the optimal size and number of reserves. *Environmental and Resource Economics*, **39**, 91–112.

Potts, M.D., Ashton, P.S., Kaufman, L.S. and Plotkin, J.B. (2002) Habitat patterns in tropical rain forests: a comparison of 105 plots in Northwest Borneo. *Ecology*, **83**, 2782–2797.

Poulsen, A.D. (1996) Species richness and density of ground herbs within a 1-ha plot of lowland rain forest in north-west Borneo. *Journal of Tropical Ecology*, **12**, 177–190.

Poulsen, A.D. and Pendry, C.A. (1995) Inventories of ground herbs at 3 altitudes on Bukit Belalong, Brunei, Borneo. *Biodiversity and Conservation*, **4**, 745–757.

Poulsen, A.D., Nielsen, I.C., Tan, S. and Balslev, H. (1996) A quantitative inventory of trees in one hectare of mixed dipterocarp forest in Temburong, Brunei Darussalam. *Tropical rainforest research—current issues* (eds D.S. Edwards, W.E. Booth and S.C. Choy), pp. 139–150. Kluwer, Dordrecht.

Prasad, S., Krishnaswamy, J., Chellam, R. and Goyal, S.P. (2006) Ruminant-mediated seed dispersal of an economically valuable tree in Indian dry forests. *Biotropica*, **38**, 679–682.

Prasad, M.S., Mahale, V.P. and Kodagali, V.N. (2007) New sites of Australasian microtektites in the central Indian Ocean: implications for the location and size of source crater. *Journal of Geophysical Research-Planets*, **112**.

Pregitzer, K.S. and Euskirchen, E.S. (2004) Carbon cycling and storage in world forests: biome patterns related to forest age. *Global Change Biology*, **10**, 2052–2077.

Prentice, I.C., Cramer, W., Harrison, S.P., Leemans, R., Monserud, R.A. and Solomon, A.M. (1992) A global biome model based on plant physiology and dominance, soil properties and climate. *Journal of Biogeography*, **19**, 117–134.

Primack, R.B. and Corlett, R.T. (2005) *Tropical rain forests: an ecological and biogeographical comparison*. Blackwell, Oxford.

Pringle, E.G., Álvarez-Loayza, P. and Terborgh, J. (2007) Seed characteristics and susceptibility to pathogen attack in tree seeds of the Peruvian Amazon. *Plant Ecology*, **193**, 211–222.

Proctor, J. (2003) Vegetation and soil and plant chemistry on ultramafic rocks in the tropical Far East. *Perspectives in Plant Ecology Evolution and Systematics*, **6**, 105–124.

Proctor, J., Anderson, J.M., Chai, P. and Vallack, H.W. (1983) Ecological studies in four contrasting lowland rain forests in Gunung Mulu National Park, Sarawak. I. Forest environment, structure and floristics. *Journal of Ecology*, **71**, 237–260.

Proctor, J., Haridasan, K. and Smith, G.W. (1998) How far north does lowland evergreen tropical rain forest go? *Global Ecology and Biogeography Letters*, **7**, 141–146.

Proctor, J., Brearley, F.Q., Dunlop, H., Proctor, K., Supramono and Taylor, D. (2001) Local wind damage in Barito Ulu, Central Kalimantan: a rare but essential event in a lowland dipterocarp forest? *Journal of Tropical Ecology*, **17**, 473–475.

Pryor, G.S., Levey, D.J. and Dierenfeld, E.S. (2001) Protein requirements of a specialized frugivore, Pesquet's Parrot (*Psittrichas fulgidus*). *Auk*, **118**, 1080–1088.

Purves, D. and Pacala, S. (2008) Predictive models of forest dynamics. *Science*, **320**, 1452–1453.

Putz, F.E. and Mooney, H.A. (1991) *The biology of vines*. Cambridge University Press, Cambridge.

Putz, F.E., Sist, P., Fredericksen, T. and Dykstra, D. (2008a) Reduced-impact logging: challenges and opportunities. *Forest Ecology and Management* **256**, 1427–1433.

Putz, F.E., Zuidema, P.A., Pinard, M.A. et al. (2008b) Improved tropical forest management for carbon retention. *PloS Biology*, **6**, e166.

Qian, H. (2007) Relationships between plant and animal species richness at a regional scale in China. *Conservation Biology*, **21**, 937–944.

Qian, H., Song, J.S., Krestov, P. et al. (2003) Large-scale phytogeographical patterns in East Asia in relation to latitudinal and climatic gradients. *Journal of Biogeography*, **30**, 129–141.

Rabineau, M., Berne, S., Olivet, J.L., Aslanian, D., Guillocheau, F. and Joseph, P. (2006) Paleo sea levels reconsidered from direct observation of paleoshoreline position during Glacial Maxima (for the last 500,000 yr). *Earth and Planetary Science Letters*, **252**, 119–137.

Rabinowitz, A.R. and Walker, S.R. (1991) The carnivore community in a dry tropical forest mosaic in Huai Kha Khaeng Wildlife Sanctuary, Thailand. *Journal of Tropical Ecology*, **7**, 37–47.

Raich, J.W., Russell, A.E., Kitayama, K., Parton, W.J. and Vitousek, P.M. (2006) Temperature influences carbon accumulation in moist tropical forests. *Ecology*, **87**, 76–87.

Ramanathan, V. and Carmichael, G. (2008) Global and regional climate changes due to black carbon. *Nature Geoscience*, **1**, 221–227.

Ranganathan, J., Chan, K.M.A. and Daily, G.C. (2007) Satellite detection of bird communities in tropical countryside. *Ecological Applications*, **17**, 1499–1510.

Rao, I.M., Borrero, V., Ricaurte, J. and Garcia, R. (1999) Adaptive attributes of tropical forage species to acid soils. V. Differences in phosphorus acquisition from inorganic and organic phosphorus sources. *Journal of Plant Nutrition*, **22**, 1175–1196.

Rasingam, L. and Parathasarathy, N. (2009) Tree species diversity and population structure across major forest formations and disturbance categories in Little Andaman Island, India. *Tropical Ecology*, **50**, 89–102.

Rasmussen, C. and Cameron, S.A. (2007) A molecular phylogeny of the Old World stingless bees (Hymenoptera: Apidae: Meliponini) and the non-monophyly of the large genus *Trigona*. *Systematic Entomology*, **32**, 26–39.

Ravenel, R.M. and Granoff, I.M.E. (2004) Illegal logging in the tropics: a synthesis of the issues. *Journal of Sustainable Forestry*, **19**, 351–366.

Rawlings, L.H., Rabosky, D.L., Donnellan, S.C. and Hutchinson, M.N. (2008) Python phylogenetics: inference from morphology and mitochondrial DNA. *Biological Journal of the Linnean Society*, **93**, 603–619.

Read, P. (2008) Biosphere carbon stock management: addressing the threat of abrupt climate change in the next few decades: an editorial essay. *Climatic Change*, **87**, 305–320.

Reaser, J.K., Meyerson, L.A. and Von Holle, B. (2008) Saving camels from straws: how propagule pressure-based prevention policies can reduce the risk of biological invasion. *Biological Invasions*, **10**, 1085–1098.

Reid, A. (1987) Low population growth and its causes in pre-colonial Southeast Asia. *Death and disease in Southeast Asia* (ed. N.G. Owen), pp. 33–47. Oxford University Press, Singapore.

Reid, M.J.C., Ursic, R., Cooper, D. et al. (2006) Transmission of human and macaque *Plasmodium* spp. to ex-captive orangutans in Kalimantan, Indonesia. *Emerging Infectious Diseases*, **12**, 1902–1908.

Ren, H., Li, Z.A., Shen, W.J. et al. (2007) Changes in biodiversity and ecosystem function during the restoration of a tropical forest in south China. *Science in China Series C-Life Sciences*, **50**, 277–284.

Rennolls, K. and Laumonier, Y. (1999) Tree species-area and species-diameter relationships at three lowland rain forest sites in Sumatra. *Journal of Tropical Forest Science*, **11**, 784–800.

Ribeiro, S.P. and Basset, Y. (2007) Gall-forming and free-feeding herbivory along vertical gradients in a lowland tropical rainforest: the importance of leaf sclerophylly. *Ecography*, **30**, 663–672.

Riley, J. (2002a) Mammals on the Sangihe and Talaud Islands, Indonesia, and the impact of hunting and habitat loss. *Oryx*, **36**, 288–296.

Riley, J. (2002b) Population sizes and the status of endemic and restricted-range bird species on Sangihe Island, Indonesia. *Bird Conservation International*, **12**, 53–78.

Riley, J. (2003) Population sizes and the conservation status of endemic and restricted-range bird species on Karakelang, Talaud Islands, Indonesia. *Bird Conservation International*, **13**, 59–74.

Riley, J.R., Greggers, U., Smith, A.D., Reynolds, D.R. and Menzel, R. (2005) The flight paths of honeybees recruited by the waggle dance. *Nature*, **435**, 205–207.

Ripley, S.D. and Beehler, B.M. (1989) Ornithogeographic affinities of the Andaman and Nicobar islands. *Journal of Biogeography*, **16**, 323–332.

Riswan, S. (1987a) Structure and floristic composition of a mixed dipterocarp forest at Lampake, East Kalimantan. *Proceedings of the third round table conference on dipterocarps* (ed. A.J. G.H. Kostermans), pp. 435–476. UNESCO, Jakarta.

Riswan, S. (1987b) Kerangas forest at Gunung Pasir, Samboja, East Kalimantan. *Proceedings of the third round table conference on dipterocarps* (ed. A.J.G.H. Kostermans), pp. 471–494. UNESCO, Jakarta.

Rivera, G., Elliott, S., Caldas, L.S., Nicolossi, G., Coradin, V.T. and Borchert, R. (2002) Increasing day-length induces spring flushing of tropical dry forest trees in the absence of rain. *Trees—Structure and Function*, **16**, 445–456.

Robbins, R.K. and Opler, P.A. (1997) Butterfly diversity and a preliminary comparison with bird and mammal diversity. *Biodiversity II. Understanding and protecting our biological resources* (eds. M.L. Reaka-Kudla, D.E. Wilson, and E.O. Wilson), pp. 69–82. Joseph Henry Press, Washington.

Roberts, R.G., Flannery, T.F., Ayliffe, L.K. et al. (2001) New ages for the last Australian megafauna: continent-wide extinction about 46,000 years ago. *Science*, **292**, 1888–1892.

Robinson, J.G. and Bennett, E.L. (2000) Carrying capacity limits to sustainable hunting in tropical forests. *Hunting for sustainability in tropical forests* (eds J.G. Robinson and E.L. Bennett), pp. 13–30. Columbia University Press, New York.

Rodrigues, A.S.L. and Brooks, T.M. (2007) Shortcuts for biodiversity conservation planning: the effectiveness of surrogates. *Annual Review of Ecology Evolution and Systematics*, **38**, 713–737.

Rodrigues, A.S.L., Pilgrim, J.D., Lamoreux, J.F., Hoffmann, M. and Brooks, T.M. (2006) The value of the IUCN Red List for conservation. *Trends in Ecology and Evolution*, **21**, 71–76.

Rodríguez, J.P., Taber, A.P., Daszak, P. et al. (2007) Globalization of conservation: a view from the South. *Science*, **317**, 755–756.

Romero, G.Q. and Benson, W.W. (2005) Biotic interactions of mites, plants and leaf domatia. *Current Opinion in Plant Biology*, **8**, 436–440.

Roos, M.C., Kessler, P.J.A., Gradstein, S.R. and Baas, P. (2004) Species diversity and endemism of five major Malesian islands: diversity-area relationships. *Journal of Biogeography*, **31**, 1893–1908.

Roubik, D.W. (1996) Wild bees of Brunei Darussalam. *Tropical rainforest research—current issues* (eds D.S. Edwards, W.E. Booth and S.C. Choy), pp. 59–66. Kluwer, Dordrecht.

Roulston, T.H. and Cane, J.H. (2000) Pollen nutritional content and digestibility for animals. *Plant Systematics and Evolution*, **222**, 187–209.

Roulston, T.H., Cane, J.H. and Buchmann, S.L. (2000) What governs protein content of pollen: pollinator preferences, pollen-pistil interactions, or phylogeny? *Ecological Monographs*, **70**, 617–643.

Round, P.D., Gale, G.A. and Brockelman, W.Y. (2006) A comparison of bird communities in mixed fruit orchards and natural forest at Khao Luang, southern Thailand. *Biodiversity and Conservation*, **15**, 2873–2891.

Ruedas, L.A. and Morales, J.C. (2005) Evolutionary relationships among genera of Phalangeridae (Metatheria: Diprotodontia) inferred from mitochondrial DNA. *Journal of Mammalogy*, **86**, 353–365.

Running, S.W., Nemani, R.R., Heinsch, F.A., Zhao, M.S., Reeves, M. and Hashimoto, H. (2004) A continuous satellite-derived measure of global terrestrial primary production. *Bioscience*, **54**, 547–560.

Russo, S.E., Davies, S.J., King, D.A. and Tan, S. (2005) Soil-related performance variation and distributions of tree species in a Bornean rain forest. *Journal of Ecology*, **93**, 879–889.

Russo, S.E., Brown, P., Tan, S. and Davies, S.J. (2008) Interspecific demographic trade-offs and soil-related habitat associations of tree species along resource gradients. *Journal of Ecology*, **96**, 192–203.

Ruxton, G.D. and Houston, D.C. (2004) Obligate vertebrate scavengers must be large soaring fliers. *Journal of Theoretical Biology*, **228**, 431–436.

Saha, S. and Howe, H.F. (2003) Species composition and fire in a dry deciduous forest. *Ecology*, **84**, 3118–3123.

Saigusa, N., Yamamoto, S., Hirata, R. et al. (2008) Temporal and spatial variations in the seasonal patterns of CO_2 flux in boreal, temperate, and tropical forests in East Asia. *Agricultural and Forest Meteorology*, **148**, 700–713.

Sakagami, S.F., Inoue, T. and Salmah, S. (1990) Stingless bees of central Sumatra. *Natural history of social wasps and bees in equatorial Sumatra*. (eds R. Ohgushi, S.F. Sakagami and D.W. Roubik), pp. 125–137. Hokkaido University Press, Sapporo.

Sakai, S. (2000) Reproductive phenology of gingers in a lowland mixed dipterocarp forest in Borneo. *Journal of Tropical Ecology*, **16**, 337–354.

Sakai, S. (2001) Phenological diversity in tropical forests. *Population Ecology*, **43**, 77–86.

Sakai, S. (2002) General flowering in lowland mixed dipterocarp forests of South-east Asia. *Biological Journal of the Linnean Society*, **75**, 233–247.

Sakai, S., Harrison, R.D., Momose, K. et al. (2006) Irregular droughts trigger mass flowering in aseasonal tropical forests in Asia. *American Journal of Botany*, **93**, 1134–1139.

Sánchez-Azofeifa, G.A., Pfaff, A., Robalino, J.A. and Boomhower, J.P. (2007) Costa Rica's payment for environmental services program: intention, implementation, and impact. *Conservation Biology*, **21**, 1165–1173.

Sandker, M., Suwarno, A. and Campbell, B.M. (2007) Will forests remain in the face of oil palm expansion? Simulating change in Malinau, Indonesia. *Ecology and Society*, **12**.

Sathiamurphy, E. and Voris, H.K. (2006) Maps of Holocene sea level transgression and submerged lakes on the Sunda Shelf. *Natural History Journal of Chulalongkorn University*, **2** (suppl.), 1–43.

Savolainen, P., Zhang, Y.P., Luo, J., Lundeberg, J. and Leitner, T. (2002) Genetic evidence for an East Asian origin of domestic dogs. *Science*, **298**, 1610–1613.

Savolainen, P., Leitner, T., Wilton, A.N., Matisoo-Smith, E. and Lundeberg, J. (2004) A detailed picture of the origin of the Australian dingo, obtained from the study of mitochondrial DNA. *Proceedings of the National Academy of Sciences of the USA*, **101**, 12387–12390.

Schaefer, H. and Renner, S.S. (2008) A phylogeny of the oil bee tribe Ctenoplectrini (Hymenoptera: Anthophila) based on mitochondrial and nuclear data: evidence for Early Eocene divergence and repeated out-of-Africa dispersal. *Molecular Phylogenetics and Evolution*, **47**, 799–811.

Schaefer, H.M., Schmidt, V. and Winkler, H. (2003) Testing the defence trade-off hypothesis: how contents of nutrients and secondary compounds affect fruit removal. *Oikos*, **102**, 318–328.

Schaefer, H.M., McGraw, K. and Catoni, C. (2008) Birds use fruit colour as honest signal of dietary antioxidant rewards. *Functional Ecology*, **22**, 303–310.

Scharlemann, J.P.W. and Laurance, W.F. (2008) Environmental science—how green are biofuels? *Science*, **319**, 43–44.

Schenkel, R. and Schenkel-Hulliger, L. (1969) The Javan rhinoceros (*Rh. sondaicus* Desm.) in Udjung Kulon Nature Reserve. Its ecology and behavior. *Acta Tropica*, **26**, 97–135.

Schepartz, L.A., Stoutamire, S. and Bekken, D.A. (2005) *Stegodon orientalis* from Panxian Dadong, a Middle

Pleistocene archaeological site in Guizhou, South China: taphonomy, population structure and evidence for human interactions. *Quaternary International*, **126**, 271–282.

Scherr, S.J. and McNeely, J.A. (2008) Biodiversity conservation and agricultural sustainability: towards a new paradigm of 'ecoagriculture' landscapes. *Philosophical Transactions of the Royal Society B: Biological Sciences*, **363**, 477–494.

Schimann, H., Ponton, S., Hattenschwiler, S. et al. (2008) Differing nitrogen use strategies of two tropical rainforest late successional tree species in French Guiana: evidence from N-15 natural abundance and microbial activities. *Soil Biology and Biochemistry*, **40**, 487–494.

Schipper, J., Chanson, J.S., Chiozza, F. et al. (2008) The status of the world's land and marine mammals: diversity, threat, and knowledge. *Science*, **322**, 225–230.

Schlesinger, W.H. (2009) On the fate of anthropogenic nitrogen. *Proceedings of the National Academy of Sciences of the USA*, **106**, 203–208.

Schlesinger, W.H., Bruijnzeel, L.A., Bush, M.B. et al. (1998) The biogeochemistry of phosphorous after the first century of soil development on Rakata Island, Krakatau, Indonesia. *Biogeochemistry*, **40**, 37–55.

Schleucher, E. (2002) Metabolism, body temperature and thermal conductance of fruit-doves (Aves: Columbidae, Treroninae). *Comparative Biochemistry and Physiology A-Molecular and Integrative Physiology*, **131**, 417–428.

Schnitzer, S.A. (2005) A mechanistic explanation for global patterns of liana abundance and distribution. *American Naturalist*, **166**, 262–276.

Schnitzer, S.A. and Bongers, F. (2002) The ecology of lianas and their role in forests. *Trends in Ecology and Evolution*, **17**, 223–230.

Scofield, D.G. and Schultz, S.T. (2006) Mitosis, stature and evolution of plant mating systems: low-Phi and high-Phi plants. *Proceedings of the Royal Society B: Biological Sciences*, **273**, 275–282.

Scotland, R.W. and Wortley, A.H. (2003) How many species of seed plants are there? *Taxon*, **52**, 101–104.

Scott, M.P. (1998) The ecology and behavior of burying beetles. *Annual Review of Entomology*, **43**, 595–618.

Searchinger, T., Heimlich, R., Houghton, R.A. et al. (2008) Use of US croplands for biofuels increases greenhouse gases through emissions from land-use change. *Science*, **319**, 1238–1240.

SEAZA (2007) SEAZA Future 2015: 8-point action plan for success. South East Asian Zoos Association.

Seddon, P.J., Soorae, P.S. and Launay, F. (2005) Taxonomic bias in reintroduction projects. *Animal Conservation*, **8**, 51–58.

Seddon, P.J., Armstrong, D.P. and Maloney, R.F. (2007) Combining the fields of reintroduction biology and restoration ecology. *Conservation Biology*, **21**, 1388–1390.

Seidler, T.G. and Plotkin, J.B. (2006) Seed dispersal and spatial pattern in tropical trees. *PLoS Biology*, **4**, 2132–2137.

Seino, T., Takyu, M., Aiba, S., Kitayama, K. and Ong, R.C. (2006) Floristic composition, stand structure, and aboveground biomass of the tropical rain forest of Deramakot and Tangkulap Forest Reserve in Malaysia under different forest managements. *Proceedings of second workshop on synergy between carbon management and biodiversity conservation in tropical rain forests*. (eds Y.F. Lee, A.Y.C. Chung and K. Kitayama), pp. 29–52. DIWPA, Kyoto.

Sekercioglu, C.H., Ehrlich, P.R., Daily, G.C., Aygen, D., Goehring, D. and Sandi, R.F. (2002) Disappearance of insectivorous birds from tropical forest fragments. *Proceedings of the National Academy of Sciences of the USA*, **99**, 263–267.

Sessions, J. (2007) *Harvesting operations in the tropics*. Springer, Berlin.

Shanahan, M., So, S., Compton, S.G. and Corlett, R.T. (2001) Fig-eating by vertebrate frugivores: a global review. *Biological Reviews*, **76**, 529–572.

Shek, C.T. (2006) *A field guide to the terrestrial mammals of Hong Kong*. AFCD, Hong Kong.

Shen, Z.-H., Tang, Y.-Y., Lu, N., Zhao, J.X., Li, D.-X. and Wang, G.-F. (2007) Community dynamics of seed rain in mixed evergreen broad-leaved and deciduous forests in a subtropical mountain of Central China. *Journal of Integrative Plant Biology*, **49**, 1294–1303.

Shepherd, C.R. and Nijman, V. (2007) An assessment of wildlife trade at Mong La Market on the Myanmar-China border. *TRAFFIC Bulletin*, **21**, 85–88.

Shi, P., Körner, C. and Hoch, G. (2008a) A test of the growth-limitation theory for alpine tree line formation in evergreen and deciduous taxa of the eastern Himalayas. *Functional Ecology*, **22**, 213–220.

Shi, P., Luo, J.-Q., Xia, N.-B., Wi, H.-W. and Song, J.-Y. (2008b) Suggestions on management measures of pine forest ecosystems invaded by *Bursaphelenchus xylophilus*. *Forestry Studies in China*, **10**, 45–48.

Shimizu, Y. (2003) The nature of Ogasawara and its conservation. *Global Environmental Research*, **7**, 3–14.

Shine, R., Ambariyanto, Harlow, P.S. and Mumpuni (1998) Ecological traits of commercially harvested water monitors, *Varanus salvator*, in northern Sumatra. *Wildlife Research*, **25**, 437–447.

Shine, R., Ambariyanto, Harlow, P.S. and Mumpuni (1999) Reticulated pythons in Sumatra: biology, harvesting and sustainability. *Biological Conservation*, **87**, 349–357.

Shivik, J.A. (2006) Are vultures birds, and do snakes have venom, because of macro- and microscavenger conflict? *Bioscience*, **56**, 819–823.

Shono, K., Cadaweng, E.A. and Durst, P.B. (2007a) Application of assisted natural regeneration to restore degraded tropical forestlands. *Restoration Ecology*, **15**, 620–626.

Shono, K., Davies, S.J. and Chua, Y.K. (2007b) Performance of 45 native tree species on degraded lands in Singapore. *Journal of Tropical Forest Science*, **19**, 25–34.

Shoo, L.P. and VanDerWal, J. (2008) No simple relationship between above-ground tree growth and fine-litter production in tropical forests. *Journal of Tropical Ecology*, **24**, 347–350.

Shoocongdej, R. (2000) Forager mobility organization in seasonal tropical environments of western Thailand. *World Archaeology*, **32**, 14–40.

Siddique, I., Engel, V.L., Parrotta, J.A. et al. (2008) Dominance of legumes alters nutrient relations in mixed species forest restoration plantings within seven years. *Biogeochemistry*, **88**, 89–101.

Sidle, R.C., Ziegler, A.D., Negishi, J.N., Nik, A.R., Siew, R. and Turkelboom, F. (2006) Erosion processes in steep terrain—Truths, myths, and uncertainties related to forest management in Southeast Asia. *Forest Ecology and Management*, **224**, 199–225.

Siebert, S.F. (2002) From shade- to sun-grown perennial crops in Sulawesi, Indonesia: implications for biodiversity conservation and soil fertility. *Biodiversity and Conservation*, **11**, 1889–1902.

Singh, S., Boonratana, R., Bezuijen, M. and Phonvisay, A. (2006a) *Trade in natural resources in Attapeu Province, Lao PDR: an assessment of the wildlife trade*. TRAFFIC Southeast Asia, Vientiane.

Singh, S., Boonratana, R., Bezuijen, M. and Phonvisay, A. (2006b) *Trade in natural resources in Stung Treng Province, Cambodia: an assessment of the wildlife trade*. TRAFFIC Southeast Asia, Vientiane.

Singleton, I. and van Schaik, C.P. (2001) Orangutan home range size and its determinants in a Sumatran swamp forest. *International Journal of Primatology*, **22**, 877–911.

Siniarovina, U. and Engardt, M. (2005) High-resolution model simulations of anthropogenic sulphate and sulphur dioxide in Southeast Asia. *Atmospheric Environment*, **39**, 2021–2034.

Sist, P., Picard, N. and Gourlet-Fleury, S. (2003a) Sustainable cutting cycle and yields in a lowland mixed dipterocarp forest of Borneo. *Annals of Forest Science*, **60**, 803–814.

Sist, P., Sheil, D., Kartawinata, K. and Priyadi, H. (2003b) Reduced-impact logging in Indonesian Borneo: some results confirming the need for new silvicultural prescriptions. *Forest Ecology and Management*, **179**, 415–427.

Slade, E.M., Mann, D.J., Villanueva, J.F. and Lewis, O.T. (2007) Experimental evidence for the effects of dung beetle functional group richness and composition on ecosystem function in a tropical forest. *Journal of Animal Ecology*, **76**, 1094–1104.

Slagsvold, T. and Sonerud, G.A. (2007) Prey size and ingestion rate in raptors: importance for sex roles and reversed sexual size dimorphism. *Journal of Avian Biology*, **38**, 650–661.

Slik, J.W.F., Keßler, P.J.A. and Van Welzen, P.C. (2003) *Macaranga* and *Mallotus* species (Euphorbiaceae) as indicators for disturbance in the mixed lowland dipterocarp forest of East Kalimantan (Indonesia). *Ecological Indicators*, **2**, 311–324.

Slik, J.W.F., Bernard, C.S., Ven Beek, M., Breman, F.C. and Eichhorn, K.A.O. (2008) Tree diversity, composition, forest structure and aboveground biomass dynamics after single and repeated fires in a Bornean rain forest. *Oecologia* **158**, 579–588.

Small, A., Martin, T.G., Kitching, R.L. and Wong, K.M. (2004) Contribution of tree species to the biodiversity of a 1 ha Old World rainforest in Brunei, Borneo. *Biodiversity and Conservation*, **13**, 2067–2088.

Smulders, M.J.M., Westende, W.P.C. van't, Diway, B. et al. (2008) Development of microsatellite markers in *Gonystylus bancanus* (Ramin) useful for tracing and tracking of wood of this protected species. *Molecular Ecology Resources*, **8**, 168–171.

Sodhi, N.S., Koh, L.P., Brook, B.W. and Ng, P.K.L. (2004) Southeast Asian biodiversity: an impending disaster. *Trends in Ecology and Evolution*, **19**, 654–660.

Sodhi, N.S., Koh, L.P., Prawiradilaga, D.M., Tinulele, I., Putra, D.D. and Tan, T.H.T. (2005a) Land use and conservation value for forest birds in Central Sulawesi (Indonesia). *Biological Conservation*, **122**, 547–558.

Sodhi, N.S., Soh, M.C.K., Prawiradilaga, D.M., Darjono and Brook, B.W. (2005b) Persistence of lowland rainforest birds in a recently logged area in central Java. *Bird Conservation International*, **15**, 173–191.

Sodhi, N.S., Brook, B.W. and Bradshaw, C.J.A. (2007) *Tropical conservation biology*. Blackwell, Oxford.

Sodhi, N.S., Acciaioli, G., Erb, M. and Tan, A.K.-J. (2008a) *Biodiversity and human livelihoods in protected areas: case studies from the Malay Archipelago*. Cambridge University Press, Cambridge.

Sodhi, N.S., Koh, L.P., Peh, K.S.H. et al. (2008b) Correlates of extinction proneness in tropical angiosperms. *Diversity and Distributions*, **14**, 1–10.

Song, Y. (1995) The essential characteristics and main types of the broad-leaved evergreen forest in China. *Phytocoenologia*, **16**, 105–123.

Sotta, E.D., Corre, M.D. and Veldkamp, E. (2008) Differing N status and N retention processes of soils under old-growth lowland forest in Eastern Amazonia, Caxiuana, Brazil. *Soil Biology and Biochemistry*, **40**, 740–750.

Spriggs, M. (2003) Chronology of the Neolithic transition in island Southeast Asia and the Western Pacific: a view from 2003. *Review of Archaeology*, **24**, 57–80.

Sri-Ngernyuang, K., Kanzaki, M., Mizuno, T. et al. (2003) Habitat differentiation of Lauraceae species in a tropical lower montane forest in northern Thailand. *Ecological Research*, **18**, 1–14.

Stamp, L.D. and Lord, L. (1923) The ecology of part of the riverine tract of Burma. *Journal of Ecology*, **11**, 129–159.

Steadman, D.W. (2006) *Extinction and biogeography of tropical Pacific birds*. University of Chicago Press, Chicago.

Stevens, M. and Cuthill, I.C. (2007) Hidden messages: are ultraviolet signals a special channel in avian communication? *Bioscience*, **57**, 501–507.

Stokstad, E. (2008) Environmental regulation: New rules on saving wetlands push the limits of the science. *Science*, **320**, 162–163.

Stone, R. (2006) The day the land tipped over. *Science*, **314**, 406–409.

Stone, R. (2007) Biodiversity crisis on tropical islands: last-gasp effort to save Borneo's tropical rainforests. *Science*, **317**, 192.

Stone, R. (2008a) Natural disasters—ecologists report huge storm losses in China's forests. *Science*, **319**, 1318–1319.

Stone, R. (2008b) Showdown looms over a biological treasure trove. *Science*, **319**, 1604.

Stork, N.E. and Brendell, M.J.D. (1990) Variation in the insect fauna of Sulawesi trees with season, altitude and forest type. *Insects and the rain forests of South East Asia (Wallacea)* (eds W.J. Knight and J.D. Holloway), pp. 173–190. Royal Entomological Society of London, London.

Stott, P. (1990) Stability and stress in the savanna forests of mainland South-East Asia. *Journal of Biogeography*, **17**, 373–383.

Stott, P. (2000) Combustion in tropical biomass fires: a critical review. *Progress in Physical Geography*, **24**, 355–377.

Streets, D.G. (2007) Dissecting future aerosol emissions: warming tendencies and mitigation opportunities. *Climatic Change*, **81**, 313–330.

Streets, D.G., Bond, T.C., Carmichael, G.R. et al. (2003). An inventory of gaseous and primary aerosol emissions in Asia in the year 2000. *Journal of Geophysical Research D: Atmospheres*, **108**.

Streicher, U. (2007) Release and re-introduction efforts in Indochina. *Re-introduction News*, **26**, 5–7.

Streicher, U. and Nadler, T. (2003) Re-introduction of pygmy lorises in Vietnam. *Re-introduction News*, **23**, 38–40.

Strickland, D.A. (1967) Ecology of the rhinoceros in Malaya. *Malayan Nature Journal*, **20**, 1–17.

Stronza, A. (2007) The economic promise of ecotourism for conservation. *Journal of Ecotourism*, **6**, 210–230.

Struck, U., Altenbach, A.V., Gaulke, M. and Glaw, F. (2002) Tracing the diet of the monitor lizard *Varanus mabitang* by stable isotope analyses (delta N-15, delta C-13). *Naturwissenschaften*, **89**, 470–473.

Struebig, M.J., Harrison, M.E., Cheyne, S.M. and Limin, S.H. (2007) Intensive hunting of large flying foxes *Pteropus vampyrus natunae* in Central Kalimantan, Indonesian Borneo. *Oryx*, **41**, 390–393.

Styrsky, J.D. and Eubanks, M.D. (2007) Ecological consequences of interactions between ants and honeydew-producing insects. *Proceedings of the Royal Society B: Biological Sciences*, **274**, 151–164.

Su, H.J. (1984) Studies on the climate and vegetation types of the natural forests in Taiwan. (II) Altitudinal vegetation zones in relation to temperature gradient. *Quarterly Journal of Chinese Forestry*, **17**, 57–73.

Su, S.H., Chang-Yang, C.H., Lu, C.L. et al. (2007) *Fushan subtropical forest dynamics plot: tree species characteristics and distribution patterns*. Taiwan Forestry Research Institute, Taipei.

Sugiura, S., Okochi, I. and Tamada, H. (2006) High predation pressure by an introduced flatworm on land snails on the oceanic Ogasawara Islands. *Biotropica*, **38**, 700–703.

Sukardjo, S., Hagihara, A., Yamakura, T. and Ogawa, H. (1990) Floristic composition of a tropical rain forest in Indonesian Borneo. *Bulletin of the Nagoya University Forests*, **10**, 1–10.

Sukumar, R. (2003) *The living elephants: evolutionary ecology, behavior, and conservation*. Oxford University Press, New York.

Sultana, F., Hu, Y.G., Toda, M.J., Takenaka, K. and Yafuso, M. (2006) Phylogeny and classification of *Colocasiomyia* (Diptera, Drosophilidae), and its evolution of pollination mutualism with aroid plants. *Systematic Entomology*, **31**, 684–702.

Sun, I.F. and Hsieh, C.F. (2004) Nanjenshan Forest Dynamics Plot. *Tropical forest diversity and dynamism: findings from a large-scale plot network* (eds E.C. Losos and E.G. Leigh), pp. 564–573. University of Chicago Press, Chicago.

Sun, X.J. and Wang, P.X. (2005) How old is the Asian monsoon system? Palaeobotanical records from China. *Palaeogeography Palaeoclimatology Palaeoecology*, **222**, 181–222.

Sun, I.F., Hsieh, C.F. and Hubbell, S.P. (1996) The structure and species composition of a subtropical monsoon forest in southern Taiwan on a steep wind-stress gradient. *Biodiversity and the dynamics of ecosystems* (eds I.M. Turner, C.H. Diong, S.S.L. Lim and P.K.L. Ng). DIWPA, Kyoto.

Sun, G., Xu, Q., Jin, K., Wang, Z. and Lang, Y. (1998) The historical withdrawal of wild *Elephas maximus* from China and its relationship with human population pressure. *Journal of the Northeast Forestry University*, **26**, 47–50.

Sun, I.F., Chen, Y.Y., Hubbell, S.P., Wright, S.J. and Noor, N. (2007) Seed predation during general flowering events of varying magnitude in a Malaysian rain forest. *Journal of Ecology*, **95**, 818–827.

Swenson, N.G., Enquist, B.J., Thompson, J. and Zimmerman, J.K. (2007) The influence of spatial and size scale on phylogenetic relatedness in tropical forest communities. *Ecology*, **88**, 1770–1780.

Symes, C.T. and Marsden, S.J. (2007) Patterns of supra-canopy flight by pigeons and parrots at a hill-forest site in Papua New Guinea. *Emu*, **107**, 115–125.

Tacconi, L., Moore, P.F. and Kaimowitz, D. (2007) Fires in tropical forests—What is really the problem? Lessons from Indonesia. *Mitigation and Adaptation Strategies for Global Change*, **12**, 55–66.

Takanose, Y. and Kamitani, T. (2003) Fruiting of fleshy-fruited plants and abundance of frugivorous birds: phenology correspondence in a temperate forest in central Japan. *Ornithological Science*, **2**, 25–32.

Takenaka, K., Yin, J.T., Wen, S.Y. and Toda, M.J. (2006) Pollination mutualism between a new species of the genus *Colocasiomyia* de Meijere (Diptera: Drosophilidae) and *Steudnera colocasiifolia* (Araceae) in Yunnan, China. *Entomological Science*, **9**, 79–91.

Takeuchi, Y. and Nakashizuka, T. (2007) Effect of distance and density on seed/seedling fate of two dipterocarp species. *Forest Ecology and Management*, **247**, 167–174.

Tan, C.L. and Drake, J.H. (2001) Evidence of tree gouging and exudate eating in pygmy slow lorises (*Nycticebus pygmaeus*). *Folia Primatologica*, **72**, 37–39.

Tanaka, S., Wasli, M.E.B., Kotegawa, T. et al. (2007) Soil properties of secondary forests under shifting cultivation by the Iban of Sarawak, Malaysia in relation to vegetation condition. *Tropics*, **16**, 385–398.

Tang, C.Q. (2006) Forest vegetation as related to climate and soil conditions at varying altitudes on a humid subtropical mountain, Mount Emei, Sichuan, China. *Ecological Research*, **21**, 174–180.

Tang, A.M.C., Corlett, R.T. and Hyde, K.D. (2005) The persistence of ripe fleshy fruits in the presence and absence of frugivores. *Oecologia*, **142**, 232–237.

Tang, Y., Cao, M. and Fu, X.F. (2006) Soil seedbank in a dipterocarp rain forest in Xishuangbanna, Southwest China. *Biotropica*, **38**, 328–333.

Taylor, A.B. (2006) Feeding behavior, diet, and the functional consequences of jaw form in orangutans, with implications for the evolution of *Pongo*. *Journal of Human Evolution*, **50**, 377–393.

Teejuntuk, S., Pongsak, S., Katsutoshi, S. and Witchaphart, S. (2003) Forest structure and tree species diversity along an altitudinal gradient in Doi Inthanon National Park, Northern Thailand. *Tropics*, **12**, 85–102.

Tella, J.L. and Carrete, M. (2008) Broadening the role of parasites in biological invasions. *Frontiers in Ecology and the Environment*, **6**, 11–12.

ten Kate, K., Bishop, J. and Bayon, R. (2004) *Biodiversity offsets: views, experience, and the business case*. IUCN and Insight Investment, London.

Teo, D.H.L., Tan, H.T.W., Corlett, R.T., Wong, C.M. and Lum, S.K.Y. (2003) Continental rain forest fragments in Singapore resist invasion by exotic plants. *Journal of Biogeography*, **30**, 305–310.

Terborgh, J., Van Schaik, C.P., Rao, M. and Davenport, I. (2002) *Making parks work: strategies for preserving tropical nature*. Island Press, Washington, DC.

Terborgh, J., Nuñez-Iturri, G., Pitman, N.C.A. et al. (2008) Tree recruitment in an empty forest. *Ecology*, **89**, 1757–1768.

Ter Steege, H., Pitman, N., Sabatier, D. et al. (2003) A spatial model of tree alpha-diversity and tree density for the Amazon. *Biodiversity and Conservation*, **12**, 2255–2277.

Tewksbury, J.J., Huey, R.B. and Deutsch, C.A. (2008) Putting the heat on tropical animals. *Science*, **320**, 1296–1297.

Theimer, T., C. and Gehring, C.A. (2007) Mycorrhizal plants and vertebrate seed and spore dispersal: incorporating mycorrhizas into the seed dispersal paradigm. *Seed dispersal: theory and its application in a changing world* (eds A.J. Dennis, E.W. Schupp, R.J. Green and D.A. Westcott), pp. 463–478. CABI International, Wallingford.

Thiollay, J.M. (1995) The role of traditional agroforests in the conservation of rain forest bird diversity in Sumatra. *Conservation Biology*, **9**, 335–353.

Thiollay, J.M. (1998) Distribution patterns and insular biogeography of South Asian raptor communities. *Journal of Biogeography*, **25**, 57–72.

Thompson, L.G., Mosley-Thompson, E., Brecher, H. et al. (2006) Abrupt tropical climate change: past and present. *Proceedings of the National Academy of Sciences of the USA*, **103**, 10536–10543.

Thornton, I.W.B., Runciman, D., Cook, S. et al. (2002) How important were stepping stones in the colonization of Krakatau? *Biological Journal of the Linnean Society*, **77**, 275–317.

Timmins, R.J. and Duckworth, J.W. (2008) Diurnal squirrels (Mammalia Rodentia Sciuridae) in Lao PDR: distribution, status and conservation. *Tropical Zoology*, **21**, 11–56.

Toda, T., Takeda, H., Tokuchi, N., Ohta, S., Wacharinrat, C. and Kaitpraneet, S. (2007) Effects of forest fire on the nitrogen cycle in a dry dipterocarp forest, Thailand. *Tropics*, **16**, 41–45.

Tong, H.W. and Liu, J.G. (2004) The Pleistocene-Holocene extinctions of mammals in China. *Proceedings of the Ninth Annual Symposium of the Chinese Society of Vertebrate Paleontology* (ed. W. Dong), pp. 111–119. China Ocean Press, Beijing.

Townsend, A.R., Cleveland, C.C., Asner, G.P. and Bustamante, M.M.C. (2007) Controls over foliar N: P ratios in tropical rain forests. *Ecology*, **88**, 107–118.

Tran, H., Uchihama, D., Ochi, S. and Yasuoka, Y. (2006) Assessment with satellite data of the urban heat island effects in Asian mega cities. *International Journal of Applied Earth Observation and Geoinformation*, **8**, 34–48.

Tsahar, E., del Rio, C.M., Izhaki, I. and Arad, Z. (2005) Can birds be ammonotelic? Nitrogen balance and excretion in two frugivores. *Journal of Experimental Biology*, **208**, 1025–1034.

Tsang, A.C.W. and Corlett, R.T. (2005) Reproductive biology of the *Ilex* species (Aquifoliaceae) in Hong Kong, China. *Canadian Journal of Botany*, **83**, 1645–1654.

Tsujino, R. and Yumoto, T. (2004) Effects of sika deer on tree seedlings in a warm temperate forest on Yakushima Island, Japan. *Ecological Research*, **19**, 291–300.

Tsujita, K., Sakai, S. and Kikuzawa, K. (2008) Does individual variation in fruit profitability override color differences in avian choice of red or white *Ilex serrata* fruits? *Ecological Research*, **23**, 445–450.

Tudhope, A.W., Chilcott, C.P., McCulloch, M.T. et al. (2001) Variability in the El Nino—Southern oscillation through a glacial-interglacial cycle. *Science*, **291**, 1511–1517.

Turner, I.M. (1996) Species loss in fragments of tropical rain forest: a review of the evidence. *Journal of Applied Ecology*, **33**, 200–209.

Turner, I.M. (2001) *The ecology of trees in the tropical rain forest*. Cambridge University Press, Cambridge.

Turner, B.L. (2008) Resource partitioning for soil phophorus: a hypothesis. *Journal of Ecology*, **96**, 698–702.

Turner, I.M. and Corlett, R.T. (1996) The conservation value of small, isolated fragments of lowland tropical rain forest. *Trends in Ecology and Evolution*, **11**, 330–333.

Turner, I.M., Wong, Y.K., Chew, P.T. and bin Ibrahim, A. (1997) Tree species richness in primary and old secondary tropical forest in Singapore. *Biodiversity and Conservation*, **6**, 537–543.

Turner, J.A., Maplesden, F. and Johnson, S. (2007) Measuring the impacts of illegal logging. *Tropical Forest Update*, **17**, 19–22.

Turton, S.M. (2008) Landscape-scale impacts of Cyclone Larry on the forests of northeast Australia, including comparisons with previous cyclones impacting the region between 1858 and 2006. *Austral Ecology*, **33**, 409–416.

Turton, S.M. and Stork, N.E. (2008) Environmental impacts of tourism and recreation in the wet tropics. *Living in a dynamic tropical forest landscape* (eds N.E. Stork and S.M. Turton), pp. 349–356. Blackwell, Malden.

Tweddle, J.C., Dickie, J.B., Baskin, C.C. and Baskin, J.M. (2003) Ecological aspects of seed desiccation sensitivity. *Journal of Ecology*, **91**, 294–304.

Tzeng, H.Y., Lu, F.Y., Ou, C.H., Lu, K.C. and Tseng, L.J. (2006) Pollinational-mutualism strategy of *Ficus erecta* var. *beecheyana* and *Blastophaga nipponica* in seasonal Guandaushi Forest Ecosystem, Taiwan. *Botanical Studies*, **47**, 307–318.

Underwood, E.C., Shaw, M.R., Wilson, K.A. et al. (2008) Protecting biodiversity when money matters: maximizing return on investment. *PLoS ONE*, **1**, e1515.

UNDP (2007) *Human development report 2007/2008. Fighting climate change: human solidarity in a divided world*. United Nations Development Program, New York.

UN Population Division (2006) *World urbanization prospects: the 2005 revision*. United Nations, Department of Economic and Social Affairs, Population Division.

Uriarte, M., Condit, R., Canham, C.D. and Hubbell, S.P. (2004) A spatially explicit model of sapling growth in a tropical forest: does the identity of neighbours matter? *Journal of Ecology*, **92**, 348–360.

Uryu, Y., Mott, C., Foead, N. et al. (2008) Deforestation, forest degradation, biodiversity loss and CO_2 emissions in Riau, Sumatra, Indonesia. WWF Indonesia, Jakarta.

USDA (2006) *Keys to Soil Taxonomy*, 10th edition. United States Department of Agriculture, Washington, DC.

van den Bergh, G.D. (1999) The Late Neogene elephantoid-bearing faunas of Indonesia and their palaeozoogeographic implications. A study of the terrestrial faunal succession of Sulawesi, Flores and Java, including evidence for early hominid dispersal east of Wallace's Line. *Scripta Geologica*, **117**, 1–419.

van den Bergh, G.D., de Vos, J. and Sondaar, P.Y. (2001) The Late Quaternary palaeogeography of mammal evolution in the Indonesian Archipelago. *Palaeogeography Palaeoclimatology Palaeoecology*, **171**, 385–408.

van den Bergh, G.D., Due Awe, R., Morwood, M.J., Sutikna, T., Jatmiko and Wahyu Saptomo, E. (2008a) The youngest stegodon remains in Southeast Asia from the Late Pleistocene archaeological site Liang Bua, Flores, Indonesia. *Quaternary International*, **182**, 16–48.

van den Bergh, G.D., Meijer, H.J.M., Due Awe, R. et al. (2008b) The Liang Bua faunal remains: a 95 kyr sequence from Flores, East Indonesia. *Journal of Human Evolution*.

van der Heijden, G.M.F. and Phillips, O.L. (2008) What controls liana success in Neotropical forests? *Global Ecology and Biogeography*, **17**, 372–383.

Vander Wall, S.B. and Longland, W.S. (2004) Diplochory: are two seed dispersers better than one? *Trends in Ecology and Evolution*, **19**, 155–161.

Van Driesche, R., Hoddle, M. and Center, T. (2008) *Control of pests and weeds by natural enemies: an introduction to biological control*. Blackwell Publishing, Malden.

Van Gulik, R.H. (1967) *The gibbon in China*. E.J. Brill, Leiden.

van Steenis, C.G.G.J. (1942) Gregarious flowering of *Strobilanthes* (Acanthaceae) in Malaysia. *Annals of the Royal Botanic Garden Calcutta*, (150th Anniversary), 91–97.

Van Welzen, P.C., Slik, J.W.F. and Alahuhta, J. (2005) Plant distribution patterns and plate tectonics in Malesia. *Biologiske Skrifter,* **55**, 199–217.

Veldman, J.W., Murray, K.G., Hull, A.L. et al. (2007) Chemical defense and the persistence of pioneer plant seeds in the soil of a tropical cloud forest. *Biotropica,* **39**, 87–93.

Venkataraman, V. (2007) *A matter of attitude: the consumption of wild animal products in Ha Noi.* TRAFFIC Southeast Asia, Hanoi.

Verdu, M. and Pausas, J.G. (2007) Fire drives phylogenetic clustering in Mediterranean Basin woody plant communities. *Journal of Ecology,* **95**, 1316–1323.

Verheij, E.W.M. and Coronel, R.E. (1991) *Plant resources of South-East Asia 2. Edible fruits and nuts.* Pudoc-DLO, Wageningen.

Vesk, P.A. and Westoby, M. (2004) Sprouting ability across diverse disturbances and vegetation types worldwide. *Journal of Ecology,* **92**, 310–320.

Vieira, D.L.M., Scariot, A., Sampaio, A.B. and Holl, K.D. (2006) Tropical dry-forest regeneration from root suckers in Central Brazil. *Journal of Tropical Ecology,* **22**, 353–357.

Voigt, C.C., Capps, K.A., Dechmann, D.K., Michener, R.H. and Kunz, T.H. (2008) Nutrition or detoxification: why bats visit mineral licks of the Amazonian rainforest. *PLoS ONE,* **3**, e2011.

Volkov, I., Banavar, J.R., He, F.L., Hubbell, S.P. and Maritan, A. (2005) Density dependence explains tree species abundance and diversity in tropical forests. *Nature,* **438**, 658–661.

Von der Lippe, M. and Kowarik, I. (2008) Do cities export biodiversity? Traffic as dispersal vector across urban-rural gradients. *Diversity and Distributions,* **14**, 18–25.

Wäckers, F.L., Romeis, J. and van Rijn, P. (2007) Nectar and pollen feeding by insect herbivores and implications for multitrophic interactions. *Annual Review of Entomology,* **52**, 301–323.

Walker, J.S. (2007) Dietary specialization and fruit availability among frugivorous birds on Sulawesi. *Ibis,* **149**, 345–356.

Wallace, A.R. (1876) *The geographical distribution of animals.* Macmillan, London.

Walsh, R.P.D. (1996) Climate. *Tropical rain forest: an ecological study* (eds P. W. Richards, R. P. D. Walsh, I. C. Baillie and P. Greig-Smith), pp. 159–205. Cambridge University Press, Cambridge.

Wang, S.J., Li, R.L., Sun, C.X. et al. (2004) How types of carbonate rock assemblages constrain the distribution of karst rocky desertified land in Guizhou Province, PR China: phenomena and mechanisms. *Land Degradation and Development,* **15**, 123–131.

Wang, S.Y., Lu, H.Y., Liu, J.Q. and Negendank, J.F.W. (2007a) The early Holocene optimum inferred from a high-resolution pollen record of Huguangyan Maar Lake in southern China. *Chinese Science Bulletin,* **52**, 2829–2836.

Wang, X.M., Sun, X.J., Wang, P.X. and Stattegger, K. (2007b) A high-resolution history of vegetation and climate history on Sunda Shelf since the last glaciation. *Science in China Series D-Earth Sciences,* **50**, 75–80.

Wang, X.H., Kent, M. and Fang, X.F. (2007c) Evergreen broad-leaved forest in Eastern China: its ecology and conservation and the importance of resprouting in forest restoration. *Forest Ecology and Management,* **245**, 76–87.

Wang, Z.H., Tang, Z.Y. and Fang, J.Y. (2007d) Altitudinal patterns of seed plant richness in the Gaoligong Mountains, south-east Tibet, China. *Diversity and Distributions,* **13**, 845–854.

Wang, X.K., Manning, W., Feng, Z.W. and Zhu, Y.G. (2007e) Ground-level ozone in China: distribution and effects on crop yields. *Environmental Pollution,* **147**, 394–400.

Wang, Y., Cheng, H., Edwards, R.L. et al. (2008) Millennial- and orbital-scale changes in the East Asian monsoon over the past 224,000 years. *Nature,* **451**, 1090–1093.

Ward, M., Dick, C.W., Gribel, R. and Lowe, A.J. (2005) To self, or not to self ... A review of outcrossing and pollen-mediated gene flow in neotropical trees. *Heredity,* **95**, 246–254.

Wardell-Johnson, G.W., Kanowski, J., Catterall, C.P., Price, M. and Lamb, D. (2008) Rainforest restoration for biodiversity and the production of timber. *Living in a dynamic tropical forest landscape* (eds N.E. Stork and S.M. Turton), pp. 494–509. Blackwell, Malden.

Wasser, S.K., Mailand, C., Booth, R. et al. (2007) Using DNA to track the origin of the largest ivory seizure since the 1989 trade ban. *Proceedings of the National Academy of Sciences of the USA,* **104**, 4228–4233.

Watanabe, T., Misawa, S., Hiradate, S. and Osaki, M. (2008) Characterization of root mucilage from *Melastoma malabathricum*, with emphasis on its roles in aluminum accumulation. *New Phytologist,* **178**, 581–589.

Watari, Y., Takatsuki, S. and Miyashita, T. (2008) Effects of exotic mongoose (*Herpestes javanicus*) on the native fauna of Amami-Oshima Island, southern Japan, estimated by distribution patterns along the historical gradient of mongoose invasion. *Biological Invasions,* **10**, 7–17.

Wattanaratchakit, N. and Srikosamatara, S. (2006) Small mammals around a Karen village in Northern Mae Hong Son Province, Thailand: abundance, distribution and human consumption. *Natural History Bulletin of the Siam Society,* **54**, 195–207.

Webb, S.D. (1997) The great American faunal interchange. *Central America: a natural and cultural history* (ed. A. G. Coates), pp. 97–122. Yale University Press, New Haven.

Webb, C.O. and Peart, D.R. (1999) Seedling density dependence promotes coexistence of Bornean rain forest trees. *Ecology,* **80**, 2006–2017.

Webb, C.O., Gilbert, G.S. and Donoghue, M.J. (2006) Phylodiversity-dependent seedling mortality, size structure, and disease in a Bornean rain forest. *Ecology,* **87**, S123-S131.

Webb, C.O., Cannon, C.H. and Davies, S.J. (2008) Ecological organization, biogeography, and the phylogenetic structure of tropical forest tree communities. *Tropical forest community ecology* (eds W.P. Carson and S.A. Schnitzer), pp. 79–97. Wiley-Blackwell, Oxford.

Wei, G.J., Deng, W.F., Liu, Y. and Li, X.H. (2007) High-resolution sea surface temperature records derived from foraminiferal Mg/Ca ratios during the last 260 ka in the northern South China Sea. *Palaeogeography Palaeoclimatology Palaeoecology,* **250**, 126–138.

Weir, J.E.S. and Corlett, R.T. (2007) How far do birds disperse seeds in the degraded tropical landscape of Hong Kong, China? *Landscape Ecology,* **22**, 131–140.

Wells, D. (2007) *Birds of the Thai-Malay Peninsula, Volume 2.* Christopher Helm, London.

Wells, K., Smales, L.R., Kalko, E.K.V. and Pfeiffer, M. (2007) Impact of rain-forest logging on helminth assemblages in small mammals (Muridae, Tupaiidae) from Borneo. *Journal of Tropical Ecology,* **23**, 35–43.

Weng, E. and Zhou, G. (2006) Modeling distribution changes of vegetation in China under future climate change. *Environmental Modeling and Assessment,* **11**, 45–58.

Westaway, K.E., Zhao, J.X., Roberts, R.G., Chivas, A.R., Morwood, M.J. and Sutikna, T. (2007) Initial speleothem results from western Flores and eastern Java, Indonesia: were climate changes from 47 to 5 ka responsible for the extinction of *Homo floresiensis*? *Journal of Quaternary Science,* **22**, 429–438.

Westerkamp, C. and Classen-Bockhoff, R. (2007) Bilabiate flowers: the ultimate response to bees? *Annals of Botany,* **100**, 361–374.

Wharton, C.H. (1966) Man, fire and wild cattle in north Cambodia. *Proceedings of the Annual Tall Timbers Fire Ecology Conference 6*, pp. 23–65.

White, J.C., Penny, D., Kealhofer, L. and Maloney, B. (2004a) Vegetation changes from the late Pleistocene through the Holocene from three areas of archaeological significance in Thailand. *Quaternary International,* **113**, 111–132.

White, E., Tucker, N., Meyers, N. and Wilson, J. (2004b) Seed dispersal to revegetated isolated rainforest patches in North Queensland. *Forest Ecology and Management,* **192**, 409–426.

Whitmore, T.C. and Burslem, D.F.R.P. (1998) Major disturbances in tropical rainforests. *Dynamics of tropical communities* (eds D.M. Newbery, H.H.T. Prins and N.D. Brown), pp. 549–565. Blackwell Science, Oxford.

Whitmore, T.C. and Sidiyasa, K. (1986) Composition and structure of a lowland rain forest at Toraut, northern Sulawesi. *Kew Bulletin,* **41**, 747–756.

Whitten, T. and Balmford, A. (2006) Who should pay for tropical forest conservation, and how could the costs be met? *Emerging threats to tropical forests* (eds W.F. Laurance and C.A. Peres), pp. 317–336. University of Chicago Press, Chicago.

Wich, S.A. and van Schaik, C.P. (2000) The impact of El Niño on mast fruiting in Sumatra and elsewhere in Malesia. *Journal of Tropical Ecology,* **16**, 563–577.

Wiens, F., Zitzmann, A. and Hussein, N.A. (2006) Fast food for slow lorises: is low metabolism related to secondary compounds in high-energy plant diet? *Journal of Mammalogy,* **87**, 790–798.

Wiens, F., Zitzmann, A., Lachance, M.A. et al. (2008) Chronic intake of fermented floral nectar by wild treeshrews. *Proceedings of the National Academy of Sciences of the USA,* **105**, 10426–10431.

Wikramanayake, E.D., Dinerstein, E., Loucks, C.J. et al. (2002) *Terrestrial ecoregions of the Indo-Pacific: a conservation assessment.* Island Press, Washington, DC.

Williams, S.E. and Hilbert, D.W. (2006) Climate change as a threat to the biodiversity of tropical rainforests in Australia. *Emerging threats to tropical forests* (eds W.F. Laurance and C.A. Peres), pp. 33–52. Chicago University Press, Chicago.

Williams, J.W., Jackson, S.T. and Kutzbacht, J.E. (2007) Projected distributions of novel and disappearing climates by 2100 AD. *Proceedings of the National Academy of Sciences of the USA,* **104**, 5738–5742.

Williams, L.J., Bunyavejchewin, S. and Baker, P.J. (2008) Deciduousness in a seasonal tropical forest in western Thailand: interannual and intraspecific variation in timing, duration and environmental cues. *Oecologia,* **155**, 571–582.

Williamson, G.B. and Ickes, K. (2002) Mast fruiting and ENSO cycles—does the cue betray a cause? *Oikos,* **97**, 459–461.

Wilson, E.O. (2005) Oribatid mite predation by small ants of the genus *Pheidole. Insectes Sociaux,* **52**, 263–265.

Wilson, D.E. and Reeder, D.M. (2005) *Mammal species of the world: a taxonomic and geographic reference.* John Hopkins University Press, Baltimore.

Wilting, A., Buckley-Beason, V.A., Feldhaar, H., Gadau, J., O'Brien, S.J. and Linsenmair, K.E. (2007) Clouded leopard phylogeny revisited: support for species recognzition and population division between Borneo and Sumatra. *Frontiers in Zoology,* **4**, 15.

Winkler, H. and Christie, D.A. (2002) Family Picidae (Woodpeckers). *Handbook of the Birds of the World,*

Volume 7. (eds J. del Hoyo, A. Elliott and J. Sargatal), pp. 296–555. Lynx Edicions, Barcelona.

Winkler, H., Christie, D.A. and Nurney, D. (1995) *Woodpeckers: an identification guide to the woodpeckers of the world*. Houghton Mifflin, Boston.

Wong, S.T., Servheen, C. and Ambu, L. (2002) Food habits of Malayan sun bears in lowland tropical forests of Borneo. *Ursus*, **13**, 127–136.

Wong, S.T., Servheen, C., Ambu, L. and Norhayati, A. (2005) Impacts of fruit production cycles on Malayan sun bears and bearded pigs in lowland tropical forest of Sabah, Malaysian Borneo. *Journal of Tropical Ecology*, **21**, 627–639.

Woodruff, D.S. (2003) Neogene marine transgressions, palaeogeography and biogeographic transitions on the Thai-Malay Peninsula. *Journal of Biogeography*, **30**, 551–567.

Woods, K. and Elliott, S. (2004) Direct seeding for forest restoration on abandoned agricultural land in northern Thailand. *Journal of Tropical Forest Science*, **16**, 248–259.

Wösten, J.H.M., Clymans, E., Page, S.E., Rieley, J.O. and Limin, S.H. (2008) Peat-water interrelationships in a tropical peatland ecosystem in Southeast Asia. *Catena*, **73**, 212–224.

Wright, S.J. (2002) Plant diversity in tropical forests: a review of mechanisms of species coexistence. *Oecologia*, **130**, 1–14.

Wright, S.J. and Muller-Landau, H.C. (2006a) The future of tropical forest species. *Biotropica*, **38**, 287–301.

Wright, S.J. and Muller-Landau, H.C. (2006b) The uncertain future of tropical forest species. *Biotropica*, **38**, 443–445.

Wright, S.J., Sanchez-Azofeifa, G.A., Portillo-Quintero, C. and Davies, D. (2007a) Poverty and corruption compromise tropical forest reserves. *Ecological Applications*, **17**, 1259–1266.

Wright, S.J., Stoner, K.E., Beckman, N. et al. (2007b) The plight of large animals in tropical forests and the consequences for plant regeneration. *Biotropica*, **39**, 289–291.

Wu, J.G. (2007) World without borders: wildlife trade on the Chinese-language Internet. *TRAFFIC Bulletin*, **21**, 75–84.

Wu, L., Shinzato, T., Chen, C. and Aramoto, M. (2008) Sprouting characteristics of a subtropical evergreen broad-leaved forest following clear-cutting in Okinawa, Japan. *New Forests*, **36**, 239–246.

Wunder, S. (2006) Are direct payments for environmental services spelling doom for sustainable forest management in the tropics? *Ecology and Society*, **11**.

Wunder, S. (2007) The efficiency of payments for environmental services in tropical conservation. *Conservation Biology*, **21**, 48–58.

Wunder, S., Engel, S. and Pagiola, S. (2008) Taking stock: a comparative analysis of payments for environmental services programs in developed and developing countries. *Ecological Economics*, **65**, 834–852.

WWF (2007) *Gone in an instant: how the trade in illegally grown coffee is driving the destruction of rhino, tiger and elephant habitat*. WWF-Indonesia, Jakarta.

Xiao, Z.S., Wang, Y.S., Harris, M. and Zhang, Z.B. (2006) Spatial and temporal variation of seed predation and removal of sympatric large-seeded species in relation to innate seed traits in a subtropical forest, Southwest China. *Forest Ecology and Management*, **222**, 46–54.

Xiao, J.Y., Lu, H.B., Zhou, W.J., Zhao, Z.J. and Hao, R.H. (2007a) Evolution of vegetation and climate since the last glacial maximum recorded at Dahu peat site, South China. *Science in China Series D-Earth Sciences*, **50**, 1209–1217.

Xiao, Z.S., Harris, M.K. and Zhang, Z.B. (2007b) Acorn defenses to herbivory from insects: implications for the joint evolution of resistance, tolerance and escape. *Forest Ecology and Management*, **238**, 302–308.

Xie, Y., Mackinnon, J. and Li, D. (2004) Study on biogeographical divisions of China. *Biodiversity and Conservation*, **13**, 1391–1417.

Xu, Y.C., Fang, S.G. and Li, Z.K. (2007) Sustainability of the South China tiger: implications of inbreeding depression and introgression. *Conservation Genetics*, **8**, 1199–1207.

Yafuso, M. (1993) Thermogenesis of *Alocasia odora* (Araceae) and the role of *Colocasiomyia* flies (Diptera, Drosophilidae) as cross-pollinators. *Environmental Entomology*, **22**, 601–606.

Yamada, I. (1975) Forest ecological studies of the montane forest of Mt Pangrango, West Java. *South East Asian Studies*, **13**, 402–426.

Yamada, T., Suzuki, E., Yamakura, T. and Tan, S. (2005a) Tap-root depth of tropical seedlings in relation to species-specific edaphic preferences. *Journal of Tropical Ecology*, **21**, 155–160.

Yamada, A., Inoue, T., Wiwatwitaya, D. et al. (2005b) Carbon mineralization by termites in tropical forests, with emphasis on fungus combs. *Ecological Research*, **20**, 453–460.

Yamada, A., Inoue, T., Wiwatwitaya, D. and Ohkuma, M. (2007) A new concept of the feeding group composition of termites (Isoptera) in tropical ecosystems: Carbon source competitions among fungus-growing termites, soil-feeding termites, litter-layer microbes, and fire. *Sociobiology*, **50**, 135–153.

Yamashita, N., Tanaka, N., Hoshi, Y., Kushima, H. and Kamo, K. (2003) Seed and seedling demography of invasive and native trees of subtropical Pacific islands. *Journal of Vegetation Science*, **14**, 15–24.

Yan, J.H., Wang, Y.P., Zhou, G.Y. and Zhang, D.Q. (2006) Estimates of soil respiration and net primary production of three forests at different succession stages in South China. *Global Change Biology*, **12**, 810–821.

Yan, E.-R., Wang, X.-H. and Zhou, W. (2008) N:P stoichiometry in secondary succession in evergreen broad-leaved forest, Tiantong, East China. *Journal of Plant Ecology*, **32**, 13–22.

Yancheva, G., Nowaczyk, N.R., Mingram, J. et al. (2007) Influence of the intertropical convergence zone on the East Asian monsoon. *Nature*, **445**, 74–77.

Yang, Y.S., Chen, G.S., Lin, P., Xie, J.S. and Guo, J.F. (2004) Fine root distribution, seasonal pattern and production in four plantations compared with a natural forest in subtropical China. *Annals of Forest Science*, **61**, 617–627.

Yang, Y.S., Chen, G.S., Guo, J.F., Xie, J.S. and Wang, X.G. (2007) Soil respiration and carbon balance in a subtropical native forest and two managed plantations. *Plant Ecology*, **193**, 71–84.

Yasuda, M., Miura, S. and Nor Azman, H. (2000) Evidence for food hoarding behaviour in terrestrial rodents in Pasoh Forest Reserve, a Malaysian lowland rain forest. *Journal of Tropical Forest Science*, **12**, 164–173.

Ye, W.-H., Cao, H.-L., Huang, Z.-L. et al. (2008) Community structure of a 20 hm-2 lower subtropical evergreen broadleaved forest plot in Dinghushan, China. *Journal of Plant Ecology*, **32**, 274–286.

Yip, J.Y., Corlett, R.T. and Dudgeon, D. (2004) A fine-scale gap analysis of the existing protected area system in Hong Kong, China. *Biodiversity and Conservation*, **13**, 943–957.

Yip, J.Y., Corlett, R.T. and Dudgeon, D. (2006) Selecting small reserves in a human-dominated landscape: a case study of Hong Kong, China. *Journal of Environmental Management*, **78**, 86–96.

Yoda, K. (1983) Community respiration in a lowland rain forest in Pasoh, Peninsular Malaysia. *Japanese Journal of Ecology*, **33**, 183–197.

Yoder, A.D. and Yang, Z.H. (2004) Divergence dates for Malagasy lemurs estimated from multiple gene loci: geological and evolutionary context. *Molecular Ecology*, **13**, 757–773.

Yokoyama, Y., Falguères, C., Sémah, F., Jacob, T. and Grün, R. (2008) Gamma-ray spectrometric dating of late *Homo erectus* skulls from Ngandong and Sambungmacan, Central Java, Indonesia. *Journal of Human Evolution*, **55**, 274–277.

Yonariza and Webb, E.L. (2007) Rural household participation in illegal timber felling in a protected area of West Sumatra, Indonesia. *Environmental Conservation*, **34**, 73–82.

Yoneda, T., Nishimura, S. and Chairul (2000) Impacts of dry and hazy weather in 1997 on a tropical rainforest ecosystem in West Sumatra, Indonesia. *Ecological Research*, **15**, 63–71.

Yu, Y., Baskin, J.M., Baskin, C.C., Tang, Y. and Cao, M. (2008a) Ecology of seed germination of eight non-pioneer tree species from a tropical seasonal rain forest in southwest China. *Plant Ecology*, **197**, 1–16.

Yu, Y., Yu, J., Shan, Q., Fang, L. and Jiang, D. (2008b) Organic acid exudation from the roots of *Cunninghamia lanceolata* and *Pinus massoniana* seedlings under low phosphorus stress. *Frontiers of Forestry in China*, **3**, 117–120.

Yuan, W.P., Liu, S., Zhou, G.S. et al. (2007) Deriving a light use efficiency model from eddy covariance flux data for predicting daily gross primary production across biomes. *Agricultural and Forest Meteorology*, **143**, 189–207.

Yumoto, T., Noma, N. and Maruhashi, T. (1998) Cheek-pouch dispersal of seeds by Japanese monkeys (*Macaca fuscata yakui*) on Yakushima Island, Japan. *Primates*, **39**, 325–338.

Zachos, J., Pagani, M., Sloan, L., Thomas, E. and Billups, K. (2001) Trends, rhythms, and aberrations in global climate 65 Ma to present. *Science*, **292**, 686–693.

Zackey, J. (2007) Peasant perspectives on deforestation in southwest China: social discontent and environmental mismanagement. *Mountain Research and Development*, **27**, 153–161.

Zahawi, R.A. (2008) Instant trees: using giant vegetative stakes in tropical forest restoration. *Forest Ecology and Management*, **255**, 3013–3016.

Zang, R.-G., Zhang, W.-Y. and Ding, Y. (2007) Seed dynamics in relation to gaps in a tropical montane rainforest of Hainan Island, South China: (1) seed rain. *Journal of Integrative Plant Biology*, **49**, 1565–1572.

Zedler, P.H. (2007) Fire effects in grasslands. *Plant disturbance ecology: the process and the response* (eds E.A. Johnson and K. Miyanishi). Elsevier, Amsterdam.

Zeng, Y., Jiang, Z.G. and Li, C.W. (2007) Genetic variability in relocated Pere David's deer (*Elaphurus davidianus*) populations—Implications to reintroduction program. *Conservation Genetics*, **8**, 1051–1059.

Zeppel, H.D. (2006) *Indigenous ecotourism: sustainable development and management*. CABI, Wallingford.

Zhang, L. and Corlett, R.T. (2003) Phytogeography of Hong Kong bryophytes. *Journal of Biogeography*, **30**, 1329–1337.

Zhang, Z.-B., Xiao, Z.-S. and Li, H.-J. (2005) Impact of small rodents on tree seeds in temperate and subtropical forests, China. *Seed fate: predation, dispersal and seedling establishment* (eds P.M. Forget, J.E. Lambert, P.E. Hulme and S.B. Vander Wall), pp. 269–282. CABI Publishing, Wallingford.

Zhang, J., Ge, Y., Chang, J. et al. (2007) Carbon storage by ecological service forests in Zhejiang Province, subtropical China. *Forest Ecology and Management*, **245**, 64–75.

Zhang, W., Mo, J., Yu, G. et al. (2008a) Emissions of nitrous oxide from three tropical forests in southern China in response to simulated nitrogen deposition. *Plant and Soil*, **306**, 221–236.

Zhang, L., Hua, N. and Sun, S. (2008b) Wildlife trade, consumption and conservation awareness in southwest China. *Biodiversity and Conservation*, **17**, 1493–1516.

Zhao, Q.K. (1999) Responses to seasonal changes in nutrient quality and patchiness of food in a multigroup community of Tibetan macaques at Mt. Emei. *International Journal of Primatology*, **20**, 511–524.

Zheng, Z., Feng, Z.L., Cao, M., Li, Z.F. and Zhang, J.H. (2006) Forest structure and biomass of a tropical seasonal rain forest in Xishuangbanna, Southwest China. *Biotropica*, **38**, 318–327.

Zhong, L., Buckley, R. and Xie, T. (2007) Chinese perspectives on tourism eco-certification. *Annals of Tourism Research*, **34**, 808–811.

Zhou, Z.H. and Jiang, Z.G. (2005) Identifying snake species threatened by economic exploitation and international trade in China. *Biodiversity and Conservation*, **14**, 3525–3536.

Zhou, H., Chen, J. and Chen, F. (2007) Ant-mediated seed dispersal contributes to the local spatial pattern and genetic structure of *Globba lancangensis* (Zingiberaceae). *Journal of Heredity*, **98**, 317–324.

Zhou, Y.B., Zhang, L., Kaneko, Y., Newman, C. and Wang, X.-M. (2008a) Frugivory and seed dispersal by a small carnivore, the Chinese ferret-badger, *Melogale moschata*, in a fragmented subtropical forest of central China. *Forest Ecology and Management*, **255**, 1595–1603.

Zhou, Y.B., Slade, E., Newman, C., Wang, X.M. and Zhang, S.Y. (2008b) Frugivory and seed dispersal by the yellow-throated marten, *Martes flavigula*, in a subtropical forest of China. *Journal of Tropical Ecology*, **24**, 219–223.

Zhou, Y.B., Zhang, J.S., Slade, E. et al. (2008c) Dietary shifts in relation to fruit availability among masked palm civets (*Paguma larvata*) in central China. *Journal of Mammalogy*, **89**, 435–447.

Zhu, H. (2008a) Advances in biogeography of the tropical rain forest in southern Yunnan, southwestern China. *Tropical Conservation Science*, **1**, 34–42.

Zhu, H. (2008b) Species composition and diversity of lianas in tropical forests of southern Yunnan (Xishuangbanna), south-western China. *Journal of Tropical Forest Science*, **20**, 111–122.

Zhu, H., Xu, Z.F., Wang, H. and Li, B.G. (2004) Tropical rain forest fragmentation and its ecological and species diversity changes in southern Yunnan. *Biodiversity and Conservation*, **13**, 1355–1372.

Zhu, H., Ma, Y.-X. and Hu, H.-B. (2007a) The relationship between geography and climate in the generic-level patterns of Chinese seed plants. *Acta Phytotaxonomic Sinica*, **45**, 134–166.

Zhu, L., Sun, O.J., Sang, W.G., Li, Z.Y. and Ma, K.P. (2007b) Predicting the spatial distribution of an invasive plant species (*Eupatorium adenophorum*) in China. *Landscape Ecology*, **22**, 1143–1154.

Zhu, Y., Zhao, G.-F., Zhang, L.-W. et al. (2008) Community composition and structure of Gutianshan Forest Dynamic Plot in a mid-subtropical evergreen broad-leaved forest. *Journal of Plant Ecology*, **32**, 262–273.

Ziegler, T., Abegg, C., Meijaard, E. et al. (2007) Molecular phylogeny and evolutionary history of Southeast Asian macaques forming the *M-silenus* group. *Molecular Phylogenetics and Evolution*, **42**, 807–816.

Zimmerman, J.K., Wright, S.J., Calderon, O., Pagan, M.A. and Paton, S. (2007) Flowering and fruiting phenologies of seasonal and aseasonal neotropical forests: the role of annual changes in irradiance. *Journal of Tropical Ecology*, **23**, 231–251.

Zong, Y., Chen, Z., Innes, J.B., Chen, C., Wang, Z. and Wang, H. (2007) Fire and flood management of coastal swamp enabled first rice paddy cultivation in east China. *Nature*, **449**, 459–462.

Zvereva, E.L., Toivonen, E. and Kozlov, M.V. (2008) Changes in species richness of vascular plants under the impact of air pollution: a global perspective. *Global Ecology and Biogeography*, **17**, 305–319.

INDEX

Page numbers in *italics* refer to figures; those in **bold** to tables.